AN OUTLINE OF
STRUCTURAL GEOLOGY

AN OUTLINE OF STRUCTURAL GEOLOGY

Bruce E. Hobbs

Department of Earth Sciences
Monash University
Melbourne, Australia

Winthrop D. Means

Department of Geological Sciences
State University of New York at Albany
Albany, New York, U.S.A.

Paul F. Williams

Institute of Geology & Mineralogy
Leiden, Netherlands

JOHN WILEY & SONS, INC.

New York · London · Sydney · Toronto

Library of Congress Cataloging in Publication Data:

Hobbs, Bruce E
  An outline of structural geology.

  Includes index.
  1. Geology, Structural. I. Means, Winthrop
Dickinson, joint author. II. Williams, Paul Frederick,
1938–   joint author. III. Title.
QE601.H6   551.8   75-20393
ISBN 0-471-40156-0

Printed in the United States of America

10 9 8 7 6 5 4 3 2 1

PREFACE

For some time now, we, and many of our colleagues, have bemoaned the fact that there exists no concise textbook in structural geology that covers modern developments in the subject and that can be given to undergraduates as basic reading in conjunction with more detailed lectures. Twenty years ago E. S. Hills' *Outlines of Structural Geology* served this purpose admirably but, unfortunately, that book is now out of print and many aspects of it are outdated. We have tried, in this book, to give a new introduction to fundamental aspects of the subject. Since the book is meant to be both compact and introductory the treatment is by no means complete, and most parts will have to be amplified either by a lecture course or by supplementary reading. Some topics are taken to an advanced level because of their importance, but many other advanced aspects are not touched upon. We have included a large number of references to provide a lead-in to advanced study and to aid students who are working independently. Although the book is elementary we hope that the treatment is rigorous and that there are no parts that will have to be relearned.

B. E. Hobbs
W. D. Means
P. F. Williams

INTRODUCTION

Structural geology deals with ways in which rocks respond to the application of deforming forces and with the structures that result from deformation. The subject is complicated by the enormous range of chemical and mineralogical compositions that rocks may have and the large range in physical and chemical environments in which deformation occurs. Each rock type possesses different mechanical properties, and these properties depend on the conditions under which deformation takes place. The result is a great variety of responses that manifest themselves in the diverse structures displayed in the crust of the earth. It is, of course, these structures — such as various types of folds, faults, foliations, lineations, and joints — that are of direct interest to the field geologist and of greatest importance for delineating the form of an ore body, the extent of an oil field, or the structural history of an area.

The ultimate aim of structural geology is to establish the history of displacements, strains, stresses, strain rates, temperatures, and pressures that the crust and associated upper mantle of the earth have experienced. Fieldwork remains the most essential method of working toward this goal, but other techniques may also be important, including use of the optical microscope and scanning electron microscope to examine the microstructure of a deformed rock, use of the optical microscope and X-ray goniometer to measure the crystallographic orientation of constituent crystals, and use of the transmission electron microscope to look at structural details within individual crystals. All such small-scale features may shed considerable light on the structural history of a particular piece of rock.

The various factors that interact with each other to produce structural changes in rocks are schematically represented in the diagram on page viii.

To understand these factors and the ways in which they can interact with each other, we might consider a body of rock, say in the form of a cube a mile on the side somewhere in the crust of the earth. Imagine further that this cube is ultimately placed in a situation where deforming forces begin to act on the boundaries of the body. At this stage conditions of temperature and of confining pressure are established and in general, there will be gradients of these two quantities from point to point across the body. Also, conditions will not be static so that at each point the temperature and the confining pressure will be changing slowly with time. The rock body itself is unlikely to be homogeneous throughout and will consist of layers of different rock types. Some aspects of

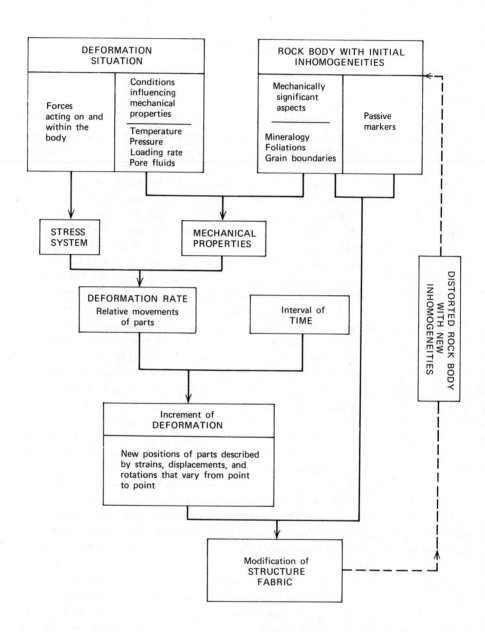

the rock body will be mechanically significant. Thus the mineralogical composition from point to point will largely govern the way in which the material responds to the forces that have been imposed on it. Again there will be structural features that are important in governing the mechanical response of the rock body. There may be foliations, schistosities, cleavages, and bedding planes present locally or throughout the material. If the confining pressure is relatively low then perhaps it will be possible for slip to occur on these foliations during a deformation. If, however, the confining pressure is high, then the frictional resistance to slip on such surfaces may be quite high, and it may be easier for deformation to occur by some other process. In addition there may be other markers present in the body, such as color bands and the like. All such structural features, whether they are mechanically significant during the deformation or quite passive, are destined to outline the structure of the rock after deformation has ceased.

We have, then, the deformation situation on one hand and the rock body on the other. The forces imposed on the boundaries of the rock body, together with gravitational forces, give rise to other systems of forces at each point within the body. This system of forces that varies from point to point defines a stress system. At the same time, the mechanically significant aspects of the rock body (such as the mineral composition, the presence of foliation, and so on) combine with the conditions imposed on the material (such as temperature and pressure) to produce a particular set of mechanical properties characteristic of the material under those particular physical conditions. The action of the stress system on the material with these particular mechanical properties gives rise to a certain rate of deformation in which the configuration of the material is changed as a whole and parts move with respect to each other. With the passing of time, new positions of parts of the body are taken up. These new positions are described by various strains, displacements, and rotations that vary from point to point in time during the deformation. The new positions of parts are reflected in changes in the structure and fabric of the deformed rock body. The structure is largely outlined by inhomogeneities that happen to be present in the rock body initially and have become distorted during the deformation. If deformation continues, then the changes in structure and fabric that have already taken place within the rock body will generally play a part in determining how the body responds to further deformation, as suggested by the dashed line up the right side of the diagram. Most rock structures evolve gradually, over long periods of time, from many small increments of deformation applied to material of changing mechanical properties.

As outlined above, structural geology is quite closely related to aspects of materials science, solid state physics, physical metallurgy and ceramics and, in fact, recent developments in the field of structural geology are closely allied to developments in these other physical sciences. However, the material that the structural geologist has to work with is on a much grander scale than that encountered by the metallurgist or solid state physicist, and furthermore the structural geologist is often presented with a complicated three-dimensional structure to work out from limited two-dimensional exposure. The structural geologist has developed specific techniques for unraveling such structures. He also relies heavily on the associated fields of stratigraphy, petrology, and mineralogy. However, many of the most fundamental problems in structural geology are similar to those encountered in other sciences dealing with the deformation of solids.

Each chapter in this book elaborates on specific parts of the scheme presented above. In Chapter 1 the mechanical aspects of rock deformation are considered and the concepts of stress, strain, and the response of rocks to stress are discussed in detail. In Chapter 2 the microfabric of deformed rocks is outlined, emphasis being placed on the microstructure that develops during deformation, including the crystal defects that are responsible for much of the deformation that takes place. The crystallographic preferred orientations that develop during deformation are also considered in this chapter. In Chapter 3 the primary structures present in rock prior to any deformation are considered. These include bedding and the primary sedimentary structures that may be present due to sedimentary processes. These primary structures enable the structural geologist to tell the facing or direction of younging of a stratigraphic sequence even after it has been heavily deformed. In Chapter 4, perhaps the most spectacular type of deformation is considered, that is, folding of originally undisturbed layers. The various types of folds that can be seen in deformed rocks are considered along with the distribution of strain likely to be present in such folds and the mechanisms by which folds develop. In Chapters 5 and 6, the features common to folded rocks are considered. These are the various kinds of foliations and lineations that are typically associated with such rocks. In Chapter 7, aspects of the deformation of brittle rocks are considered, including joints and faults. Chapter 8 explains methods used to work out the structural geometry of deformed areas. In Chapter 9 the various structural associations that are widely distributed over the surface of the earth are considered, and in Chapter 10 an attempt is made to briefly summarize some of the deformational situations that have been proposed as responsible for the structures seen in rocks. A large part of this chapter is devoted to the new and currently very popular concept of plate tectonics.

ACKNOWLEDGMENTS

We thank our wives, Robyn, Anna and Pam and many friends and colleagues who helped us prepare this book. We especially thank the following people, who read various parts of the manuscript and suggested improvements: M. B. Bayly, T. H. Bell, J. W. A. Bodenhausen, G. J. Borradaile, J. D. Byerlee, K. A. W. Crook, G. A. Davis, J. F. Dewey, M. A. Etheridge, P. J. Fox, N. C. Gay, J. Glass, J. Granath, R. J. Holcombe, R. J. Lisle, G. S. Lister, J. F. Savage, A. Senior, T. Thyrsted, R. Trouw, R. H. Vernon, C. J. L. Wilson, D. S. Wood, and H. J. Zwart. Others who helped in various ways are J. Boland, B. Bryant, J. R. Conolly, D. D. Graham, O. J. Groot, W. C. Laurijssen, D. Marshall, A. C. McLaren, D. Morrison-Smith, C. D. van Panhuys, and A. P. S. Reymer. Miss D. Kelly, Mrs. M. Leicester, Mrs. F. G. M. Madjlessi-Leenen, and Miss T. W. Terpstra did the typing.

B. E. H.
W. D. M.
P. F. W.

CONTENTS

AN OUTLINE OF
STRUCTURAL GEOLOGY

1

MECHANICAL
ASPECTS

1.1 INTRODUCTION

Any portion of the earth will generally be acted upon by forces that tend to displace and distort the rocks within that region. Some of these forces arise solely from the weight of the overlying rocks; others arise from large-scale motions of material composing adjacent parts of the crust or of the mantle of the earth. In addition, gravity acts on each element of rock. In some instances, these forces are small or act only for short periods of time so that no significant deformation results. In other instances, the forces act for relatively long periods of time and spectacular permanent deformations, such as large-scale folding, result. In still other instances, the fracture strength of the rocks may be exceeded and faulting is then the most conspicuous mode of deformation. Whether the rocks composing a region deform permanently or not and whether any deformation is predominantly by folding, by faulting, or by yet other modes, depends on the interplay of a number of physical and chemical factors including the temperature, the hydrostatic pressure, the pressure of any pore fluids, the rate at which the deforming forces are applied, the rate at which deformation proceeds, and the composition (including the fluid content) of the rocks. The main purpose of this chapter is to examine these factors in order to gain a clearer physical insight into the manner in which rocks deform in nature. A secondary purpose is to introduce a number of concepts that will be useful in subsequent chapters.

This aspect of the study of the deformation of rocks is conveniently divided into three parts: *the analysis of stress* (Section 1.2), *the analysis of deformation* (Section 1.3), and *the response of rocks to stress* (Section 1.4).

1

1.2 **ANALYSIS OF STRESS**

Consider a small cube of rock within a large volume of rock undergoing deformation. Figure 1.1*a* shows such a small cube within a large fold as an example. All six faces of this cube will be pressed on by adjacent parts of the rock, and there will be corresponding reactions from the material within the cube. In addition, each particle within the cube is acted on by gravity. A system of forces, therefore, exists throughout the cube, and the resultant forces acting upon the faces of the cube are illustrated in Figure 1.1*c*. Some components of these forces tend to accelerate the cube relative to adjacent parts of the rock mass whereas others, acting as moments, tend to change the angular velocity of the cube. Under natural conditions, however, these accelerations are extremely small in magnitude. At the same time, the imposed forces tend to distort the shape of the cube so that, as the fold grows, each small region changes its shape as it rotates and translates relative to adjacent parts (Fig. 1.1*a* and *b*).

The magnitudes of the forces that act on the faces of the small cube in order to produce a given deformation depend on the areas of these faces: the larger the cube, the larger the force required to produce a given change of shape. It is convenient to have a measure of the deforming agents that is independent of the size of cube considered. The measure commonly used is the *force per unit area* or the *stress*. Using such a measure, it becomes easier to compare the agents responsible for the deformation of different-sized cubes or those responsible for the deformation of the same region of rock as it progressively changes shape. Stress has the dimensions $[ML^{-1}T^{-2}]$, and some units commonly used in geology are given in Table 1.1.

1.2.1 **Components of Stress. Stress at a Point**

The force on each of the cube faces in Figure 1.1*c* may be resolved into three orthogonal components, one normal to the face and two parallel to the face (Fig. 1.1*d*). If the magnitudes of each of these three components is divided by the area of the cube face then the magnitudes of three *components of stress* are obtained (Fig. 1.1*e*). Using the edges of the cube as a system of Cartesian coordinates (x_1, x_2, x_3) and employing the symbol σ_{ij} to denote that component of stress that acts on the face normal to x_i and in the direction of x_j (here, since there are three coordinate axes, i and j range over the values 1, 2, and 3), the various components of stress may be labeled as in Figure 1.1*e*, and written down in a systematic manner in the following array:

$$\begin{bmatrix} \sigma_{11} & \sigma_{12} & \sigma_{13} \\ \sigma_{21} & \sigma_{22} & \sigma_{23} \\ \sigma_{31} & \sigma_{32} & \sigma_{33} \end{bmatrix} \tag{1.1}$$

TABLE 1.1 Conversion Table for Units of Stress Commonly Used in Geology

	Bar	Kbar	Dynes cm^{-2}	Atmosphere	Kg cm^{-2}	Newton-m^{-2}	Pascal (Pa)	Gigapascal (GPa)	Pounds ins^{-2} (psi)
Bar	1.0	10^{-3}	10^6	0.9869	1.0197	10^5	10^5	10^{-4}	14.503
Kbar	10^3	1.0	10^9	0.9869×10^3	1.0197×10^3	10^8	10^8	10^{-1}	14.503×10^3
Dynes cm^{-2}	10^{-6}	10^{-9}	1.0	0.9869×10^{-6}	1.0197×10^{-6}	10^{-1}	10^{-1}	10^{-10}	14.503×10^{-6}
Atmosphere	1.0133	1.0133×10^{-3}	1.0133×10^6	1.0	1.0333	1.0133×10^5	1.0133×10^5	1.0133×10^{-4}	14.695
Kg cm^{-2}	0.9807	0.9807×10^{-3}	0.9807×10^6	0.9678	1.0	0.9807×10^5	0.9807×10^5	0.9807×10^{-4}	14.223
Newton-meter^{-2}	10^{-5}	10^{-8}	10	0.9869×10^{-5}	1.0197×10^{-5}	1.0	1.0	10^{-9}	14.503×10^{-5}
Pascal (Pa)	10^{-5}	10^{-8}	10	0.9869×10^{-5}	1.0197×10^{-5}	1.0	1.0	10^{-9}	14.503×10^{-5}
Gigapascal (GPa)	10^4	10	10^{10}	0.9869×10^4	1.0197×10^4	10^9	10^9	1.0	14.503×10^4
Pounds ins^{-2} (psi)	6.895×10^{-2}	6.895×10^{-5}	6.895×10^4	6.805×10^{-2}	7.031×10^{-2}	6.895×10^3	6.895×10^3	6.895×10^{-6}	1.0

To use this table start in the left-hand column and read along to the column for which a conversion is required. Thus, 1 atmosphere = 1.013×10^5 N-m^{-2} whereas 1 N-m^{-2} = 1.02×10^{-5} kg cm^{-2}.

Sources: Baumeister and Marks, 1967; The Symbols Committee of the Royal Society (1971).

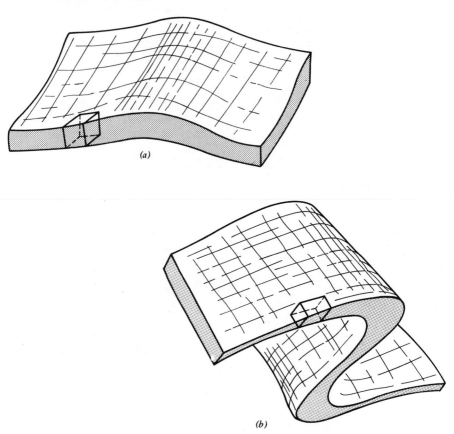

FIGURE 1.1 Components of stress. (a) Small cube in a layer of rock undergoing folding. (b) The same cube distorted after the layer has become folded. (c) Forces acting on the faces of a deforming cube. (d) Resolution of these forces, both normal and parallel to the faces of the cube. (e) Stress system acting upon the faces of the cube referred to the coordinate system Ox_1, Ox_2, Ox_3. (f) Stress system acting on the faces of an infinitesimal cube.

Thus, σ_{13} is a stress component that acts on the face normal to x_1 and in the direction of x_3, whereas σ_{31} acts on the face normal to x_3 and in the direction of x_1. Stress components represented by σ_{ij} where the two subscripts are the same ($i = j$), and which act normal to a cube face, are known as *normal stresses*. Those stress components represented by σ_{ij} where the two subscripts are different ($i \neq j$), and which act parallel to a cube face, are known as *shear stresses*. In some textbooks normal stresses are represented by symbols such as σ_{ij} ($i = j$), whereas a different symbol τ_{ij} with $i \neq j$ is used to represent the shear stresses.

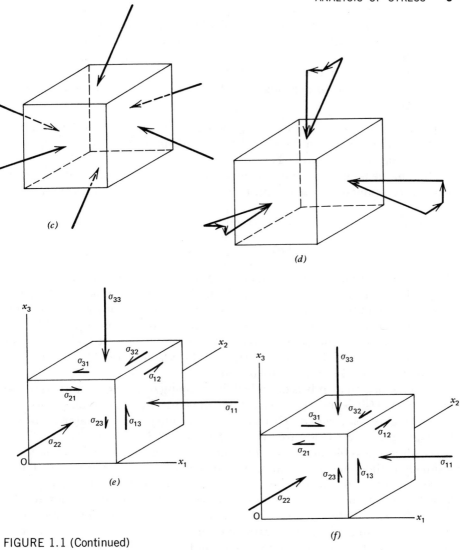

FIGURE 1.1 (Continued)

The situation illustrated in Figure 1.1*a* is likely to be complicated by variation in the magnitude and direction of force over each cube face, and it becomes convenient in such inhomogeneous situations to consider the *state of stress at a point*. This is achieved by imagining the cube of Figure 1.1*e* to shrink to a point, the stress at a point being defined as the limiting ratio of force to area as the area of the face approaches zero. Three important features arise as this limiting situation is approached:

1. The distribution of forces over each face approaches uniformity.
2. Forces on opposite faces approach each other in magnitude and direction.
3. Unless the angular acceleration of the cube is to become infinite, forces that are capable of exerting a resultant torque on the cube must tend to balance out.

This implies that $\sigma_{12} = \sigma_{21}$, $\sigma_{23} = \sigma_{32}$, and $\sigma_{31} = \sigma_{13}$ in the limit [Nye (1964, pp. 83–87)] and that an infinitesimal cube would have stresses acting upon its faces such as those shown in Figure 1.1*f*.

Thus, in the limit, when the state of stress at a point is considered, the array of stress components at that point may be written in a symmetrical manner:

$$\begin{bmatrix} \sigma_{11} & \sigma_{12} & \sigma_{13} \\ \sigma_{12} & \sigma_{22} & \sigma_{23} \\ \sigma_{13} & \sigma_{23} & \sigma_{33} \end{bmatrix} \tag{1.2}$$

Thus, there are just six independent components of stress at a point in any material. This is true whether or not the body is at rest or is accelerating and whether or not the distribution of forces throughout the body is uniform. It is not true, however, if there exists a body torque. Such a torque is probably rare in geological bodies, but an example may arise in a magnetic body due to an external magnetic field [Nye (1964, pp. 86–87). See also McClintock and Argon (1966, pp. 54, 55) for discussion].

The *stress field* in a body is described by mapping out the components of the array (1.2) at all points. If these components are the same at all points, the stress field is *homogeneous*. Otherwise it is *inhomogeneous*. Examples of inhomogeneous stress fields are given in Section 1.2.3.

It is always possible, at any point in a homogeneous stress field, to find three mutually orthogonal planes upon which the shear stresses are zero. These three planes are known as the *principal planes of stress* and their normals are the *principal axes of stress*. The normal stresses across the principal planes are the *principal stresses*, often denoted by σ_1, σ_2 and σ_3 with the convention that $\sigma_1 > \sigma_2 > \sigma_3$. In geological literature compressive stresses are commonly taken as positive. The *maximum, intermediate,* and *minimum principal stresses* are σ_1, σ_2, and σ_3, respectively. The state of stress at a point may, therefore, be characterized by giving these *three* principal stresses and their directions, or by giving the *six* independent stress components contained in the array (1.2) when the faces of the reference cube are not parallel to the principal planes of stress.

It is important to note that although stress has many of the physical characteristics of force associated with it, the concept always contains the additional physical association with area. Hence, the value of a stress not only varies with the orientation and magnitude of the imposed force but varies also as the area

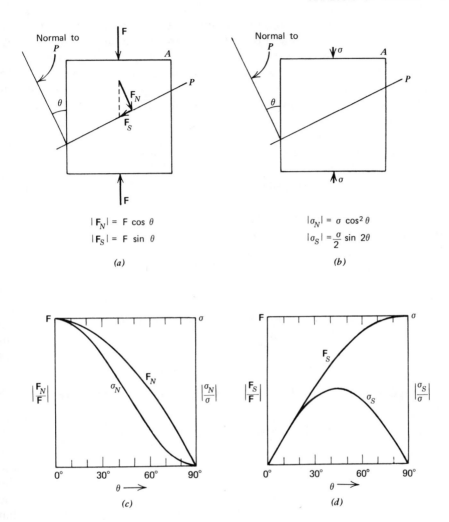

$$|\mathbf{F}_N| = F \cos\theta$$
$$|\mathbf{F}_S| = F \sin\theta$$

(a)

$$|\sigma_N| = \sigma \cos^2\theta$$
$$|\sigma_S| = \frac{\sigma}{2} \sin 2\theta$$

(b)

(c)

(d)

FIGURE 1.2   An illustration that stresses may not be resolved as though they were vector quantities. (a) Cross section through a cube with area of each face A. $\mathbf{F}$ is a force inclined at angle θ to the normal to the plane P. The component of force normal to the plane is $\mathbf{F}_N$, whereas that parallel to the plane is $\mathbf{F}_S$. The magnitudes of $\mathbf{F}_N$ and $\mathbf{F}_S$ may be obtained by resolving $\mathbf{F}$ normal and parallel to P, respectively. (b) Cross section through the same cube showing the stress σ due to the force $\mathbf{F}$. The magnitudes of σ_N and of σ_S, the normal and shear stresses, respectively, on the plane P may not be found by resolving σ normal and parallel to P. (c) Variation of $\mathbf{F}_N$ and σ_N with θ. (d) Variation of $\mathbf{F}_S$ and σ_S with θ.

of action changes orientation and magnitude. This may be seen more clearly in Figure 1.2, where cross sections of cubes are shown with a force of magnitude F acting normally to one cube face of area A (Fig. 1.2a). Cutting across the cube is another plane, P, whose normal is inclined at angle θ to $\mathbf{F}$. We ask: What are the normal and shear components of *force* across the plane P, and how do they differ in magnitude from the normal and shear components of *stress* across P?

In Figure 1.2a the force $\mathbf{F}$ has been resolved into components normal and parallel to the plane P. The components have magnitudes

$$\left|\mathbf{F}_N\right| = F\cos\theta \qquad \left|\mathbf{F}_S\right| = F\sin\theta \qquad (1.3)$$

respectively. Now, in Figure 1.2b, the stress σ on the cube face has the magnitude F/A whereas the area of the plane P is

$$A_P = \frac{A}{\cos\theta} \qquad (1.4)$$

Hence,

$$\left|\mathbf{F}_N\right| = F\cos\theta = A\sigma\cos\theta = A_P\sigma\cos^2\theta \qquad (1.5)$$

and

$$\left|\mathbf{F}_S\right| = F\sin\theta = A\sigma\sin\theta = A_P\sin\theta\cos\theta$$

Thus, the magnitudes of the normal and shear components of stress across P are

$$\left|\sigma_N\right| = \frac{\left|\mathbf{F}_N\right|}{A_P} = \sigma\cos^2\theta = \frac{F}{A}\cos^2\theta$$

and $\qquad\qquad\qquad\qquad\qquad\qquad\qquad\qquad\qquad\qquad$ (1.6)

$$\left|\sigma_S\right| = \frac{\left|\mathbf{F}_S\right|}{A_P} = \frac{\sigma}{2}\sin2\theta = \frac{F}{A}\sin\theta\cos\theta$$

The manner in which the magnitudes of $\mathbf{F}_N$ and σ_N, and of $\mathbf{F}_S$ and σ_S vary with θ is shown in Figures 1.2c and d, respectively. A comparison of Equations 1.3 and 1.6 shows that stresses may not be resolved as though they were forces and that the change in the magnitude of the area of action must be considered also.

Where the principal stresses are σ_1 and σ_2 the equations for the normal and shear stresses across a plane whose normal is inclined at θ to σ_1 are

$$\sigma_N = \tfrac{1}{2}(\sigma_1 + \sigma_2) + \tfrac{1}{2}(\sigma_1 - \sigma_2)\cos2\theta$$

$$\sigma_S = \tfrac{1}{2}(\sigma_2 - \sigma_1)\sin2\theta \qquad (1.7)$$

respectively, and reference should be made to Jaeger (1969, pp. 5–20) where the three-dimensional case is also considered. Note that (1.7) reduces to (1.6) when σ_2 is zero.

Comparison of Equations 1.3 and 1.6 shows that the stresses may not be calculated as though they were forces. Stress is an example of another type of quantity known as a *second-order tensor* [Nye (1964)].

Equations 1.7 lead to a convenient representation of stress known as the *Mohr diagram* (Fig. 1.3). Two orthogonal coordinate axes, along which the

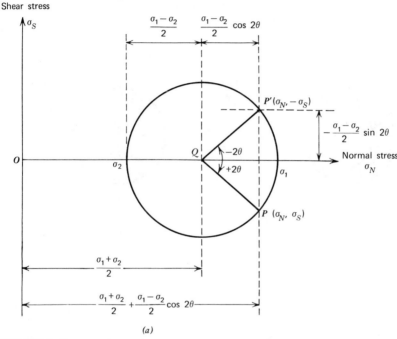

(a)

FIGURE 1.3   (a) The Mohr diagram for a two-dimensional stress state in which the principal stresses are σ_1 and σ_2. Normal stresses are plotted along one axis and shear stresses along the other. If a circle is drawn as shown with diameter $(\sigma_1 - \sigma_2)$ and center at

$$\left[\frac{\sigma_1 + \sigma_2}{2}, O\right] \text{ then at any point } P \text{ on the circle the coordinates are}$$

$$\left[\frac{\sigma_1 + \sigma_2}{2} + \frac{\sigma_1 - \sigma_2}{2}\cos 2\theta, \ \frac{\sigma_2 - \sigma_1}{2}\sin 2\theta\right].$$

Hence, such a point has coordinates (σ_N, σ_S). The angle 2θ is as shown. (b) Mohr diagram for a uniaxial stress. (c) Mohr diagram for a biaxial stress. In this part and (d), (e) all possible stress states lie in the hatched area. (d) Mohr diagram for a triaxial stress. (e) Mohr diagram for a pure shear stress. (f) Mohr diagram for a hydrostatic pressure. (g) Mohr diagram for a hydrostatic tension.

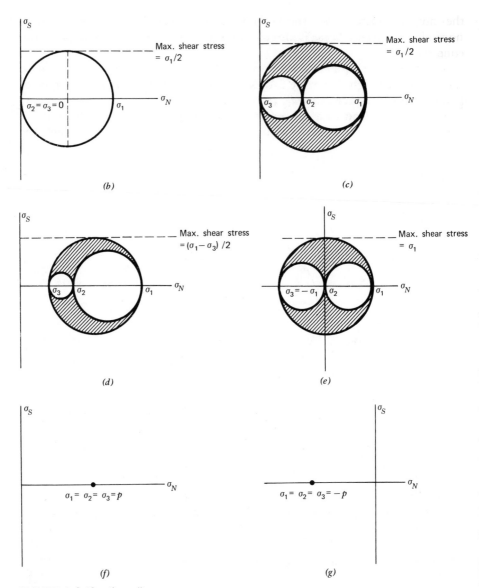

FIGURE 1.3 (Continued)

normal and shear stresses are plotted, are shown in Figure 1.3*a* and, for a two-dimensional stress in which the principal stresses are σ_1 and σ_2, a circle is constructed (as shown) with diameter $(\sigma_1 - \sigma_2)$ and center at $Q = \left[(\sigma_1 + \sigma_2)/2, 0\right]$. Then any point P on the circle has coordinates (σ_N, σ_S) where σ_N and σ_S are given by (1.7) and 2θ is the angle between the σ_N axis and the line PQ, measured in the sense shown. The coordinates of any point P on the circle, then, give the normal and shear stresses across a plane whose normal is inclined at θ to σ_1 where the principal stresses are σ_1 and σ_2. This construction may also be used to find σ_1, σ_2, and θ given σ_N and σ_S on two planes at right angles [Jaeger (1969, pp. 9–10)]. Three-dimensional stress states may also be represented by a slightly more complicated Mohr diagram [Figure 1.3c, see Jaeger (1969, pp.18–20)]. Mohr diagrams are used extensively in discussions of the fracturing of rock masses (see Chapter 7).

Some examples of special stress states are:

1. *Uniaxial stress* in which one principal stress (σ_1 or σ_3) is nonzero and the other two are equal to zero. The Mohr diagram for this state of stress is given in Figure 1.3b.
2. *Biaxial stress* in which two of the principal stresses are nonzero and the other is zero. An example of a Mohr diagram for this state of stress is shown in Figure 1.3c. The Mohr diagram is more complicated now than in the two-dimensional situation. [See Jaeger (1969, pp.18–20) for a discussion of this diagram.]
3. *Triaxial stress* in which σ_1, σ_2, and σ_3 have nonzero values. This is the most general stress system and is the one probably developed most commonly in nature. The Mohr diagram is illustrated in Figure 1.3d.
4. *Pure shear stress* in which $\sigma_1 = -\sigma_3$ and is nonzero. $\sigma_2 = 0$. This is an example of biaxial stress. The Mohr diagram is shown in Figure 1.3e.
5. *Hydrostatic pressure* in which $\sigma_1 = \sigma_2 = \sigma_3 = p$. The Mohr diagram is shown in Figure 1.3f.

One can also talk about the situation where all three principal stresses are equal but tensile. This stress state is referred to as a *hydrostatic tension*. The Mohr diagram is shown in Figure 1.3g. In much geological literature, another term — the *lithostatic pressure* — is used. This is the hydrostatic pressure generated at a depth h below the ground surface due solely to the weight of rocks, of mean density ρ, in that interval. Naturally, this is equal to $\rho g h$ where g is the acceleration due to gravity. Such a statement, however, requires some qualification because it assumes that the stress state at depth h has become truly hydrostatic due to relaxation of all shearing stresses by some creep process. If the stress state has not been allowed to become hydrostatic, and we are talking about the stress state due solely to a pile of rocks of height h, then this is usually taken to be $\sigma_1 = \rho g h$, $\sigma_2 = \sigma_3 = [\nu/(1 - \nu)]\,\sigma_1$ where ν is Poisson's ratio [see Jaeger (1969, p. 120) and Section 1.4.2]. Another commonly used term is the *mean stress*, which is equal to $(\sigma_1 + \sigma_2 + \sigma_3)/3$. (See Section 1.2.2.)

Notice that in each of the examples given above except the last, the maximum shear stress occurs on planes inclined at 45° to σ_1. In all cases where $\sigma_1 > \sigma_2 > \sigma_3$ the planes of maximum shear stress are only two in number and intersect in σ_2. In the special situation where $\sigma_2 = \sigma_3$ or $\sigma_1 = \sigma_2$ there are an infinite number of such planes inclined at 45° to σ_1 or to σ_3, respectively. The value of the maximum shear stress is indicated in each diagram of Figures 1.3*b* to *e*. In all cases the maximum shear stress has the value $(\sigma_1 - \sigma_3)/2$.

1.2.2.   The Stress Ellipsoid

We now consider the components σ_x, σ_y, σ_z of a stress σ, acting across a plane *P*, where the coordinate directions *Ox*, *Oy* and *Oz* are taken parallel to the principal stresses σ_1, σ_2, σ_3, respectively.

Figure 1.4*a* shows a small segment of a plane *P*. The normal to *P* makes angles α, β, and γ with the coordinate axes; the cosines of these angles, *l*, *m*, and *n* are known as the direction cosines of the normal to *P*. We are interested in finding the *x* component of σ, σ_x, on *P* where the principal stresses are known and the direction cosines *l*, *m*, *n* are known. We do this by considering the necessary equilibrium or balance of forces acting in the direction *Ox* on the

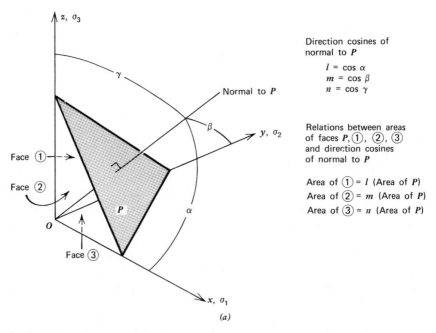

Direction cosines of
normal to **P**

$l = \cos \alpha$
$m = \cos \beta$
$n = \cos \gamma$

Relations between areas
of faces **P**, ①, ②, ③
and direction cosines
of normal to **P**

Area of ① = *l* (Area of *P*)
Area of ② = *m* (Area of *P*)
Area of ③ = *n* (Area of *P*)

(a)

FIGURE 1.4   Illustration of a situation where a stress may be represented as a vector. The principal stresses σ_1, σ_2, σ_3 are directed along the coordinate axes Ox_1, Ox_2, Ox_3, respectively. If σ is the stress acting across the plane *P* then this stress may be represented as a vector with components given by Equation (1.8). See text for discussion.

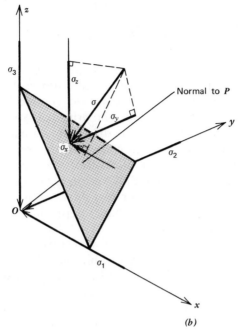

(b)

FIGURE 1.4 (Continued)

small tetrahedron bounded by P and by the three other triangular faces labeled 1, 2, and 3 in Figure 1.4a. Notice that each of these faces is normal to a principal stress, so that the total force across each is exactly perpendicular to it. Thus there is no component of force in the Ox direction on faces 2 or 3. The only forces acting on the tetrahedron in the Ox direction are $\sigma_x \times$ (area of P) and $\sigma_1 \times$ (area of 1). These forces must balance for equilibrium of the tetrahedron, so that

$$\sigma_x = \sigma_1 \left| \frac{\text{area of 1}}{\text{area of } P} \right|$$

Now, since

$$\text{Area of } 1 = l \times (\text{area of } P),$$

we have

$$\sigma_x = l\sigma_1$$

and, by similar arguments,

$$\sigma_y = m\sigma_2 \tag{1.8}$$
$$\sigma_z = n\sigma_3$$

It then follows, since $l^2 + m^2 + n^2 = 1$, that

$$\frac{\sigma_x^{\,2}}{\sigma_1^{\,2}} + \frac{\sigma_y^{\,2}}{\sigma_2^{\,2}} + \frac{\sigma_z^{\,2}}{\sigma_3^{\,2}} = 1 \tag{1.9}$$

Equation 1.9 is the equation of an ellipsoid whose principal semiaxes are in the same direction, and have the same magnitudes, as the principal stresses. This ellipsoid is one of several commonly used graphical representations of stress and is known as the *stress ellipsoid*. The principal axes of this ellipsoid are known as the *principal axes of stress*. The direction and magnitude of a radius vector of the stress ellipsoid gives a complete representation of the stress across the plane *conjugate* to that radius vector. Notice that, in general, the plane corresponding to a given radius vector is not normal to the radius vector. The relationship is shown for two dimensions in Figure 1.5*b*. The stress ellipsoid may have a variety of shapes depending on the relative magnitudes of the principal stresses.

Figure 1.5*c* shows a vector drawn in the same direction as a stress σ acting across a plane *P*. This vector is drawn with a magnitude proportional to σ. The above discussion may be summarized for two dimensions thus:

If coordinate axes Ox and Oy are chosen parallel to the maximum and least principal axes of stress, respectively, then the components of σ resolved respectively parallel to Ox and Oy are (cf. Equation 1.8)

$$\sigma_x = l\sigma_1$$
$$\text{and} \quad \sigma_y = m\sigma_2$$

where l and m are the direction cosines of the normal to *P*.

Figure 1.5*c* shows a stress vector σ in two dimensions, across a plane *P* resolved into components σ_x and σ_y, and also into components σ_N and σ_S. It is important to understand that such resolution of stresses as vectors can only be done when one is considering the stress and its components across a particular plane *P*, in a specified orientation. Otherwise, stress cannot be resolved as vectors, as illustrated in Figure 1.2.

These conclusions may also be reached for the three-dimensional situation.

It is always possible to divide a stress represented by the array (1.2) into two parts; one part consisting of only normal stresses with zero shear stresses, whereas the other part has nonzero normal and shear stresses;

$$\begin{bmatrix} \sigma_{11} & \sigma_{12} & \sigma_{13} \\ \sigma_{12} & \sigma_{22} & \sigma_{23} \\ \sigma_{13} & \sigma_{23} & \sigma_{33} \end{bmatrix} = \begin{bmatrix} p & 0 & 0 \\ 0 & p & 0 \\ 0 & 0 & p \end{bmatrix} + \begin{bmatrix} (\sigma_{11} - p) & \sigma_{12} & \sigma_{13} \\ \sigma_{12} & (\sigma_{22} - p) & \sigma_{23} \\ \sigma_{13} & \sigma_{23} & (\sigma_{33} - p) \end{bmatrix} \tag{1.10}$$

where p is the mean stress or hydrostatic pressure and is given by

$$p = (\sigma_{11} + \sigma_{22} + \sigma_{33}) / 3$$

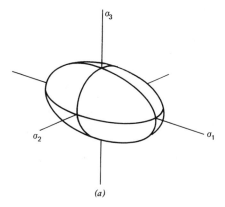

(a)

FIGURE 1.5   (a) The stress ellipsoid, with principle axes σ_1, σ_2, and σ_3. (b) Construction used to find the plane across which a given stress vector σ acts. The ellipse shown in this diagram is a cross section through the stress ellipsoid, and σ, represented by the line OP, is the stress vector. OX is the major principal axis of the ellipse, and a circle is drawn with center O and radius OX. A perpendicular to OX through the point P meets this circle in A. If α is the angle between OA and OX, then $OY = \sigma_1 \cos \alpha$. In the same way, an dentical construction may be made for the minimum principal axis OZ of the ellipse. Then, $OQ = \sigma_2 \cos \gamma$ where γ is the angle between OZ and OB in this case [right-hand side of (b)]. But since P lies on an ellipse, $OY^2/\sigma_1{}^2 + OQ^2/\sigma_2{}^2 = 1$. Hence, $\cos^2\alpha + \cos^2\gamma = 1$ and O, B, and A are colinear. Thus, comparing this situation with Equation 1.8, the line OA ($= OB$) has the geometrical significance of representing the normal to the plane across which σ acts. Either of the constructions [left or right of (b)] serve to identify the normal to this plane. Similar constructions can be carried out in three dimensions.

(c) A plane P across which a stress σ is represented by the vector σ. The coordinate axes Ox, Oz are chosen parallel to the principal stresses, σ_1 and σ_2, respectively. For this situation, and for this situation only, the stress may be resolved as though it was a vector (and not a second-order tensor) as shown. (d) and (e) An illustration of Equation (1.10) for a two-dimensional stress state. Figure (d) represents a two-dimensional stress state in which the principal stresses are σ_1 and σ_2. The center of this Mohr circle is distant $(\sigma_1 + \sigma_2)/2$ from the origin. If the circle is moved through this distance, as in (e), then the Mohr diagram is that representing a pure shear stress. The normal stresses have changed their magnitudes by an amount equal to the mean stress; the shear stresses remain unchanged.

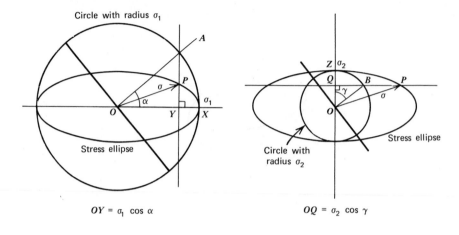

$$OY = \sigma_1 \cos \alpha \qquad\qquad OQ = \sigma_2 \cos \gamma$$

(b)

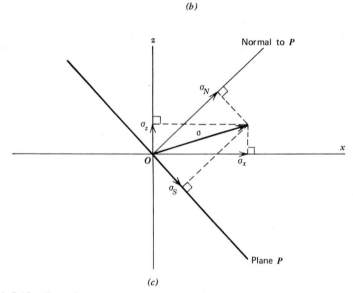

(c)

FIGURE 1.5 (Continued)

The part of the stress that consists only of normal stresses equal to p and zero shear stresses is called the *nondeviatoric stress;* the remaining part of the stress is called the *deviatoric stress.* This is illustrated for a two-dimensional stress state in Figure 1.5d,e. The state of stress with principal stresses σ_1 and σ_2 is illustrated by a Mohr diagram in Figure 1.5d. Shifting the center of the Mohr circle by the mean stress, $(\sigma_1 + \sigma_2)/2$, changes the stress to a pure shear as in

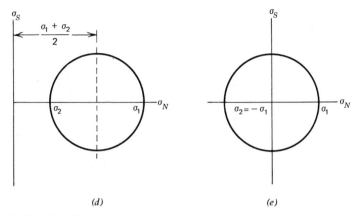

(d)   (e)

FIGURE 1.5 (Continued)

Figure 1.5*e*. The values of all shear stresses remain the same, indicating that they are independent of the hydrostatic component. The values of all normal stresses are, however, reduced by the value of the mean stress. This same argument may be extended to three dimensions.

1.2.3 Examples of Stress States in Rocks

The state of knowledge in structural geology is such that very little is known concerning the stress fields that exist in rocks during deformation, although it

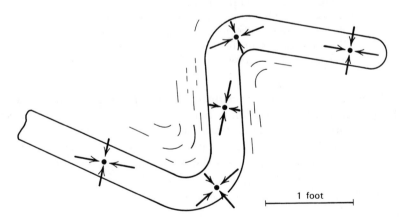

FIGURE 1.6   Sketch showing a folded quartzite layer studied by Scott, Hansen, and Twiss (1965). At each point the deformation lamellae in quartz grains were analyzed and the arrows indicate the derived orientation of σ_1 (long arrow) and σ_3 (short arrow. [From Dieterich and Carter (1969). *Amer. Jour. Sci.,* v. 267.]

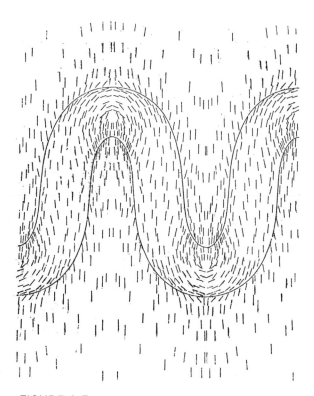

FIGURE 1.7   Computer-derived orientation of
stress in a folded layer. Each of the short lines are
drawn perpendicular to σ_1 at that point in the fold.
The fold represents 63.2 percent overall shortening
[From Dieterich (1969), *Amer. Jour., Sci.*, v. 267.]

is one of the prime goals of the subject to define these fields as closely as
possible. The lack of knowledge is, in part, due to the complexity of the stress
fields that exist in deforming bodies but mostly results from an overall lack of
information concerning the mechanical properties of rocks (see Section
1.4.2.). Four interpretations of stress states existing in rocks undergoing deformation are presented in Figures 1.6, 1.7, 1.8, and 1.9 to illustrate the
variety and complexity of states that are possible.

1.3   ANALYSIS OF DEFORMATION

When a body of rock is stressed the individual particles that make up the body
are displaced to new positions. Some parts of this motion may contribute

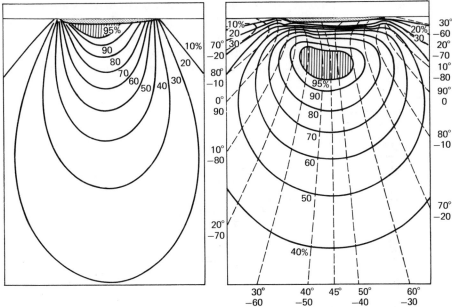

FIGURE 1.8 Stress distribution underneath a dam represented by the stippled slab at the top of the figures. Left, the vertical normal stress as percentage of the maximum value 1.16×10^7 dynes/cm^2; right, maximum shear stress as percentage of maximum value 3.08×10^6 dyne/cm^2 with inclinations of planes across which it acts. [From Gough (1969), reproduced by permission of the National Research Council of Canada from the *Candian Journal of Earth Sciences,* Vol. 6, No. 5, pp. 1067–1075.]

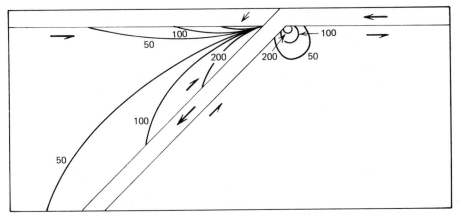

FIGURE 1.9 Shear stresses in bars calculated around a downgoing slab. The half arrows show the direction of the forces exerted by the fluid on the plates and slab. The stresses on the plate behind the island arc exceed those on the plate in front. The lithosphere is 50 km thick. [From McKenzie (1969).]

toward an overall *translation* of the body whereas other parts lead to local *distortions* and *rotations*. This is illustrated in Figure 1.10 where a sequence of stages in the progressive deformation of a layered specimen is shown. The individual squares in the undeformed specimen (Fig. 1.10*a*) are progressively *displaced* from their initial positions as they are *distorted* and *rotated*. During this

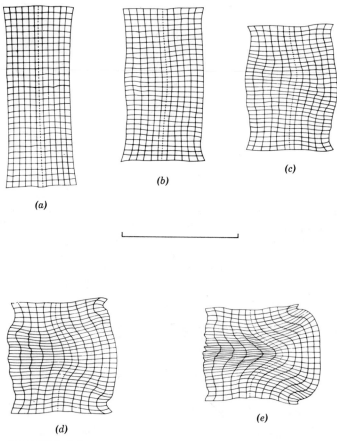

FIGURE 1.10   Gradual development of an experimental fold. The specimen is multilayered; the layers run parallel to the long axis of the figure in *(a)* and their boundaries are represented by the vertical lines. The horizontal lines complete a grid on the specimen that becomes distorted throughout the rest of the deformation. It is instructive to follow the gradual deformation and rotation of particular squares selected from various parts of the specimen. The scale bar is 5.0 cm. Mean shortenings are *(b)* 17 percent, *(c)* 32 percent, *(d)* 41 percent, *(e)* 52 percent.

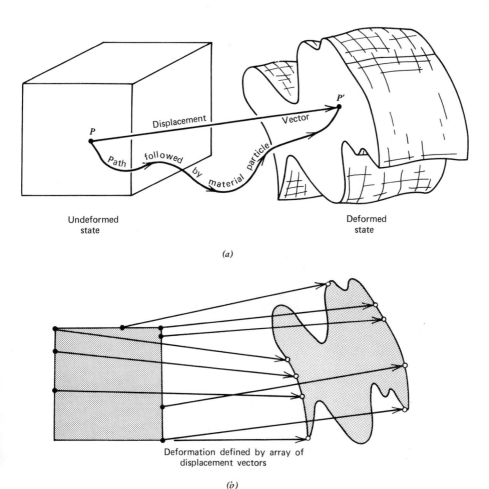

FIGURE 1.11   Homogeneous and inhomogeneous deformations. (a) An inhomogeneous deformation in which the point P in the undeformed state becomes P' in the deformed state. The actual path followed by P' during the deformation is the curved line. The vector **PP'** is the displacement vector that defines the displacement of P. (b) The array of displacement vectors that defines a deformation. (c) An inhomogeneous deformation in which flat sheets in the undeformed state become folded in the deformed state. A small cube with a circle inscribed in the undeformed state becomes a parallelopiped with an inscribed ellipse in the deformed state. The deformation of this cube may be expressed as a displacement together with a distortion and a rotation. (d) Displacement vectors for an inhomogeneous deformation. AB and CD are equal and parallel in the undeformed state but this is not true for A'B' and C'D' in the deformed state. (e) Displacement vectors for a homogeneous deformation. AB and CD are equal and parallel in the undeformed state and this is also true for A'B' and C'D' in the deformed state.

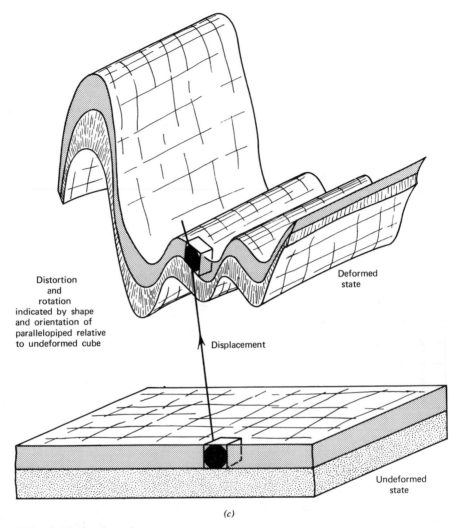

Distortion
and
rotation
indicated by shape
and orientation of
parallelopiped relative
to undeformed cube

Deformed
state

Displacement

Undeformed
state

(c)

FIGURE 1.11 (Continued)

particular progressive deformation a fold grows from initially planar layers.

The motion initiated in a body due to stress continues until the material reaches a configuration that is an equilibrium (although not necessarily unstressed) state. During this motion the path traced out by a material particle may be very complicated, but it is always possible, in principle, to construct a vector that describes the displacement of the material particle from the undeformed to the deformed state (Fig. 1.11*a*). We are concerned in this section with this displacement vector, not the actual path followed by the material

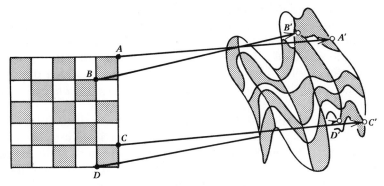

Inhomogeneous deformation

(d)

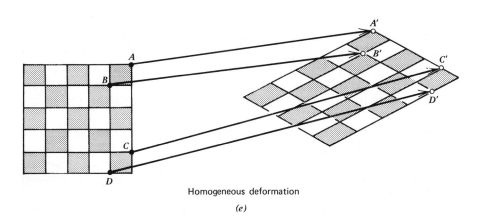

Homogeneous deformation

(e)

FIGURE 1.11 (Continued)

particle. The actual path is of concern when we discuss the *deformation history* of the body.

Displacement Fields. Deformation. Displacement Gradients.   As indicated in Figure 1.11*b,* each material point in the undeformed body may be connected to the same material point in the deformed body by an array of displacement vectors. This array of displacement vectors defines a *displacement field.* Such a displacement field may be described by a series of mathematical equations or *transformations,* examples of which are found in Jaeger (1969, pp. 24–25), Ramsay (1967, pp. 56–57), Ramsay and Graham (1970), and Hobbs (1971). This displacement field is commonly called the *deformation.* As such, the study of deformation is part of the study of geometry; the deformation is defined solely by comparing the deformed and undeformed states and is independent of the movement history of material particles.

This definition of deformation as a displacement field is the strict meaning of the term. This meaning is quite different from, and should not be confused with, the more general usage of the term deformation. In the latter usage, which is commonly convenient in geology and is also used in this book, deformation refers — in a broad way — to the process by which the displacements are accomplished, for example, in the phrase *recrystallization during deformation*.

It has already been indicated that a deformation may consist of an array of local translations, distortions, and rotations. Figure 1.11c illustrates this again where the translation component at one point is indicated by the arrow, and the distortion and rotation are indicated by changes in the shape and orientation of objects of known initial shape and orientations, such as the box and circle in Figure 1.11c. The ways in which these distortions and rotations are specified are indicated in Section 1.3.1.

The manner in which the displacement vectors vary from one point to another in a body is called the *displacement gradient*. If the displacement gradient is a constant throughout the deformed body then the deformation is *homogeneous**. Otherwise it is *inhomogeneous**. An example of an inhomogeneous deformation is illustrated in Figure 1.11d. The pairs of displacement vectors **AA'**, **CC'** and **BB'**, **DD'** relate similar points in the undeformed state to points in the deformed state. There is no consistent relationship between these vectors or other similar displacement vectors. The displacements are such that lines A'B' and C'D' in the deformed state have different lengths and orientations even though the corresponding lines AB and CD in the undeformed state had the same lengths and orientations. This means that the displacement gradients are not constant and the deformation is inhomogeneous. On the other hand, in Figure 1.11e the displacement vectors are such that lines A'B' and C'D' have the same lengths and orientation in the deformed state, just as AB and CD had in the undeformed state. In the deformation described by Figure 1.11e the displacement gradient is constant and the deformation is homogeneous. In a homogeneous deformation the translational component may vary from point to point, but the distortional and rotational components are the same from point to point. *For a homogeneous deformation, initially straight lines remain straight after deformation.*

---

*This usage corresponds to that of Sander (1930) who used the terms *affine Umformung* and *nicht affine Umformung* to mean *affine transformation* and *nonaffine transformation*, respectively. The phrases were translated by Hills (1953, p. 24) as *homogeneous strain* and *nonhomogeneous strain*, respectively, whereas a better translation of *Umformung* would have been *deformation* rather than *strain* [see translation of Sander (1948) by Phillips and Windsor (1970)]. Unfortunately, usage of such terms is very loose. The important point is that a distinction should be made between the *mathematical transformation*, which describes the overall *deformation*, and the *distortional* part of this, which is commonly called the *strain*.

1.3.1 **Strain. The Strain Ellipsoid**

When a body is distorted there are changes in the relative configuration of particles. To describe these changes we focus attention on a point in the undeformed body and imagine a small sphere centered there. In the deformed body this sphere becomes an ellipsoid. The *strain* is defined by comparing the shape and size of the ellipsoid with the shape and size of the initial sphere. The ellipsoid is called the *strain ellipsoid*.

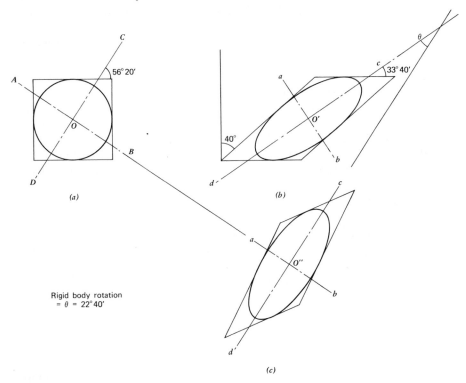

FIGURE 1.12 *(a)* An initial square with the circle center O inscribed within it. *(b)* The square in *(a)* has been distorted by simple shear through 40° to produce a parallelogram. The circle has been distorted to form an ellipse with center O'. The principal axes of this ellipse are ab and cd and are the principal axes of strain in the deformed state. In Figure *(a)*, these were the lines AB and CD that were initially at right angles. They represent the principal axes of strain in the undeformed state. *(c)* The square of Figure *(a)* has been distorted to form a parallelogram and the circle has been distorted to form an ellipse with principal axes ab and cd and center O''. In this case the principal axes of strain in the deformed state are parallel to those in the undeformed state, and the strain is a pure shear. Notice that the figure in *(c)* is identical to the figure in *(b)* except for a translation through the distance $O'\,O''$ and a rotation through the angle θ. [See Jaeger (1969, p. 33).]

The preceding paragraph applies strictly to homogeneous deformations. In such cases the strain ellipsoid has the same shape and orientation in all parts of the deformed body, and spheres of any size become perfect ellipsoids. Where deformation is inhomogeneous, spheres become ellipsoids to a close approximation if they are very small, and they become perfect ellipsoids if they are infinitesimally small. For every *point* in any deformed body therefore, there is an ellipsoid that represents the strain at that point. The ellipsoids are actually drawn with finite dimensions, but the strain they represent applies only to a single point if the deformation is inhomogeneous. Figure 4.29*a* exemplifies this.

The strain ellipsoid is a concept applicable to any deformation no matter how large in magnitude, in any class of material. The strain ellipsoid is not a concept restricted to strains of small magnitude in elastic bodies as is suggested by some geological literature.

Figure 1.12 illustrates, in two dimensions, a number of important concepts associated with the strain ellipsoid. In Figure 1.12 a square is deformed to produce a parallelogram (Fig. 1.12*b*). If a circle is inscribed within the undeformed square it becomes an ellipse in the deformed state with principal axes *ab* and *cd*. In the undeformed state those material lines are *AB* and *CD*, respectively; they also are perpendicular and, during the deformation, have been rotated through the angle θ. This is the two-dimensional equivalent of the case discussed above for three dimensions, in which a sphere becomes an ellipsoid. In the three-dimensional case, three mutually perpendicular lines may always be found in the undeformed state that remain mutually perpendicular in the strained state. These lines in the deformed state are parallel to the principal axes of the strain ellipsoid and are known as the *principal axes of strain* in the deformed state. The principal planes of the strain ellipsoid are the *principal planes of strain*. In Figure 1.12, *AB* and *CD* are the principal axes of strain in the undeformed state whereas *ab* and *cd* are these axes in the deformed state.

In this book we designate the principal axes of the strain ellipsoid λ_1, λ_2, and λ_3, corresponding to the maximum, intermediate, and minimum principal axes, respectively. λ_1, λ_2, λ_3 are the principal quadratic elongations and are defined below under "measures of strain." Note, however, that although the principal axes of the strain ellipsoid are parallel to λ_1, λ_2, and λ_3 they have magnitudes $\sqrt{\lambda_1}$, $\sqrt{\lambda_2}$, and $\sqrt{\lambda_3}$ (see Equation 1.14).

Strains Associated with Homogeneous Deformations. Homogeneous deformations at constant volume may involve several different types of strain illustrated in Figure 1.13.

1. *Axially symmetric extension* (Fig. 1.13*a*). This involves extension in one principal direction and equal shortening in all directions at right angles. The symmetry of strain is axial. The strain ellipsoid is a prolate spheroid.

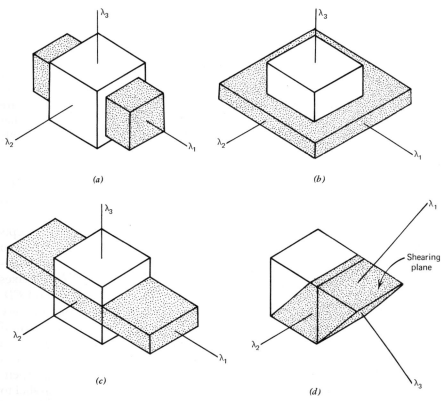

FIGURE 1.13   Simple types of homogeneous deformation. In each case the cube
represents the undeformed state and the stippled figure represents the deformed
state. (a) Simple or uniform extension: $\lambda_1 > \lambda_2 = \lambda_3$. (b) Simple or uniform
flattening: $\lambda_1 = \lambda_2 > \lambda_3$. (c) A plane strain: $\lambda_1 > \lambda_2 = 1 > \lambda_3$. (d) Simple shear
through an angle of 40°: $\lambda_1 > \lambda_2 = 1 > \lambda_3$. (e) Strain ellipsoid illustrating the
circular sections. (f) A general strain that is also a pure shear: $\lambda_1 > \lambda_2 > \lambda_3$.

(Notice that the words *extension* and *shortening* are associated with *strain* in
this book, whereas the words *tension* and *compression* are associated with
stress.)

2. *Axially symmetric shortening* (Fig. 1.13b). This involves shortening in one
principal direction and equal extension in all directions at right angles.
The strain ellipsoid is an oblate spheroid.

3. *Plane strain* (Fig. 1.13c). In plane strain the intermediate principal axis of
the ellipsoid has the same length as the diameter of the initial sphere.
Shortening and extension, respectively, occur parallel to the other two
principal directions. The strain ellipsoid is a *triaxial* ellipsoid, in the sense
that all three principal axes of the ellipsoid have different lengths.

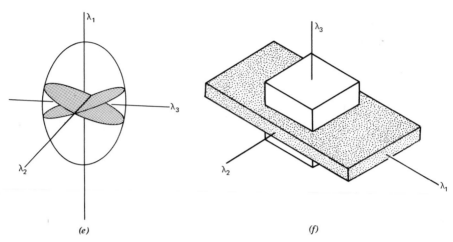

(e)   (f)

FIGURE 1.13 (Continued)

4. *General strain.* This involves extension or shortening in each of the principal directions of strain, and the strain ellipsoid is triaxial (Fig. 1.13*f*).

In describing these four classes of strain, we have looked at the different *shapes* of strain ellipsoids that are possible. We have not specified any particular *orientations* relative to the undeformed state. For example, we have not specified whether a line of particles lying along the λ_1 axis in the deformed state is parallel or not to the same line of particles in the undeformed state. For convenience Figures 1.13*a*, *b*, and *c* have been drawn as if no rotation of material lines along principal axes has occurred, but this is not necessarily a part of the definition of an axially symmetric extension, for example.

Pure Shear and Simple Shear. These are two types of homogeneous deformation for which it is usual to specify both the type of strain and the relative orientations of certain material lines in the deformed and undeformed states.

Pure shear is a homogeneous deformation involving either a plane strain or a general strain, in which lines of particles that are parallel to the principal axes of the strain ellipsoid have the same orientation before and after deformation. Examples are shown in Figure 1.12*a* to *c* and Figure 1.13*a*, *b*, *c*, and *f*. Because there is no rotation of material lines along principal directions, a pure shear is referred to as an *irrotational deformation* or, loosely, as an *irrotational strain* (see also Section 1.3.3).

It is important to emphasize that since any deformation is defined solely by the array of displacement vectors that join the same material points in the undeformed and deformed states, and is independent of the actual path traveled by these particles, any deformation is defined only by comparing the final and initial configurations. Thus a pure shear does not imply coincidence of principal axes of strain at all times in the history of straining. Many geologists use the term pure shear to imply just such a deformation history in which the principal axes of strain remain parallel and attached to the same lines of material particles from one increment of strain to the next. We prefer to use the phrase *progressive pure shear* to describe such a *deformation history* and to restrict the phrase *pure shear* to the specific relationship between the deformed and undeformed states described above and illustrated by Figures 1.12*a* and *c*. Some writers [e.g., Nye (1964, p. 103)] restrict pure shear to being a plane strain. There appears to be no advantage in this restricted usage so the term is also used here for deformations in which $\lambda_2 \neq 1$.

Simple shear is a constant volume, homogeneous deformation involving plane strain, in which a single family of parallel material planes is undistorted in the deformed state and parallel to the same family of planes in the undeformed state. Examples are shown in Figures 1.12*a* and *b* and 1.13*d*. Simple shear involves a change in orientation of material lines along two of the principal axes (λ_1 and λ_3). Simple shear is, therefore, referred to as a *rotational deformation* or, loosely, as a *rotational strain*.

Because of the geometrical relationship of the deformed and undeformed states (see Fig. 1.12*a* and *b*) in a simple shear deformation there is a natural tendency to draw an analogy between this kind of deformation and that which results from homogeneous shearing of a pack of thin cards, the undistorted planes being equivalent to the cards. This mechanism of deformation carries with it the implication of a specific history of straining involving a constant orientation of the undistorted planes and progressive rotation of the principal axes of strain. In fact, for a large number of geologists, the term *simple shear* implies just such a deformation history in which the principal axes of strain, λ_1 and λ_3, rotate during the history of deformation and are attached to different lines of material particles from one increment of deformation to the next. We prefer to use the phrase *progressive simple shear* to describe such a *deformation history* and to restrict the phrase *simple shear* to describe the specific relationship between the deformed and undeformed states defined above.

The definitions given above for pure shear and simple shear, which involve orientations of material lines, are always somewhat ambiguous in application, because whether or not an orientation change occurs for a given line of particles depends on the reference frame selected. This is illustrated in Figure 1.14*a*. Region A has suffered a pure shear if the reference frame (dotted) is fixed in orientation relative to bedding; the principal directions of strain are parallel and perpendicular to bedding before and after deformation. Region A has not suffered a pure shear, however, if the reference frame is fixed in orientation relative to the plane of the paper. Similarly, region B has suffered a simple shear if the reference frame is fixed in orientation relative to bedding, but has not suffered a simple shear if the reference frame is fixed to the paper. In any given situation the reference frame should be selected for convenience and the terms pure shear and simple shear applied accordingly.

In geology it is often impossible to know the original orientation of a material line, and it is therefore often desirable to fix the reference frame in some manner to an internal marker like bedding in Figure 1.14a. When this is done,

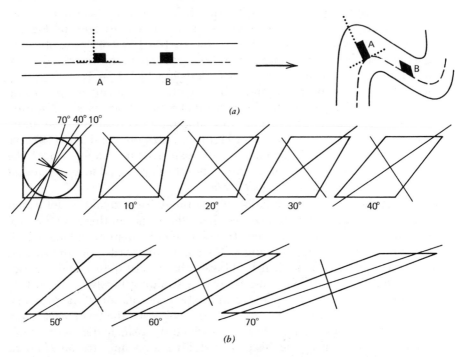

(a)

(b)

FIGURE 1.14   (a) Pure shear of region A and simple shear of region B referred to reference axes attached to the folding layer. (b) Progressive simple shear. The square is progressively sheared into the various parallelograms keeping the base of constant orientation and length. The orientations of the principal axes of strain in the deformed states are as shown and these correspond to progressively *different* principal axes of strain in the undeformed state. These axes for the undeformed state for angles of shear of 10, 40, and 70° are shown. (c) Progressive pure shears. Each of the parallelograms (A), (B), and (C) may be obtained from the initial square by progressive uniform shortening normal to the directions AA', BB', CC'. These directions correspond to maximum principal axes of strain in the undeformed state and remain constant in orientation throughout the progressive deformation to ultimately produce the axes in the deformed state aa', bb', and cc', respectively. An important point to note is that although various lines of material particles have suffered different deformation histories from Figure (b) to (c), (A), (B), and (C) in Figure (c) are identical in shape, respectively, to the 10°, 40°, and 70° figures in Figure (b). They differ only by a rigid body translation and a rigid body rotation. Thus, although a progressive simple shear may be distinguished from a progressive pure shear by examining the strain history of particles, a simple shear cannot be distinguished from a pure shear except for *orientation* and *position*.

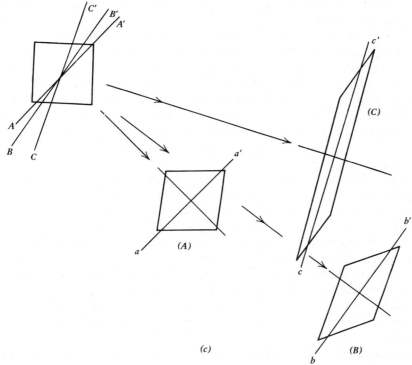

FIGURE 1.14 (Continued)

deformations such as those suffered by regions A and B are correctly termed pure shear and simple shear respectively, even though large rotations of all material lines may have occurred relative to the geographic reference frame.

GENERAL THEOREM ASSOCIATED WITH HOMOGENEOUS DEFORMATION

An important aspect of a general homogeneous deformation has been illustrated for the two dimensional case in Figure 1.12. In Figure 1.12*b* the square and circle of Figure 1.12*a* are related by a simple shear, whereas in Figure 1.12*c* the deformation is a pure shear so that the principal axes of strain are parallel before and after deformation. It is important to note that Figures 1.12*b* and *c* are identical in all respects and the two figures may be brought into coincidence by a rigid body rotation through the angle θ together with a rigid body translation through the distance O'O''. Figure 1.12 illustrates a general theorem that states that *any homogeneous constant volume deformation can be expressed as a pure shear together with a rigid body rotation and a rigid body* *translation.* The rigid body rotation is equal to the rotation undergone by the principal axes of strain.

This theorem is illustrated again in Figure 1.14. In Figure 1.14*b* a series of parallelograms all related to an initial square by a simple shear is shown. The principal axes of strain in the deformed state are shown for each parallelogram and those for the undeformed state are shown for simple shears through 10°, 40°, and 70°. If the sequence of shapes in Figure 1.14*b* is supposed to originate by a progressive simple shear then the noncoincidence of principal axes of strain in the undeformed state indicates that the principal axes of strain in each successive deformed state are attached to progressively different lines of material particles.

On the other hand, in Figure 1.14*c* the same initial square as in Figure 1.14*b* is shown but a series of three different progressive pure shears leads to the parallelograms (*A*), (*B*) and (*C*). This means that the principal axes of strain AA', BB', CC' in the undeformed state are parallel, respectively, to the principal axes of strain aa', bb', cc' in the deformed state and throughout each progressive deformation. In each progressive deformation the principal axes of strain are attached to the same lines of material particles throughout. However — and this is the important point — each of the parallelograms (*A*), (*B*), and (*C*) are identical to those that result from 10°, 40°, and 70° simple shears, respectively, in Figure 1.14*b* except for a rigid body rotation and a rigid body translation.

Any simple shear through an angle ψ can be represented by a pure shear together with a rigid body rotation through $\tan^{-1}$ (½tanψ). Notice that if ψ is small, this rotation is approximately equal to $\psi/2$, a result commonly quoted for small strains. These results are discussed in more detail in Jaeger (1969, pp. 32–33).

Thus, although the *strain histories* are different for a progressive simple shear and for a progressive pure shear, the *strain* that results from a simple shear can always be produced by a suitably oriented pure shear. *The only differences between such a pure shear and a simple shear are a rigid body rotation and a rigid body translation.*

This discussion also indicates why it is important to distinguish between the concepts of simple shear, progressive simple shear, pure shear, and progressive pure shear. Progressive simple shear and progressive pure shear are terms implying a specific history of *straining,* and the two produce entirely different end products as indicated by features such as crystallographic-preferred orientations (see Section 2.3.1) that are capabie of recording strain history. However, the *strains* produced by these different histories are identical in all respects. The terms simple shear and pure shear are simply terms that distinguish between different *orientations* of the strain ellipsoid.

Measures of Strain. In order to discuss the strain at a point, two different kinds of quantities are required. One kind measures *the relative changes of lengths of lines* whereas the other measures the *changes in the angles between pairs of lines*.

The *elongation* is a commonly used measure of the relative change of length. If l_o is the initial length of a line and l the length after straining, then the elongation ϵ is

$$\epsilon = \frac{l - l_o}{l_o} \tag{1.12}$$

Shortening is represented by a negative elongation. Another commonly used measure is the *quadratic elongation*, λ, given by

$$\lambda = \left(\frac{l}{l_o}\right)^2 = (1 + \epsilon)^2 \tag{1.13}$$

This quantity is frequently used since it is the most convenient measure of strain in the mathematical theory of large strains. Thus, for instance, the equation of the strain ellipsoid relative to Cartesian coordinates (x_1, x_2, x_3) drawn in the deformed state is

$$\frac{x_1^2}{\lambda_1} + \frac{x_2^2}{\lambda_2} + \frac{x_3^2}{\lambda_3} = 1 \tag{1.14}$$

where the undeformed sphere had unit radius and $\lambda_1, \lambda_2, \lambda_3$ are the quadratic elongations in the directions of the three principal axes of strain. The square root of the quadratic elongation, l/l_o, is the *stretch*.

The *shear strain* is commonly used as a measure of the change of angle between lines. If two initially perpendicular lines Ox_1 and Ox_2 are deformed so that the angle between them becomes $(\pi/2) - \psi$, then the shear strain is

$$\epsilon_{12} = \tan\psi \tag{1.15}$$

the symbol ϵ_{12} denoting the shear strain of Ox_1 with respect to Ox_2. Clearly $\epsilon_{12} = \epsilon_{21}$ since the change in angle between Ox_1 and Ox_2 is precisely that between Ox_2 and Ox_1. In some textbooks the symbol γ_{ij} is used instead of ϵ_{ij} for the cases where $i \neq j$. Notice that all of these quantities used to measure strain are physically dimensionless.

In three dimensions, nine quantities are required to specify a homogeneous strain. Three of these quantities measure the relative changes in lengths of three initially orthogonal lines whereas the other six are dependent upon the changes in the angles between these lines. If the three initially orthogonal lines are Ox_1, Ox_2 and Ox_3 (Fig. 1.15) then these nine components may be written in a systematic array:

$$\begin{bmatrix} \epsilon_{11} & \epsilon_{12} & \epsilon_{13} \\ \epsilon_{21} & \epsilon_{22} & \epsilon_{23} \\ \epsilon_{31} & \epsilon_{32} & \epsilon_{33} \end{bmatrix} \tag{1.16}$$

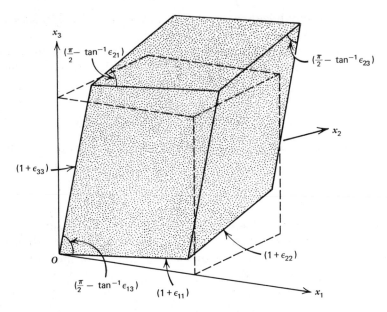

FIGURE 1.15   The components of infinitesimal strain. The unit cube with edges parallel to Ox_1, Ox_2, Ox_3 is distorted to form a parallelepiped. The edges of the cube are changed to $(1 + \epsilon_{11})$, $(1 + \epsilon_{22})$ and $(1 + \epsilon_{33})$. The original right angles are changed to

$$\left(\frac{\pi}{2} - \tan^{-1}\epsilon_{12}\right), \left(\frac{\pi}{2} - \tan^{-1}\epsilon_{23}\right), \text{ and } \left(\frac{\pi}{2} - \tan^{-1}\epsilon_{31}\right).$$

The observation that $\epsilon_{12} = \epsilon_{21}$, $\epsilon_{23} = \epsilon_{32}$, and $\epsilon_{31} = \epsilon_{13}$ means that there are just six independent components of strain in the array (1.16). In this array, quantities such as ϵ_{ii}, where the two subscripts are the same, represent the elongation in the direction of Ox_i ($i = 1, 2, 3$). Quantities such as ϵ_{ij}, where the two subscripts are different, represent the shear strain of Ox_i with respect to Ox_j ($i, j = 1, 2, 3$). These six quantities are known as the *components of strain*.

Another measure of strain used in some situations (see, for instance, Section 1.3.3) is the *natural strain* defined by

$$\bar{\epsilon} = \log_e (1 + \epsilon) \tag{1.17}$$

where e is the base of the natural logarithms and ϵ is the elongation defined by (1.12).

To give some feeling for the meaning of $\bar{\epsilon}$, the values of the elongation are given for various values of the natural strain in Table 1.2. A natural shortening of 100 percent, for instance, corresponds to a shortening of 63.2 percent.

The natural strain is useful in examining the history of a deforming body [see Jaeger (1969, pp. 68–69); Ramsay and Wood (1973)].

TABLE 1.2

Natural Strain, $\bar{\epsilon}$ Shortening (percent)	Strain, ϵ Shortening (percent)
10	9.5
20	18.1
30	25.9
40	33.0
50	39.3
60	45.1
70	50.3
80	55.1
90	59.3
100	63.2
200	86.5
300	95.0
400	98.0
500	99.3

The six independent components of strain quoted in (1.16) conveniently define a homogeneous strain only if the strain is small (this leads to the so-called theory of *infinitesimal* strain). If, however, the strains are large as is the situation in most deformations of interest in structural geology, the simple measures quoted in (1.16) are not convenient in any mathematical discussion of a homogeneous strain unless the lines Ox_1, Ox_2 and Ox_3 are parallel to principal strains. The quantities such as ϵ_{ij} $(i = j)$ need to be replaced by the quadratic elongations λ_{ij} $(i = j)$ and the shear strains by more complicated measures of strain. The subject rapidly becomes very complicated and the interested reader is referred to Brace (1961) and Jaeger (1969). The mathematical theory of large strains is called *the theory of finite strain*.

Graphical Representations of Homogeneous Strains. All types of homogeneous strain may be represented graphically on a diagram first presented by Zingg (1935) and used in structural geology by Flinn (1962). In Figure 1.16a the ratio a of major and intermediate axes of the strain ellipsoid is plotted against the ratio b of the intermediate and minor axes. The original of the graph is at $(1,1)$. This figure is generally known in the geological literature as a *Flinn diagram*. If the number k is defined by

$$k = \frac{a - 1}{b - 1} \tag{1.18}$$

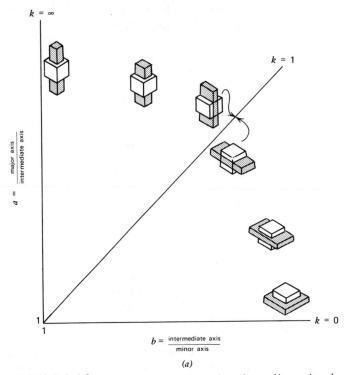

$k = \infty$

$k = 1$

$a = \dfrac{\text{major axis}}{\text{intermediate axis}}$

$k = 0$

$b = \dfrac{\text{intermediate axis}}{\text{minor axis}}$

(a)

FIGURE 1.16   Methods of representing three-dimensional strain states in two dimensions. *(a)* A Flinn diagram in which *a* is plotted against *b*. The small inset diagrams illustrate a type of strain at each point in this diagram. *(b)* A Hsu diagram. *(c)* Another form of the Flinn diagram in which the natural logarithm of *a* is plotted against the natural logirithm of *b*. *(d)* A Flinn diagram illustrating where various states of plane strain plot for deformations involving a volume change.

then Figure 1.16*a* is divided into a number of fields for different values of *k*. For *k* = 0 all strain ellipsoids are uniaxial oblate ellipsoids or "pancake" in shape. For 0 > *k* > 1, the strain ellipsoid is oblate and the deformation is of a flattening type. For *k* = 1 the deformation is a plane strain if the deformation is constant volume; all deformations by simple shear lie along this line. For 1 > *k* > ∞, the strain ellipsoid is prolate and the deformation is of a constriction type. For *k* = ∞ the strain ellipsoids are uniaxial prolate ellipsoids or "cigarlike" in shape.

Another type of graphical illustration of strain, employing natural strain, was suggested by Hsu (1966) and employed by Hossack (1968) in a study of

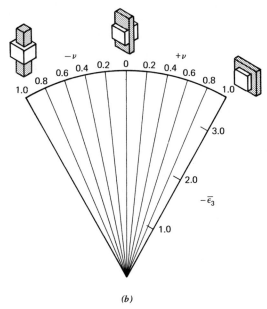

(b)

FIGURE 1.16 (Continued)

deformed conglomerates. This type of graphical illustration may prove to be more useful in geological problems than the Flinn diagram [see also Elliott (1972)].

In the Hsu diagram a quantity ν is defined by

$$\nu = \frac{2\bar{\epsilon}_2 - \bar{\epsilon}_1 - \bar{\epsilon}_3}{\bar{\epsilon}_1 - \bar{\epsilon}_3} \qquad (1.19)$$

where $\bar{\epsilon}_1, \bar{\epsilon}_2, \bar{\epsilon}_3$ are the principal natural strains defined by (1.17). The number ν compares with the k value of Flinn (1.18) as follows:

For uniform extension	$k = \infty$	$\nu = -1.0$
For plane strain at constant volume	$k = +1$	$\nu = 0.0$
For uniform shortening	$k = 0.0$	$\nu = +1.0$

Hsu (1966) has shown that all strain states may be represented on a 60° segment of a circle in which $-\bar{\epsilon}_3$ is measured along a radius and ν is measured around the circumference as in Figure 1.16*b*. On this diagram all uniform extensions plot along the radius $\nu = -1.0$, all plane strains plot along the radius $\nu = 0$, and all uniform shortenings plot along the radius $\nu = +1.0$.

Ramsay (1967, p. 138) has suggested a modified form of Flinn diagram where $(\log_e a)$ is plotted against $(\log_e b)$, a and b having the same significance as in the Flinn diagram (Fig. 1.16*a*). This is equivalent to plotting $(\bar{\epsilon}_2 - \bar{\epsilon}_3)$

against $(\bar{\epsilon}_1 - \bar{\epsilon}_2)$. All states of strain are again represented by points on this diagram (Fig. 1.16c); the line $k = 1$ where $k = (\bar{\epsilon}_1 - \bar{\epsilon}_2)/(\bar{\epsilon}_2 - \bar{\epsilon}_3)$ again represents all states of plane strain. The importance of this modified form of Flinn diagram lies in the convenience of representation of volume changes and of the history of incremental deformations in a deformed rock (see Section 1.3.3).

The diagrams used for graphical representations of strain (Figs. 1.16a, b, c) have been described so far for deformations involving no volume change. As such, the line $k = 1$ (Figs. 1.16c) represents all states of plane strain, that is, where one principal elongation is zero. The fractional change in volume, Δ, accompanying any homogeneous strain is given by

$$\Delta = \frac{V - V_o}{V_o} \tag{1.20}$$

where V and V_o are the volumes in the deformed and undeformed states, respectively. The fractional volume change, Δ, is commonly known as the *dilation*.

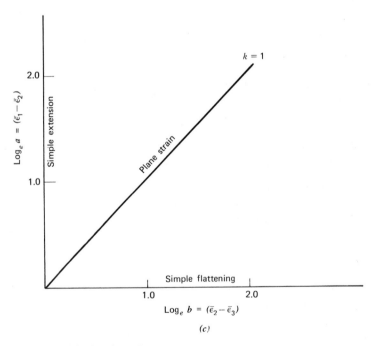

FIGURE 1.16 (Continued)

Now, in terms of the natural strains,

$$\log_e (1 + \Delta) = \bar{\epsilon}_1 + \bar{\epsilon}_2 + \bar{\epsilon}_3 \tag{1.21}$$

[see Jaeger (1969, p. 68)]

Equation 1.21 may be rewritten:

$$(\bar{\epsilon}_1 - \bar{\epsilon}_2) = (\bar{\epsilon}_2 - \bar{\epsilon}_3) - 3\bar{\epsilon}_2 + \log_e (1 + \Delta) \tag{1.22}$$

which forms the basis for representing volume changes on the modified Flinn diagram (Fig. 1.16c). For plane strains $\bar{\epsilon}_2 = 0$, and so the lines that represent all states of plane strain for different dilations are represented as in Figure 1.16d. The situation has been considered in some detail by Ramsay and Wood (1973) and is examined again in Section 1.3.3.

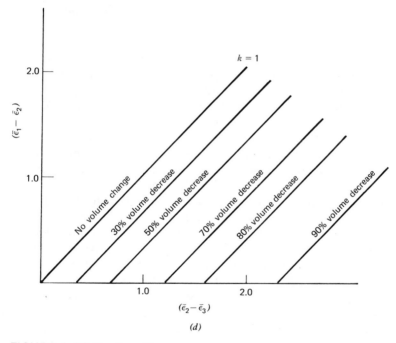

FIGURE 1.16 (Continued)

1.3.2 Inhomogeneous Deformations

Most geological deformations are inhomogeneous, that is, lines that were originally straight become curved; Figure 1.11d is an example of an inhomogeneous deformation. Provided the deformation is smooth enough, it is possible to talk about the strain at each point and to represent the variation of strain throughout an inhomogeneously deformed body by an array of strain ellipsoids [Jaeger (1969, pp. 245–251), Ramsay and Graham (1970), and Hobbs (1971)].

In some inhomogeneous deformations it may be convenient to smooth out the inhomogeneities and consider the deformation as approximating to a homogeneous one. This, for instance, might be the situation in a large body of rock folded on a very small scale; here the folded layers might be treated as approximating planar layers. In these situations the body is considered as being homogeneously deformed and a strain ellipsoid could be drawn for the deformed body. Such a strain ellipsoid is commonly referred to as the *mean strain ellipsoid.*

In some rocks, such as those with well-developed slaty cleavage or layering due to metamorphic differentiation, there may have been local changes in volume associated with the concentration of minerals in planar zones (see Sections 5.3 and 5.4.1). In such instances the dilation is inhomogeneous and this represents a particular type of inhomogeneous deformation, perhaps common in rocks, but that so far has received little attention in the geological literature. Figure 1.17 illustrates a particular example of an inhomogeneous deformation in which inhomogeneous dilation plays an important role. The undeformed state is illustrated in Figure 1.17*a* and consists of a pelite with a well-developed slaty cleavage, *S*. In Figure 1.17*b* a crenulation cleavage has been developed so that the specimen is subdivided into planar zones defined by two different mean orientations of *S*; these two types of zones are labeled A and B. The deformation depicted in Figure 1.17*b* involves a 20 percent shortening normal to the crenulation cleavage planes together with a shear strain parallel to the crenulation cleavage. Up until this stage there has been no change in volume, the total deformation is accomplished by straining and rotating individual grains. In Figure 1.17*c* metamorphic differentiation has taken place with B layers becoming rich in mica relative to A layers. This differentiation has been accomplished by preferentially removing quartz from B layers leaving A layers unchanged so that the specimen undergoes an inhomogeneous volume change and is shortened even further normal to the crenulation cleavage planes. The total shortening normal to the crenulation cleavage in Figure 1.17*c* is 30 percent and the volume change is 12.5 percent. In Figure 1.17*d* the deformation has continued in the same manner, the total shortening normal to the crenulation cleavage being 40 percent and the volume change 25 percent. The quartz removed from the B layers accumulates nearby as quartz veins. In Figure 1.17*e* it is shown comprising a vein normal to the crenulation axes where it contributes toward an overall extension of the rock mass parallel to these axes. This example, although hypothetical, illustrates the important role that inhomogeneous dilation may play in the deformation of some rocks (see also Section 5.4.4).

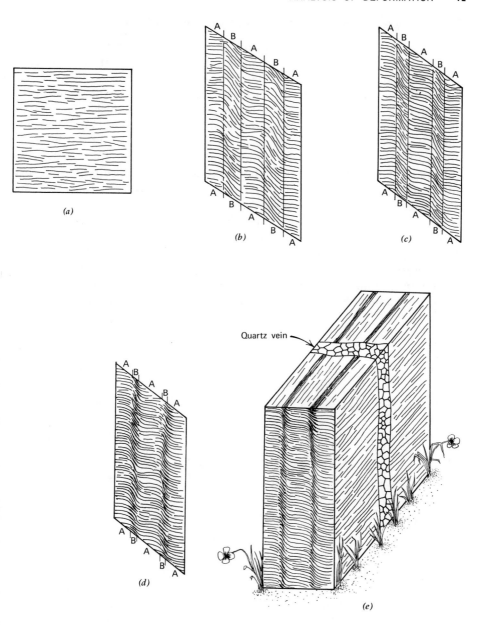

FIGURE 1.17   An inhomogeneous deformation involving inhomogeneous volume change. See text for discussion.

1.3.3 Progressive Strains. Strain Paths. Deformation Paths

During the deformation of a piece of rock, a particular material line (i.e., a line through the same material points) may experience different kinds of strain at different stages in the deformation history. This is illustrated in Figure 1.18a, where initially planar layers are shown at various stages in the development of a tight fold. The layers are first shown thickening as they undergo layer parallel shortening, but a fold finally nucleates [Fig. 1.18a (ii)] after 31 percent overall shortening of the specimen. This fold slowly grows until, at 60 percent, overall shortening of the specimen the fold has become quite tight [Fig. 1.18a (v)]. Throughout this deformation history there is very little extension parallel to the fold axis (normal to the page), so that the strain is approximately plane. Two small regions, A and B, have been delineated in Figure 1.18a and their histories may be followed as the deformation continues. One region, A, is destined to occupy the hinge region of the fold, and it starts by undergoing considerable shortening parallel to the layering (A2) even before any appreciable folding has developed. Later in the folding (A4) this region actually is extended parallel to the bedding until it is almost the initial unstrained shape. It is then shortened parallel to the bedding (A5)

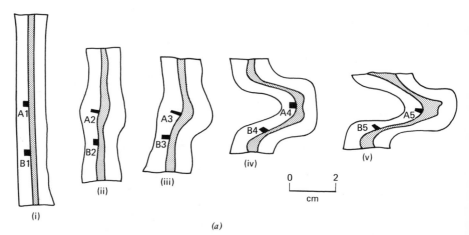

FIGURE 1.18 (a) A sequence of experiments in which a multilayered specimen is progressively shortened. The layers consist of mixtures of NaCl and mica (blank layers in diagrams) or of KCl and mica (shaded layers). Regions A and B delineated by markers so that their deformation may be followed. (i) No shortening, (ii) 31 percent, (iii) 40 percent, (iv) 51 percent, and (v) 60 percent overall shortening of specimen. (b) Flinn diagrams showing strain paths for regions A and B of Figure 1.18a. (c) Modified Flinn diagram showing deformation paths for regions A and B of Figure 1.18a. The angle between the layering and the major principal axis of strain at any point at each state in the deformation is θ.

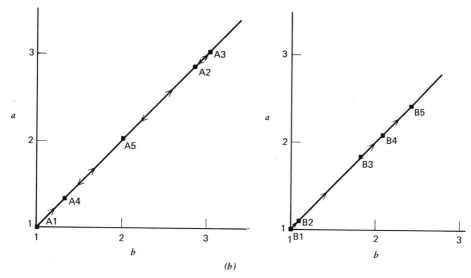

FIGURE 1.18 (Continued)

during the next increments of strain. The other region (B), destined to occupy the limb region of the fold, remains almost unstrained for a while (B2) but is finally shortened parallel to the bedding (B3) as folding begins. The region is then rotated as it undergoes considerable shear strain parallel to the layering (B4 and B5). Since the strain is close to being a plane strain throughout the folding shown in Figure 18a, each state of strain is represented by a point on the line $k = 1$ on a conventional Flinn diagram (see Fig. 1.16). This is shown in Figure 18b. The continuous lines drawn through the points for A and B are called *strain paths*. The contrast in strain history for these two regions A and B can be better illustrated by the modified form of Flinn diagram shown in Figure 18c in which an extra coordinate axis has been added normal to the a and b coordinate axes of Figure 1.16a. Along this third axis, the angle θ between the layering and the major principal axis of strain is measured. This diagram now not only portrays states of strain but also shows the relationships between a particular line of material particles (parallel to the layering) and the principal axes of strain. The continuous lines drawn through the points for A and B in Figure 18c could be conveniently called *deformation paths* to distinguish them from the strain paths of the conventional Flinn diagram. The important point is that the deformation path gives information on the history of rotation of the principal axes of strain; the strain path does not [see Hsu (1966)].

Because of their contrasted deformation histories, the two regions A and B can be expected to exhibit different kinds of small-scale structures. The sequence of development of small-scale structures for region A is shown in

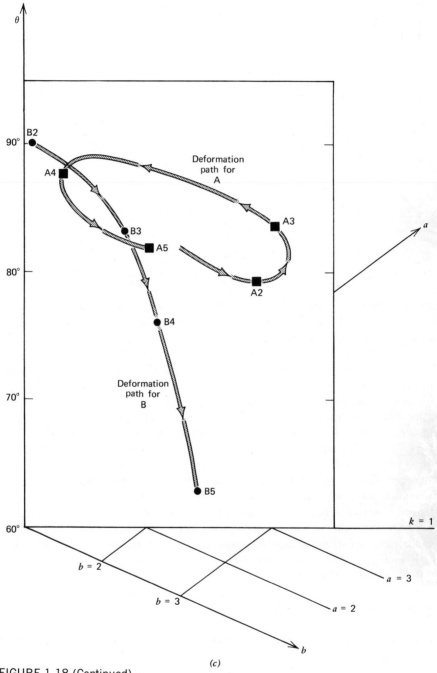

FIGURE 1.18 (Continued)

Figure 1.19. Region *A* 1 consists of a strongly foliated material and this acts as the initial undeformed state. In A2 (Fig. 1.19*a*) a number of symmetrical folds have formed in the foliation and these become slightly tighter in A3 (Fig. 1.19*b*). In A4 extension parallel to the layering has occurred and the old folds in the foliation have more or less vanished (Fig. 1.19*c*). In A5 shortening parallel to the layering has taken place again and open symmetrical folds develop in the foliation (Fig. 1.19*d*).

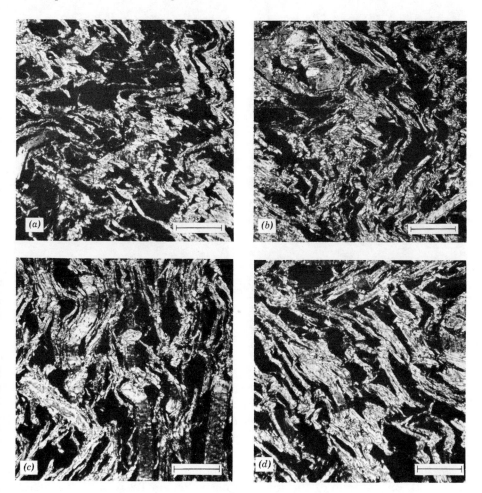

FIGURE 1.19   Photomicrographs (crossed polars) showing microstructures developed in region A of Figure 1.18*a*. (*a*) 31 percent, (*b*) 40 percent, (*c*) 51 percent, (*d*) 60 percent overall shortening of specimen. Axis of specimen shortening is parallel to long axis of page. Black areas are NaC1 aggregates. Light areas are mica. Scale marks 0.3mm.

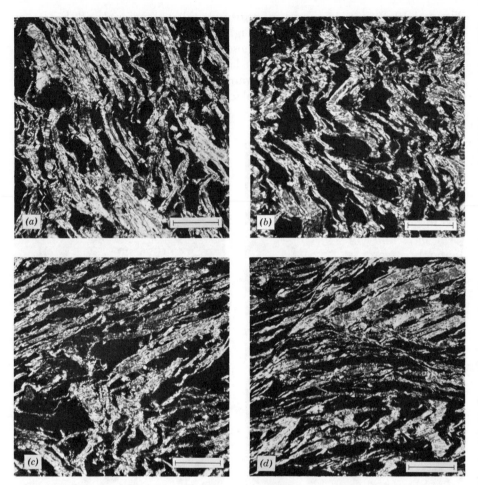

FIGURE 1.20 Photomicrographs (crossed polars) showing microstructures developed in region B of Figure 1.18a. (a) 31 percent, (b) 40 percent, (c) 51 percent, (d) 60 percent overall shortening of specimen. Axis of specimen shortening is parallel to long axis of page. Black areas are NaCl aggregates. Light areas are mica. Scale marks 0.3 mm.

In Figure 1.20 the sequence of development of small-scale structures in region B, on the limb of the fold, is shown. Region B2 is only slightly deformed and the initial foliation shows slight warping (Fig. 1.20a). In B3 some shortening parallel to the layering has occurred, and more or less symmetrical folds in the foliation have developed (Fig. 1.20b). In B4 these folds have been obliterated (Fig. 1.20c), to be replaced by a very strong foliation at a low angle to the layering. In B5 this foliation is itself deformed by a widely spaced crenulation cleavage oblique to both the layering and the axial plane of the fold (Fig. 1.20*d*). This sequence should be contrasted with that developed in region A in the hinge of the fold.

The detailed examination of small-scale structures within the context of the larger structural environment can, potentially, supply considerable information on the strain history of a particular deformed rock, and some examples of this kind of approach are given by Ramsay (1967, pp. 114–120) and by Paterson and Weiss (1968).

The general principles underlying the strain history of a particular line of material particles may be illustrated by noting that, at any stage in time during a deformation, the strain may be represented by two strain ellipsoids: one strain ellipsoid describes the strain up to that time, and another strain ellipsoid describes the small amount of strain that is about to be imposed in the next instant. The first of these is called the *finite strain ellipsoid* and the second is the *incremental strain ellipsoid*. In each of these ellipsoids it is possible to define three fields: a field in which lines have their initial length, one in which all lines have been shortened, and one in which all lines have been extended. This is illustrated in Figure 1.21. Since in most deformations of geological interest, the principal axes of strain rotate progressively with respect to material points, the principal axes of the finite and incremental strain ellipsoids at any instant will not coincide (Fig. 1.21) so that the superposition of these six fields defines four asymmetrically arranged fields labeled I, II, III, and IV in Figure 1.21.

In field I all lines have been extended and will continue to be extended in the next strain increment. In field II all lines have been extended but will be shortened. In field III all lines have been shortened but will be extended, and in field IV all lines have been shortened and will continue to be shortened in the next strain increment. It is the history of a particular layer with respect to the finite and incremental strain ellipsoids throughout the deformation that defines what kind of structures are expected to develop in that layer. For the most part, this history is likely to be very complicated so that a particular layer may pass from one field through all four fields or may progressively move backwards and forwards through two or three of them. However, a layer that has spent most of its strain history in field I is expected to display boudinage along one or more directions; one that has moved from IV into III may show boudinaged folds whereas one that has remained in field IV may show only folds. Some of these relationships are illustrated in Figure 1.21.

Extended discussions of progressive strain are given by Ramberg (1959), Flinn (1962), Ramsay (1967, p. 114), Jaeger (1969, pp. 235–245), and Elliott (1972).

If the principal axes of the incremental strain ellipsoids always have the same orientation as the principal axes of the finite strain ellipsoid then the strain history is said to be *coaxial*. Otherwise, if these two sets of axes rotate

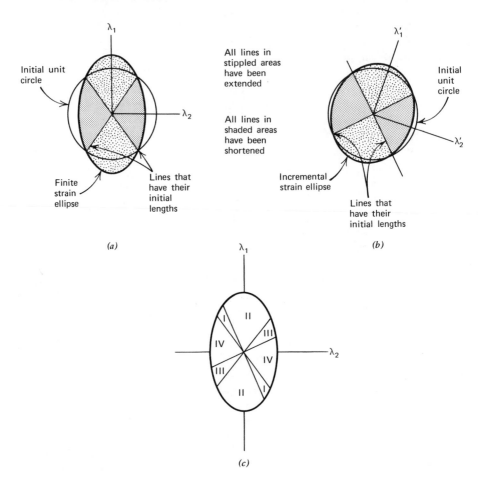

FIGURE 1.21   Incremental deformation involving constant volume. *(a)* The finite
strain ellipsoid with principal axes of strain λ_1 and λ_2. An initial unit circle intersects
this ellipse along two lines that have their initial lengths. The stippled area includes
all lines that have been extended and the shaded area includes all lines that have been
shortened. *(b)* The incremental strain ellipsoid; this is almost a circle and represents
the increment of strain that will be added to the finite strain ellipsoid in *(a)* during the
next increment of deformation. Notice that the principal axes of the incremental
strain ellipsoid λ_1' and λ_2' are not parallel to the finite axes of strain. Again the
stippled areas and shaded areas have the same significance as in *(a)*. *(c)* The four
different fields that are marked out by the intersection of the four fields in *(a)* and *(b)*.
In field I, all lines have been extended but will be shortened in the next increment. In
II, all lines have been extended and will continue to be extended in the next
increment. In III, all lines have been shortened but will be extended in the next
increment, and in IV all lines have been shortened and will continue to be shortened
in the next increment.

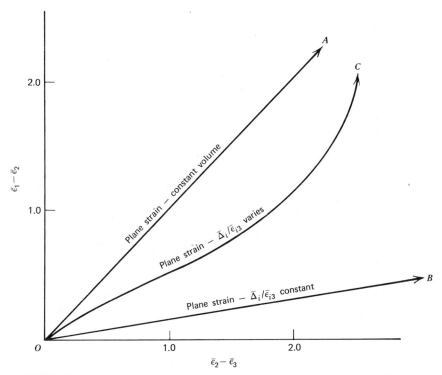

FIGURE 1.22   Flinn diagram showing plane strain coaxial strain paths. *OB* represents a coaxial, plane progressive strain in which the ratio of the natural incremental volume loss to the natural incremental shortening is constant. *OC* represents a coaxial, plane progressive strain in which the ratio of the natural incremental volume loss to the natural incremental shortening varies throughout the strain history.

with respect to each other during the deformation, then the strain history is *noncoaxial* [see Hsu (1966), Hossack (1968), and Elliott (1972)]. Thus, for a strain described as a pure shear, if the principal axes of strain have maintained their orientations relative to material points throughout the deformation, the strain history is coaxial. For a simple shear, if the undistorted surfaces have maintained their orientation relative to the unstrained state throughout the deformation, the strain history is noncoaxial. In the geological literature the terms *irrotational* and *rotational strains* are commonly used when *coaxial* and *noncoaxial strain histories,* respectively, are meant. As pointed out in Section 1.3.2 the common geological usage is loose since strains are defined only with respect to the initial and final configurations of particles and are independent of the history of straining. The terms *coaxial* and *noncoaxial strain histories* are preferred since they refer specifically to the history of strain. Note that the Flinn diagram (Fig. 1.16) is incapable of distinguishing between co-axial and noncoaxial strain histories because Flinn diagrams only show successive *shapes* of strain ellipsoids and not successive *orientations* of principal axes of strain relative to material particles [see, however, Hsu (1966) and Fig. 1.18c].

Strain paths are commonly discussed for deformations that involve no volume change. The general case of a noncoaxial strain history involving incremental volume changes is difficult to discuss, but Ramsay and Wood (1973) have considered the situation for coaxial strain histories involving only plane strain. If Δ_i is the incremental volume change in the ith strain increment, and the natural dilation is defined as $\bar{\Delta}_i = \log_e (1 + \Delta_i)$, Ramsay and Wood have shown that the slope of the strain path on a Flinn diagram using natural strain measures, in that ith increment, is $\tan^{-1}\left[1 - \bar{\Delta}_i/\bar{\epsilon}_{i3}\right]$ where $\bar{\epsilon}_{i3}$ is the principal increment of natural shortening. Thus, when there is no incremental volume change, the slope of the strain path is constant at 45°. If, on the other hand, there is a constant ratio between the natural incremental volume loss and the natural incremental shortening, then the strain path will be a straight line of slope other than 45°. If this ratio changes from one increment to the next then the strain path will be curved. Some of the strain paths are illustrated in Figure 1.22. More complicated effects of volume change on the strain path for these deformations are illustrated by Ramsay and Wood (1973). They point out that the general trend is to produce strain paths that quickly trend into the region $1 < k < 0$ if there is a volume loss during deformation.

1.3.4 Determination of Strain in Deformed Rocks

Determination of strain in deformed rocks depends on the presence of suitable markers that supply enough information to determine either the six components of strain or the three principal strains directly. As may be expected, this is possible only rarely, as is attested by the fact that although the theory for the analysis of finite strain has been known since the work of Kelvin and Tait (1883) only a handful of complete determinations of strain exist in the literature. Nevertheless, it is possible to determine strain in rocks given suitable markers and enough ingenuity in the manipulation of the mathematical theory. Some examples are Cloos (1947, 1971), Ramsay (1963), Gay (1969), Ramsay and Graham (1970), Badoux (1970), Siddans (1972), Coward (1973), Borradaile and Johnson (1973), and Wood (1974). Some of these papers assume a specific model of deformation in order to arrive at the state of strain; others assume that a principal plane of strain or a principal axis of strain is already established by a foliation or lineation.

The range of objects that serve as useful markers in rocks may be divided into two groups: (1) initially spherical or ellipsoidal objects, and (2) other objects of known initial shape.

1. Objects that were initially spherical or ellipsoidal include pebbles in conglomerates, rock fragments in sandstones, vesicules in lavas, reduction spots in slates, oolites, and radiolaria. If the object was initially spherical then the shape of the strain ellipsoid may be determined directly although nothing is known concerning volume changes. However, the usual situation is that the object is known to have been ellipsoidal rather than spherical, and this introduces errors into any calculations of strain. Dunnet (1969) and Elliott (1970)

Deformed

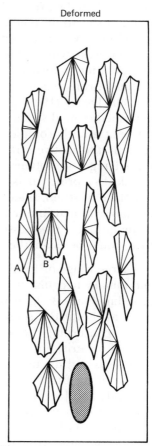

Undeformed

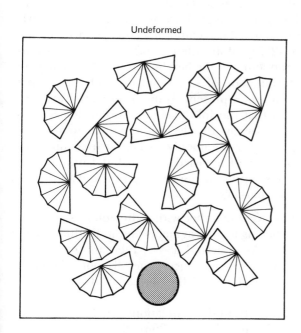

FIGURE 1.23   The shape changes in an assemblage of brachiopods due to a homogeneous strain. The principal axes of this strain may be determined from the two fossils A and B in which two lines originally at right angles have remained at right angles. There is enough information in the remaining distorted fossil shapes to calculate the shape of the section through the strain ellipsoid parallel to this plane. (From *Folding and Fracturing of Rocks* by J. G. Ramsay. Copyright © 1967 by McGraw-Hill, Inc. Used with permission of McGraw-Hill Book Company.)

have considered the deformation of initially ellipsoidal objects and present graphical methods of determining the strain. The major limitation of the use of these, and most other markers, is that because they have different mechanical properties to the rest of the rock, the strain that they record may be different from that of the bulk of the rock.

2. Initially nonspherical objects generally only give a semiquantitative indication of the magnitude of strain, although they may precisely determine the orientation of the principal axes of strain. Thus, the deformed brachiopods illustrated in Figure 1.23 allow the principal axes of an elliptical section through the strain ellipsoid (parallel to the bedding) to be determined by establishing two material lines (the hinge line and trace of the symmetry plane of the valves), which were initially perpendicular and which are still perpendicular after deformation. There is enough information in the variously oriented fossils to determine the shape of this elliptical section [Ramsay (1967, p. 239); Jaeger (1969, pp. 233–235)]. With other suitable markers, such as deformed crinoid stems lying randomly in the bedding, it may be possible to arrive at a fairly complete determination of strain, but it is only rarely that such an informative array of markers is present. If one is prepared to make simplifying assumptions, such as that the fold axis or a certain lineation is parallel to a principal axis of strain or that the cleavage is parallel to a principal plane of strain, then the problem is commonly trivial. Such assumptions are unwarranted in our present state of knowledge of rock deformation; in many instances these assumptions represent the very point about the strain that should be established by the markers.

A large range of fossils may be used together with sedimentary structures and initially irregular markers such as fossil fragments and individual grains in sedimentary or igneous rocks. As indicated above, their use commonly leads to useful but semiquantitative results, since generally not enough is known concerning the precise initial shapes of the objects or of the orientations of material lines.

Reviews and discussions of finite strain determination in deformed rocks are given by Brace (1961), Hobbs and Talbot (1966), Ramsay (1967, pp. 185–251) and Jaeger (1969, pp. 228–235).

A different approach to the study of strain in rocks makes use of the fact that the strain involved in folding is inhomogeneous and that, in order to ensure continuity from one part of a fold to another, only a distinct range of strain states is possible in order to produce the fold from initially planar and parallel layers. This approach enables the shapes of folds and of distorted lineations to be used as markers along with those mentioned above. The theory again becomes complicated and reference should be made to Ramsay and Graham (1970) and to Hobbs (1971).

1.3.5 **Examples of Strain Distributions**

In addition to the strain distributions illustrated in Figures 1.16 and 1.18 two other examples are presented here in order to emphasize the variation in strain states that may exist in deformed rocks.

The first example (Fig. 1.24) is taken from Dieterich (1969) and shows the distribution of strain in the same folded layer used to illustrate a distribution

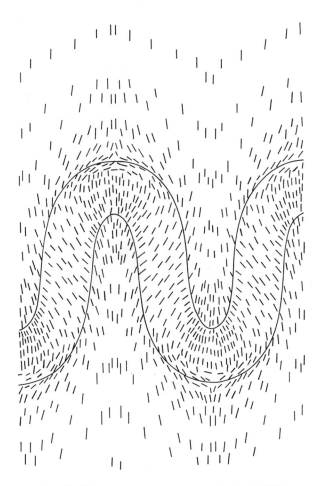

FIGURE 1.24   Distribution of strain in the fold drawn in Figure 1.7. Each line is parallel to λ_1 at each point in the fold and surrounding material. [From Dieterich (1969), *Amer. Jour. Sci.*, v. 267.]

of stress in Figure 1.7. The materials comprising the layers are assumed to be ideally viscous and the strain distribution has been calculated numerically. This example illustrates two important points. First, the orientation of the minor principal axis of the strain ellipsoid may be parallel to a folded layer at one point in the layer but normal to this same layer nearby. All variations between these extremes exist. However, the distribution of strain is quite systematic with respect to the fold hinge, the precise pattern being a function of whether a particular layer has high or low viscosity relative to adjacent layers. Second, comparison of the distribution of strain in this fold (Fig. 1.24) with the distribution of stress (Fig. 1.7) shows that *the principal axes of stress and of strain rarely coincide.* The noncoincidence of the principal axes of stress and of the finite strain ellipsoid is general and arises because the *strain* is determined from the *displacements* whereas the stress depends on the *rates* at which the strains occur instead of on the strains themselves (see Equation 1.26). Thus, the principal axes of stress depend on the principal axes of *strain rate.* A given imposed strain rate can lead to a progressive variety of strains so it is clear that, in general, the principal axes of strain do not coincide with the principal axes of strain rate. Hence the principal axes of stress and of strain do not coincide except in special symmetrical situations.

FIGURE 1.25   Distribution of strain as illustrated by distorted oolites in a large anticline. [After Cloos (1971).]

Thus, in general, identical shape changes will lead to different stresses if they are imposed on different materials or at different rates on the same material. The strain, however, is always the same. This points to the difficulties associated with determining the stresses in deformed rocks from a knowledge of the strains alone.

The second example is that of Cloos (1971), who determined the strain throughout a large fold using deformed ooids (Fig. 1.25). The principal planes of strain fan across the structure in a manner somewhat similar to that calculated by Dieterich (1969).

These and other distributions of strain in folded rocks will be considered in more detail in Sections 4.7 and 4.8.

1.4 THE RESPONSE OF ROCKS TO STRESS

The response of materials to an imposed stress system varies enormously and, even for a particular material, may be extremely variable depending on the physical conditions of deformation. A unified theory of the response of materials to stress is still in the earliest stages of development, and most progress is made by distinguishing a number of ideal classes of response (such as elastic, viscous, plastic, and so on) which some materials display to various degrees of approximation over various ranges of physical conditions. In this section some of the simpler of these ideal classes of response are considered, and an attempt is made to outline some of the mechanisms of deformation that are thought to give rise to such responses in geological materials.

Before proceeding it is necessary to distinguish between classes of *materials* and classes of *response*.

1.4.1 Classes of Materials

Materials may be mechanically *homogeneous* or *inhomogeneous,* a strictly homogeneous material being one in which all samples have identical mechanical properties. Statistical homogeneity means that all samples have the same mechanical properties within certain narrow limits. Statistical homogeneity is, however, dependent on the scale of the samples investigated; a massive sandstone bed is inhomogeneous if the samples are only the size of a few grains but is statistically homogeneous if the samples contain thousands of grains. A thinly laminated quartzite-pelite sequence is inhomogeneous if the samples only contain a few layers but is statistically homogeneous if the samples contain hundreds of layers. In what follows it is assumed that the material is statistically homogeneous on the scale of consideration and the word "statistically" is dropped.

Homogeneous materials may be mechanically *isotropic* or *anisotropic,* an isotropic material being one in which the mechanical properties are independent of the *direction* in which they are measured. An example of anisotropic mechanical behavior is given in Figure 1.26.

1.4.2 Classes of Response

When a sound wave from an earthquake or an explosion travels through a body of rock, the particles comprising the rock are locally displaced from their

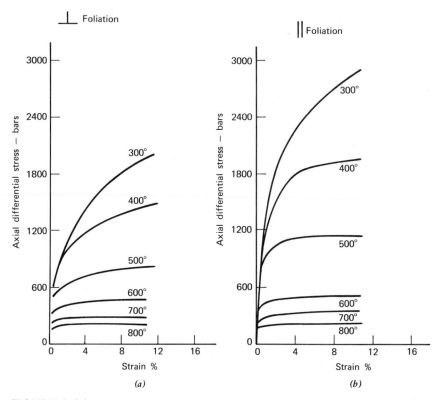

FIGURE 1.26   Stress-strain curves at various temperatures for Yule marble extended in *(a)* normal to the foliation and in *(b)* parallel to the foliation. The anisotropy in strength, especially at lower temperatures, is clear. [After Heard and Raleigh (1972).] Strain rates for these experiments are 3.3×10^{-6} sec^{-1} at 500°C and below, 1.95×10^{-6} sec^{-1} at higher temperatures.

normal positions but return to these positions once the disturbance has passed. Under such circumstances the stresses involved in the disturbance are small and the strain rates are relatively large (at least of the order of 10^{-3}sec^{-1}); there is no *permanent distortion* of the rock. The same kind of nonpermanent or *recoverable* deformation occurs if a rock or mineral specimen is loaded at relatively large strain rates in the laboratory using the techniques described in Chapter 7. Figure 1.27*a* illustrates the behavior of a number of rocks and minerals loaded axially at low hydrostatic pressures and temperatures and at strain rates of 10^{-4}sec^{-1}. The instant the load is applied, the specimen begins to deform, the relationship between the axial stress and the longitudinal strain being linear. Provided the specimen does not break, the specimen immediately returns to its prestraining size and shape once the

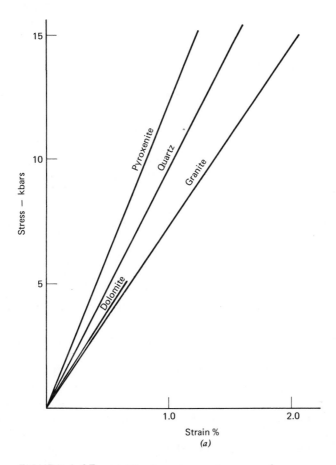

FIGURE 1.27   (a) Elastic stress-strain curves for
pyroxenite, quartz, dolomite, and granite. Rock curves
based on data in Griggs, Turner, and Heard (1960). Quartz
curve based on data in Hobbs, McLaren, and Paterson
(1972). (b) Anelastic behavior of an axially loaded specimen.

load is removed. This type of behavior, where the strain takes place instantaneously once the load is applied or removed, is called *elastic*. The important
features of this type of deformation are that the *response to stress is instantaneous*
and that the *strain is recoverable*. For a large number of materials the relationship between stress and strain is linear as in Figure 1.27a and the behavior is
then called *Hookean elasticity*. The relationship between stress and strain may
be written

$$\sigma = E\epsilon \qquad\qquad (1.23)$$

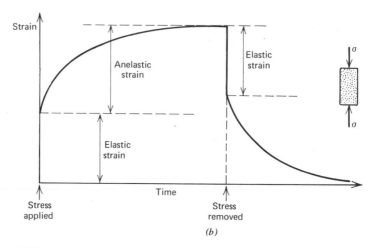

FIGURE 1.27 (Continued)

where σ is the stress, ϵ is the strain, and E is a constant known as *Young's modulus* if the deformation is a simple extension of shortening. A number of other constants are defined for an isotropic elastic material: if the deformation is by simple shear then another constant — G, the *modulus of rigidity* — is defined as the ratio of shear stress to shear strain. For a uniform hydrostatic pressure producing a uniform dilation, the *bulk modulus* or *incompressibility* K is the ratio of the hydrostatic pressure to that dilation. Another number, *Poisson's ratio*, ν, may be defined in a simple extension experiment; ν is the ratio of the lateral contraction of a rod to the longitudinal extension. These four quantities are related by the expressions:

$$G = \frac{E}{2(1 + \nu)} = \frac{3K(1 - 2\nu)}{2(1 + \nu)} \tag{1.24}$$

For further discussion see Jaeger (1969, pp. 54–58).

Equation 1.23 is quite simple in form but is only true for an isotropic material. For anisotropic materials the single linear equation is replaced by nine linear equations, one for each of the components of stress. These equations state that each stress component is a linear function of the nine components of strain. The system of equations may be written

$$\sigma_{11} = C_{1111}\epsilon_{11} + C_{1112}\epsilon_{12} + C_{1113}\epsilon_{13} + C_{1121}\epsilon_{21} + C_{1122}\epsilon_{22}$$
$$+ C_{1123}\epsilon_{23} + C_{1131}\epsilon_{31} + C_{1132}\epsilon_{32} + C_{1133}\epsilon_{33} \tag{1.25}$$

with eight other similar equations relating the stresses to the strains. In these nine equations there are 81 constants, C_{ijkl}, known as the *elastic stiffnesses*. Symmetry arguments coupled with other arguments based on elastic strain energy reduce this number so that 21 independent constants are required to relate stress to strain in the elastic distortion of a triclinic crystal. Symmetry arguments again reduce this number to 13 for the monoclinic system, 9 for the orthorhombic system, and 3 for the cubic system [Nye (1964, pp. 137–142)]. Two independent constants are required to relate stress to strain in an elastic isotropic material. Commonly used ones are Young's modulus and the rigidity modulus.

Rocks generally exhibit another kind of behavior where, although the strain is recoverable, the complete response is not instantaneous. This is illustrated in Figure 1.27*b* where the strain is plotted against time for a specimen loaded axially. When the stress is first applied there is an instantaneous elastic response, but the straining continues on after this, decreasing exponentially with time. When the stress is removed, there is again an instantaneous elastic response and then the strain slowly decays exponentially back to zero. This type of behavior — where the *strain is recoverable* but is also *time dependent* — is called *anelastic* and is of great importance in many rock mechanics problems associated with mining, tunneling and quarrying [Jaeger and Cook (1969)]. Anelastic behavior of some undetermined origin also leads to the high attenuation of seismic waves·as they pass through parts of the upper mantle of the earth [Gordon and Nelson (1966)].

The above two classes of response involve strains that are either instantaneously or ultimately recoverable. However, most of the deformations of interest to structural geologists involve strains that are *permanent*. That is, the rock remains in a strained state after the stress has been removed. Moreover, the intricately folded rocks that exist in some tectonic environments indicate that under certain conditions rocks can undergo large permanent strains without obvious faulting or loss of continuity.

One type of material that is capable of undergoing large permanent strains is the ideally *viscous* or *Newtonian* material in which the stress is related in a linear manner to the strain-rate (Fig. 1.28*a*).

$$\sigma = \eta\dot{\epsilon} \tag{1.26}$$

where $\dot{\epsilon}$ is the strain rate. Equation 1.26 says that the stress is proportional to the strain rate. The constant η is the *viscosity*. Materials with little internal structure tend to behave in an ideally viscous manner. Thus, water at room temperature and pressure behaves in this way whereas polymers with long chain molecules and crystalline materials rarely do. It was thought for some time that the deformation of polycrystalline aggregates (such as rocks) by diffusion alone could lead to steady Newtonian behavior, and it has been postulated that at the high temperatures (1000–1500°C) and slow strain rates (10^{-12} to 10^{-14} sec^{-1}) expected in the upper mantle of the earth that such behavior would be widespread [Orowan (1967), Weertman (1968)]. However, recent discussions have indicated that this is improbable [Weertman

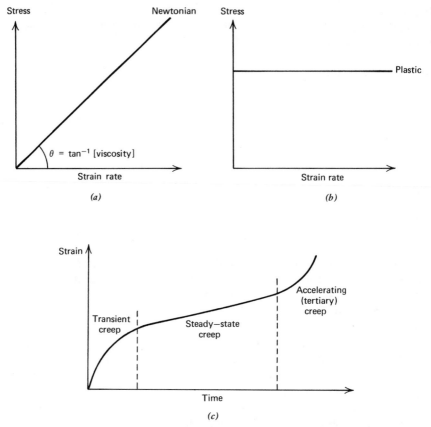

FIGURE 1.28   *(a)* Newtonian mechanical behavior. *(b)* Perfectly plastic mechanical behavior. *(c)* Creep behavior.

(1970), Green (1970)], and there now exists no adequate mechanism to explain why rocks could behave steadily as Newtonian materials. Nevertheless, a wide variety of problems have been treated with the assumption of Newtonian behavior as a basis [e.g., McKenzie (1969), Crittenden (1967), Dieterich and Carter (1969), and Torrance and Turcotte (1971)]. It may be that Newtonian behavior is a close approximation to that of real rocks but more work is needed to establish this. For anisotropic materials (1.26) is replaced by a system of nine linear equations similar in form to (1.25).

The *ideally plastic* material is incapable of maintaining a stress greater than some critical value σ_c (Fig. 1.28*b*) at which it continuously deforms in a permanent manner. Below this critical value of stress no deformation takes place. The ideally plastic material displays a characteristic kind of deformation in that strain only takes place in localized regions where the critical value of stress is reached. By contrast, ideally viscous materials show deformation throughout wherever a deviatoric stress is present.

In the laboratory, the response of rocks to stress is more complicated than that of either ideal viscosity or ideal plasticity. For the most part, rocks and minerals only suffer large permanent deformations if the hydrostatic stress is large; at atmospheric pressure, most rocks and minerals deform by breaking into fragments either parallel to the axis of loading or on faults inclined to that axis. This type of deformation, where rocks deform by developing marked discontinuities across which there is often a break in cohesion, is called *brittle* behavior and is discussed in Chapter 7. The opposite type of deformation, where rocks deform by distributing the strain in a smoothly varying manner throughout the deforming mass, is called *ductile* behavior. As the hydrostatic pressure is increased the behavior of a particular type of rock passes through a transition between what is definitely brittle behavior to what is definitely ductile behavior. This is called the *brittle-ductile transition* and is a function not only of hydrostatic pressure, but of temperature and of strain rate as well [Paterson (1958), Heard (1960)]. In general, the lower the temperature and hydrostatic pressure, and the higher the strain rate, the more likely is a particular rock to behave in a brittle manner. On the other hand, the higher the temperature and hydrostatic pressure and the lower the strain rate, the more likely is a particular rock to behave in a ductile manner. The mechanisms of brittle behavior commonly involve faulting, microfracturing, and associated cataclastic effects; the mechanisms of ductile behavior commonly involve deformation of individual grains by crystallographic slip, twinning, or other processes in which atomic diffusion plays a part. Thus, the effect of increased hydrostatic pressure on the processes operating during deformation is to inhibit fracturing and cracking of individual grains. The effect of increased temperature and decreased strain rate is to promote *thermally activated* processes such as crystal slip and atomic diffusion.

Characteristic changes associated with the brittle-ductile transition in rocks are summarized in Figure 1.29.

From this discussion it is expected that rocks deformed at low hydrostatic pressures and temperatures and at high strain rates will commonly show deformation by faulting and jointing or even by developing *tectonic breccias*. Therefore, this type of deformation is to be expected at shallow levels in the earth's crust because of the generally low temperature and pressure. On the other hand, rocks deformed at high temperatures and hydrostatic pressures and at low strain rates should typically show little signs of faulting and should exhibit large permanent deformation without evidence of cataclastic behavior. This type of deformation is to be expected at deep levels of the earth's crust and in the mantle.

Brittle behavior of rocks is discussed in detail in Chapter 7.

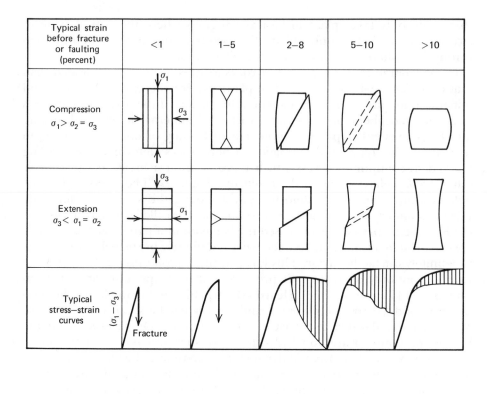

Typical strain before fracture or faulting (percent)	<1	1–5	2–8	5–10	>10
Compression $\sigma_1 > \sigma_2 = \sigma_3$					
Extension $\sigma_3 < \sigma_1 = \sigma_2$					
Typical stress–strain curves					

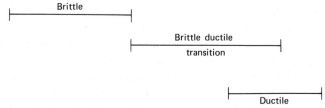

FIGURE 1.29   Spectrum of behavior illustrating the transition from perfectly brittle behavior to perfectly ductile behavior. Figure illustrates behavior for typical compression experiments and for typical extension experiments. The shape of the specimen is indicated along with the manner in which it deforms and the shape of the stress-strain curve. [After Griggs and Handin (1960).]

1.4.3 Ductile Behavior of Rocks

In the laboratory, the ductile behavior of rocks is a function of temperature and strain rate. In addition, the presence of trace amounts of impurities such as (OH) in the crystal structure of silicates has a profound effect on the mechanical properties [Griggs (1967)].

Two types of experiments are commonly conducted, both at initial hydrostatic pressures sufficient to ensure ductility. One is the *constant strain rate experiment,* where the specimen is loaded in such a manner as to ensure that straining occurs at a constant rate and the stress is allowed to vary in order to maintain this rate. The other is the *creep* or *constant stress experiment,* where the stress is kept constant and the strain rate is allowed to vary in order to maintain this prescribed stress. If, during some part of either of these two types of experiment, the material begins to deform at constant strain rate and at a constant stress, conditions are said to be steady state. During the deformation of rocks in the earth's crust conditions may approximate steady state for long periods of time, but both the stress and the strain rate will normally be changing as, for instance, an initially planar layer deforms to produce a fold.

Constant Strain-Rate Experiments.   As examples of these, the experiments of Heard (1963) and of Heard and Raleigh (1972) are presented. Specimens of Yule marble were deformed at a variety of temperatures and strain rates. The stress plotted against the strain for these specimens over a range of temperature and strain rates is shown in Figure 1.30a. The specimens first deform elastically and then pass through an *elastic limit* where the stress is no longer proportional to strain. At low temperatures, the stress gradually rises from this point onward as the strain increases. This gradual rise in the stress required to produce further deformation is called *strain hardening* and the specimens are now undergoing permanent deformation. At high temperatures or slow strain rates, no strain hardening accompanies permanent deformation, and conditions approach steady state.

A measure of the strength of these specimens is taken as the stress they are able to support at 10 percent strain, and if this is plotted against the strain rate (actually minus the natural logarithm of the strain rate) for various temperatures as in Figure 1.30b, it can be seen that the strength σ is related to the strain rate $\dot\epsilon$ by

$$\dot\epsilon = A \exp -\left(\frac{Q}{RT}\right)\sigma^N \qquad (1.27)$$

where T is the absolute temperature. In these experiments $N = 8$. A and Q are constants known, respectively, as the *frequency factor* and the *activation energy.* The activation energy is related to the energy required to perform the slowest step in the deformation mechanism and, for these experiments, has a value of 62 kcal/mole. By comparison, the activation energy for diffusion of CO_2 through the calcite structure is measured as 89 kcal/mole.

Extrapolation of the data of Figure 1.30b to geological strain rates of approximately 10^{-14} sec^{-1} indicates that these specimens should have strengths of approximately 10^3 bars at 250°C and 20 bars at 800°C.

Examples of the stress-strain curves obtained from single crystals of quartz at various temperatures are given in Figures 1.31a and b. Figure 1.31a shows

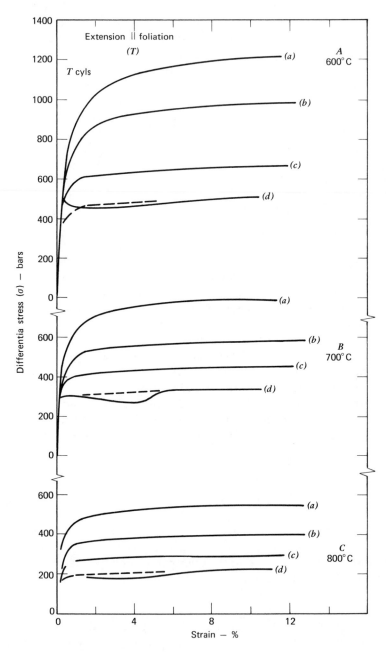

FIGURE 1.30 *(a)* Stress-strain curves at various strain rates for Yule marble extended at 600°C in *(A)*, 700°C in *(B)* and 800°C in *(C)*. The strain rates are *(a)* 2×10^{-3} sec^{-1}, *(b)* 2×10^{-4} sec^{-1}, *(c)* $2 \times$

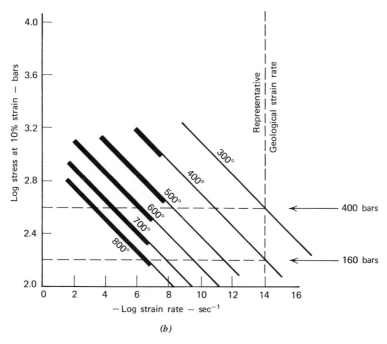

(b)

FIGURE 1.30 (Continued)

10^{-5} sec^{-1} *(d)* 2×10^{-6} sec^{-1}. [From Heard and Raleigh (1972), *Geol. Soc. Amer. Bull.*] *(b)* The data of Figure 1.30*a* plotted as the logarithm of the strength against the logarithm of the strain rate. The straight lines indicate that the behavior follows Equation 1.27. The thinner parts of the lines are extrapolations to geologically realistic strain rates. [From Heard and Raleigh (1972), *Geol. Soc. Amer. Bull.*]

the effect of adding trace amounts of (OH) to the quartz structure. Similar effects have been recorded for some other silicates [Griggs and Blacic (1965), Griggs (1967), and Blacic (1972)].

Creep Experiments.   When specimens deform at constant stress the deformation begins at a relatively high strain rate but steadily decreases until at last the strain rate becomes constant. After a still-longer period of time the strain rate increases again. If the strain is plotted against time then this behavior divides the resulting curve into three regions (Fig. 1.28*c*). The first part, where the strain rate decreases, is called *primary* or *transient creep;* the second part, where the strain rate is constant, is called *secondary* or *steady-state creep;* whereas the third part is called *tertiary* or *accelerating creep.* This third stage of the creep curve is commonly associated with fracture of the specimen although in some instances it can develop because of recrystallization of the material undergoing deformation.

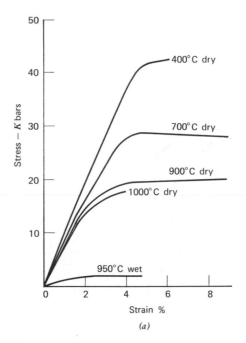

FIGURE 1.31   (a) Stress-strain curves for natural crystals of quartz at various temperatures and strain rates of 10^{-5} sec^{-1}. All of them are for a dry environment except for one at 950°C where trace amounts of water were introduced into the quartz structure. [From Griggs (1967), *Geophys. J.R. Astr. Soc.*, Vol. 14.] (b) Stress-strain curves for synthetic crystals of quartz, containing 6100 parts per million H:Si. Temperature of straining is indicated and the strain rate is 10^{-5} sec^{-1}. Orientation is normal to *r*. [From Hobbs, McLaren, and Paterson (1972), *Geophys. Mono, Ser.*, Vol. 16, p. 29–53. Copyright © by American Geophysical Union.]

During steady-state creep, the stress is related to the strain rate by a power law of the form (1.27). In that equation, if $N = 1$ then the material is behaving as a perfectly viscous fluid. Crystalline materials behave this way if diffusion of vacancies (see Section 2.2.1) through the grains from one grain boundary to another is the mechanism of deformation. Such a process is called *Herring-Nabarro* creep and this, with other more complicated diffusion-controlled creep processes, are discussed by Weertman (1970) and by Elliott (1972). Green (1970) has proposed that steady-state creep cannot be maintained by the Herring-Nabarro mechanism. Herring-Nabarro creep is observed in materials only at temperatures close to the melting point.

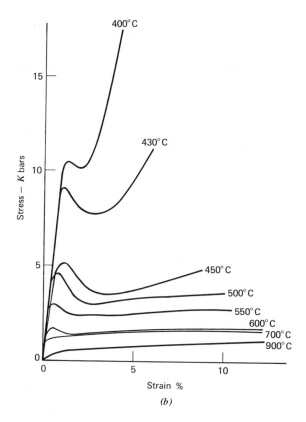

FIGURE 1.31 (Continued)

Most crystalline materials, at moderate temperatures (e.g., half the melting point) display steady state creep for which $N > 1$. Commonly $N = 3$ or 5 and such creep apparently is controlled by dislocation motion (see Section 2.2.1). The various mechanisms have been reviewed by Weertman (1970).

Two terms, commonly used in the geological literature, can now be discussed. The first of these is the term *competence*, used to denote a contrast in strength. Thus, in a sequence of interlayered quartzite and schist, the quartzite may have deformed to produce parallel folds with much jointing and faulting, whereas the schists are strongly deformed in a ductile manner and have the appearance of having flowed to occupy the spaces between the bending quartzite layers. In such a sequence the common interpretation is that the quartzites are strong and are controlling the deformation. On the other hand, the schists are pictured as being quite weak. The quartzites in this example are referred to as *competent* whereas the schists are *incompetent*. This usage is useful but quite loose since the manner in which the term strength is used is rarely, if ever, defined. Presumably the usage is meant to imply that a competent material is able to support a higher deviatoric stress level than an incompetent material.

The second term is *ductility* and is used in two quite-different manners. It is defined quite independently of the strength of a rock. In the context of this book one rock is more ductile than another if the first can be deformed to higher strains than the second before fracture ensues. In this sense, the ductility of a rock measures the amount of strain before brittle behavior commences. This is the sense in which the term is used by Griggs and Handin (1960) and by Heard (1960). Highly ductile rocks may be strong or weak; all that is implied by high ductility is the ability to undergo large strains without exhibiting brittle behavior. Some writers [e.g., Donath and Parker (1964)] use the term ductility (and the associated phrase *ductility contrast*) inappropriately to refer to the strength (or relative strength) of rocks.

PROBLEMS*

1. Show that stress (force per unit area) has the dimensions $[ML^{-1}T^{-2}]$ and that strain (fractional change of length or angle) is a dimensionless quantity. Quote some commonly used units for the measurement of stress. Are there any units for the measurement of strain? What are the units for the measurement of strain rate?

2. A plane has an area of 10 sq cm and across this plane a force of 10^{12} dynes acts, the direction of action being inclined at 72° to the normal to the plane. From first principals answer the following questions:

   (i) What is the normal stress on the plane?
   (ii) What is the maximum shear stress across the plane?
   Your answers should be expressed in dynes cm^{-2}, kbars, pounds ins^{-2}, and Gigapascals.

3. Across a particular plane, the state of stress is measured as:
Normal stress	1.2 kbar
Shear stress	−0.6 kbar
   On a plane at right angles to this plane the state of stress is measured as:
Normal stress	0.6 kbar
Shear stress	0.6 kbar

*Answers are given in Appendix B.

Assuming that one principal stress lies parallel to the intersection of the planes,

(i) What are the values of σ_1 and σ_3?
(ii) What is the angle between σ_1 and the normal to the first plane?
What other assumptions do you make in order to arrive at these answers?

4. Maximum and least principal stresses in a homogeneous rock body are measured as 1.5 kbars and 0.8 kbars, respectively. What are the normal and maximum shear stresses on the following planes, parallel to the intermediate axis of stress and whose normals are inclined at

   (i) 25°
   (ii) 45°
   (iii) 60°
   (iv) 65°
   (v) 80°
   to σ_1?

5. Assume that in both Problems 3 and 4 the stress states are two-dimensional.

   (i) What is the value of the mean stress in each case?
   (ii) What is the value of the deviatoric normal and shear stress in each case?

6. A line initially 10 cm long is shortened progressively to (i) 5 cm, (ii) 4 cm, (iii) 3 cm, (iv) 2 cm, (v) 1 cm. In each case, what is the percentage shortening and what is the natural strain? If the time taken to accomplish these strains is 10^3, 10^4, 10^5, 10^6, and 10^7 sec, respectively, what is the *average* strain rate in each case?

7. An initial angle of 90° is homogeneously distorted until it is (i) 85°, (ii) 70°, (iii) 45°, (iv) 20°, (v) 5°. In each case, what is the percentage shear strain? If the time taken to accomplish these strains is 10^2, 10^3, 10^4, 10^5, 10^6 sec, respectively, what is the *average* strain rate in each case?

8. The deformation illustrated in the accompanying diagram is a pure shear in which the initial right angle ABC undergoes a shear strain of 0.5774. Show graphically that this deformation corresponds to a simple shear through 30° together with an anticlockwise rotation through 16.1° and a small translation. Confirm that this is compatible with the general

result that a pure shear corresponds to a simple shear together with a rotation through $\tan^{-1}(\frac{1}{2} \tan \psi)$ and a translation. ψ is the angle of shear.

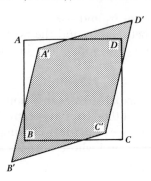

9. For the simple shear shown in the accompanying diagram, we write

$$\tan \psi = 2S$$

Jaeger (1969, p. 33) gives the following results:
(i) For a simple shear the principal axes of strain in the undeformed state make angles

$$\tfrac{1}{2}\tan^{-1}S \pm 45° \text{ with } Ox$$

(ii) For a simple shear the principal axes of strain in the deformed state make angles

$$\pm 45° - \tfrac{1}{2}\tan^{-1}S \text{ with } Ox$$

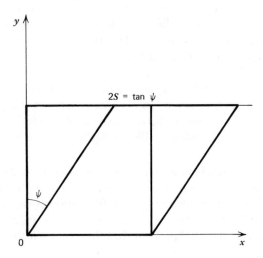

(iii) For a simple shear the major principal axis of strain has magnitude $\sqrt{S^2 + 1} + S$, and the least principal axis of strain has magnitude $\sqrt{S^2 + 1} - S$.

(a) Calculate the orientations of the principal axes of strain in the undeformed state.

(b) Calculate the orientations of the principal axes of strain in the deformed state.

(c) Calculate the magnitude of the principal strains.

(d) Plot the strain state on a Flinn diagram.

10. Construct a sequence of eight diagrams representing a progressive simple shear, the values of ψ being $0°$, $10°$, $20°$, $30°$, $40°$, $50°$, $60°$, and $70°$, respectively. Given the information in Problem 9, for each diagram in the sequence:

(i) Construct the principal axes of strain in the undeformed state.

(ii) Construct the principal axes of strain in the deformed state.

(iii) Calculate the maximum and minimum strains for each step in the sequence and plot these results on a Flinn diagram.

(iv) Construct a graph of the angle between the direction BC and the major principal axis of strain against the ratio of the major and intermediate principal strains.

(v) Consider the lines AB, BC, AC, and BD throughout the progressive simple shear. Describe their strain histories and see if you can arrive at some general conclusions regarding the strain histories of lines during a progressive simple shear.

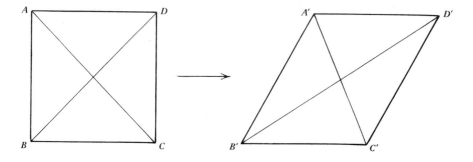

2
MICROFABRIC

2.1 INTRODUCTION

The term *fabric,* when applied to rocks, includes the complete spatial and geometrical configuration of all those components that make up the rock. It covers terms such as *texture, structure,* and *preferred orientation* and so is an all-encompassing term that describes the shapes and characters of individual parts of a rock mass and the manner in which these parts are distributed and oriented in space. As the term is currently used [Paterson and Weiss (1961); Turner and Weiss (1963, pp. 19–34)] the individual parts referred to are only considered as contributing to a fabric if they occur over and over again in a reproducible manner from one sample of the rock to another. Thus if, in any particular sample, a bedding surface occurs only once (e.g., in a massive sandstone unit), then this surface is not considered to contribute to the fabric. If, however, the sample contains many hundreds of bedding surfaces (e.g., in a shale or in the area as a whole), all similarly oriented, then these contribute a planar feature to the fabric. Similarly, five or six joint surfaces in a particular outcrop would not be considered as contributing to the fabric of the rock. On the other hand, if a region of several tens of square miles is considered these joints may occur over and over again, with much the same orientations and spacings. On this scale, the joints would contribute to the fabric of the rock.

Features that do contribute to a fabric are said to be *penetrative* [Turner and Weiss (1963, p. 22)] and are known as *fabric elements.* Otherwise, they are *nonpenetrative.* A feature may be penetrative on one scale of investigation but nonpenetrative on another.

It is generally useful to divide a body of rock into regions that are homogeneous (within prescribed limits) with respect to the orientation or pattern of orientation of a given fabric element. Such regions are *fabric domains* or simply *domains*. For example, if the bedding in a given area has the same orientation everywhere, then the area constitutes a domain on the basis of bedding orientation. If, however, the bedding is folded, then the area may be subdivided into a number of domains, within each of which the bedding has constant orientation. On the other hand, if the fold axis has constant orientation throughout the area, then the area as a whole constitutes a domain on the basis of fold-axis orientation. This concept is used at all scales, so that it is customary, for example, to divide areas of regional extent into fold domains (see Section 8.5.2), for instance, and hand specimens into quartz c axis domains for instance. A domain is not necessarily characterized by a single orientation of a fabric element but by a simple orientation pattern. For example, quartz c axis domains are commonly characterized by a girdle distribution (see Fig. 2.22k).

This chapter is concerned with those penetrative features of deformed rocks that may be observed or measured with an optical microscope or an electron microscope. Such features contribute to the *microfabric* of a rock and include grain boundaries, deformation bands, deformation lamellae, aggregates of grains of similar size, shape or composition, and crystallographic directions such as $[0001]$ of quartz or $\langle 111 \rangle$ of galena.

The overall aim in considering the microfabric of a deformed rock is to establish features on this scale that supply information on the mechanical and thermal history of the rock.

It is convenient to divide the discussion of microfabric into two overlapping parts: the *microstructure* and the *crystallographic preferred orientation*. In discussing the microstructure of a rock the concern is with the shapes of grains, the ways in which grains are arranged together, and the internal substructures within grains such as twins, deformation lamellae, and the like. The term microstructure is what is referred to as *texture* in many geological texts, but since there is a natural tendency to treat the study of rock fabric as related to material science, and since the term texture is used to mean crystallographic preferred orientation in studies of metals and ceramics, the term microstructure is preferred here. Crystallographic preferred orientations are the orientation patterns of crystallographic directions in certain phases in the rock. Common patterns are a parallel alignment of $[0001]$ of calcite normal to the foliation in marbles and of $[001]$ of olivine in peridotites.

2.2 THE MICROSTRUCTURES OF DEFORMED ROCKS

The microstructures of deformed rocks have been described since the early work of Sorby (1853), Sander (1911, 1930), and Harker (1932), but still today there is very little understanding of the precise meanings of many of the common microstructures. Since 1960 there has been a considerable amount of experimental work published on minerals and rocks that has helped to clarify the situation. In addition, the application of the transmission electron microscope since 1965, but particularly since 1970, has just begun to establish the detailed structure of some features. This section is based primarily on this geological work although prior work, particularly on metals, provides an overall framework for the interpretation of rock microfabrics.

Any understanding of microstructure must be based largely on the role that *crystal defects* play in the deformation process and on the ways in which the arrangements of crystal defects may be influenced by the thermal history of the rock. Accordingly, the following is a brief discussion of defects in crystals that is necessary before we consider the principles that control microstructural development.

2.2.1 Crystal Defects

The thermal and deformational history of a crystal, including whether it grew from a melt or recrystallized in the solid state, generally introduces a number of defects into what is commonly thought of as a perfect crystal structure. The nature of these defects is controlled by the crystal structure and by the chemical elements that make up the particular material. In cubic metals, these defects are relatively simple because of the simple structure and chemistry. In minerals, however, the defects have a very complicated structure that changes continuously if the defect moves through the crystal.

Defects in crystals may be broadly subdivided into three groups on the basis of their shapes: *zero dimensional* or *point defects, one dimensional* or *line defects,* and *two dimensional* or *planar defects.*

1. *Point Defects.* Point defects include missing atoms or *vacancies,* interstitial atoms, clusters of atoms, and combinations of vacancies with atoms or electrons. Some point defects are responsible for the color of crystals. Others contribute to the electrical conductivity. An extremely important point defect in silicates is that formed by the diffusion of (OH) ions through the structure. The (OH) ions are incorporated into the crystal structure converting an Si-O-Si bridge to Si-OH:HO-Si, so that the Si-O bond is replaced by the weaker OH-HO bond. This results in a great decrease in the mechanical strength of silicates [Griggs (1967)], in greater ease of recrystallization [Hobbs (1968)] and in a great increase in the rate of chemical reactions in most silicate systems. Some point defects such as vacancies and interstitials are created spontaneously if the temperature is increased, and for these types of defects there exists an equilibrium concentration for a particular temperature [Hirth and Lothe (1968, pp. 454–459)].

2. *Line Defects.* Line defects are generally referred to as dislocations, and

TABLE 2.1. Well-Established Slip and Twinning Systems in Common Minerals.*

Mineral	Slip System		Twin System		Reference
Quartz	**a**	$\{0001\}$			Christie et al. (1964)
	a	$\{10\bar{1}0\}$			Christie and Green (1964)
	c	$\{10\bar{1}0\}$			Christie et al. (1966)
	$\langle\mathbf{c + a}\rangle$	$\{10\bar{1}0\}$			
	$\langle\mathbf{c + a}\rangle$	$\{10\bar{1}1\}$			For crystallography of
	$\langle\mathbf{c + a}\rangle$	$\{01\bar{1}1\}$			quartz see Carter et
	$\langle\mathbf{c + a}\rangle$	$\{11\bar{2}2\}$			al. (1964)
	$\langle\mathbf{c + a}\rangle$	$\{2\bar{1}\bar{1}2\}$			
		$\{10\bar{1}2\}$			
		$\{01\bar{1}2\}$			
			Dauphiné twin		McLaren et al. (1967)
			Brazil twin		
Olivine	$[100]$	$\{okl\}$			Raleigh (1965)
					Raleigh (1968)
					Raleigh and Kirby (1970)
	$[001]$	$\{110\}$			Carter and Avé
	$[100]$	(010)			Lallemant (1970)
Plagioclase	$\{010\}$				Seifert (1965)
			Albite twin		Borg and Handin (1966)
			Pericline twin		Borg and Heard (1970)
Mica	$[100]$	(001)			Etheridge, Hobbs, and
	$[110]$	(001)			Paterson (1973)
	$[1\bar{1}0]$	(001)			
Calcite			$[e_1{:}r_2]$	$\{01\bar{1}2\}$	Turner, Griggs, and Heard (1954)
	$[r_1{:}f_2]$	$\{10\bar{1}1\}$			Griggs, Turner, and
	$[f_1{:}r_3]$	$\{02\bar{2}1\}$			Heard (1960)
	$[f_3{:}r_2]$	$\{02\bar{2}1\}$			
			$[r_1{:}f_2]$	$\{10\bar{1}1\}$	Borg and Handin (1967)
	$[r_1{:}f_2]$	$\{11\bar{2}0\}$			Paterson and Turner (1970)
			$[f_1{:}a_2]$	$\{02\bar{2}1\}$	For crystallography of calcite see Paterson and Turner (1970)

*For a discussion of other minerals see Spry (1969, pp. 60–61) and Carter (1971).

these are the defects that largely influence the mechanical behavior of crystals. They are, therefore, the defects of greatest interest to the structural geologist. When a single crystal of a mineral is deformed at a pressure high enough to prevent brittle fracture, and at temperatures low enough that solid state diffusion is not important, the deformation process is commonly one of *slip* on crystallographic planes with low Miller indices in crystallographic directions with low directional indices within those planes. Slip is referred to as translation gliding in older geological literature. The combination of a specific crystallographic plane with a crystallographic direction in that plane is called a *slip system*. All symmetrically equivalent combinations are included in that system. For instance, quartz commonly deforms by slip on the basal plane (0001) in the direction of one or more of the *a* axes [Christie et al. (1964); McLaren et al. (1967)]. The family of three slip systems denoted by **a** (0001) is a system common in quartz. Slip systems observed in some common minerals are given in Table 2.1 The ease with which a given slip system operates in a particular mineral is a function of temperature and of strain rate. In some minerals, such as calcite, *mechanical twinning* is common as well as slip [Paterson and Turner (1970)]. Both slip and twinning enable a grain to change its shape by allowing one part of the crystal to undergo shear with respect to a neighboring part.

Microscopically, slip may be compared to the gliding of cards past each other as a deck of cards is sheared. On the atomic scale, however, slip does not take place simultaneously over an entire slip plane in a crystal, but nucleates in a small region of high stress concentration. This region of slip then spreads out in an expanding loop across the slip plane until it ultimately intersects a grain boundary where it produces a small step. The line that separates the slipped region from the unslipped region at any instant is a *dislocation* (Fig. 2.1).

The propagation of a dislocation may be compared in many respects to the problem presented to a hypothetical person in removing a very large carpet that is loaded down with large pieces of furniture such as the odd piano and table. He could simply try and drag the carpet out of the room. This clearly would take a lot of energy and may not be very successful anyway. The analogy here is with wholesale slip along one plane in the crystal structure. Such an effort requires the simultaneous movement of a large number of atoms and is associated with relatively large amounts of energy; if it does occur in crystals it can lead to disruption or at least local melting of the crystal [Christie and Ardel (1974)]. Another way for the person to remove the carpet is to ruck up the carpet at one end and slowly propagate that ruck across the room, lifting up the legs of pieces of furniture when the need arises but ultimately moving the ruck to the other side of the carpet. By then the carpet would have been moved through an inch or so. He could repeat the process until the carpet was moved out of the room. The process requires only a small

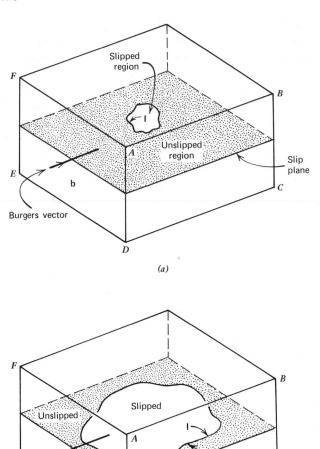

FIGURE 2.1   Propagation of a dislocation across a slip plane. In *(a)* a small region of material that has slipped relative to that around it is outlined by the line **l** lying in the slip plane. In *(b)* this region has expanded under the influence of a shear stress. The atomic arrangement around such a loop is shown in *(c)*. Atoms below the slip plane are shown as open circles whereas those on and above the slip plane are shown as dots. The Burgers vector, **b**, is in the direction of displacement of atoms and is defined in Figure 2.4. Notice that the segments of the dislocation loop lying normal to the Burgers vector represent the ends of an extra plane of atoms inserted above or below the slip

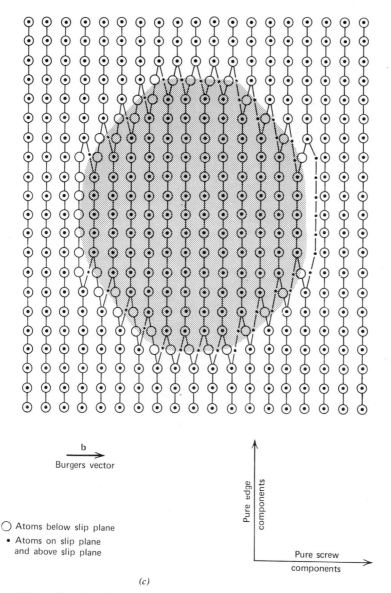

b
⟶
Burgers vector

Pure edge components

○ Atoms below slip plane
• Atoms on slip plane
and above slip plane

Pure screw components

(c)

FIGURE 2.1 (Continued)

plane. The shaded region represents the area that has slipped relative to the structure around it. The reader should analyze this figure and follow the individual atom movements in detail. In *(d)* the dislocation loop has intersected the edges of the crystal producing a slight displacement. The distortion of a cubic lattice about such a dislocation line is shown in *(e)*.

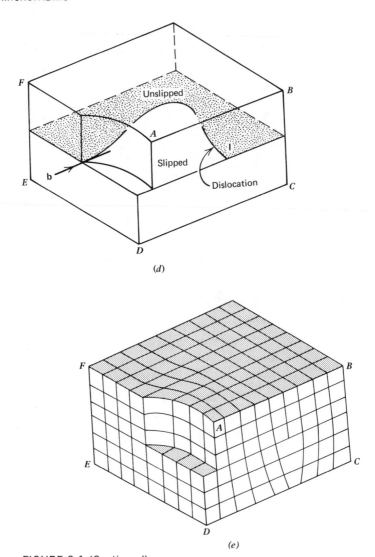

FIGURE 2.1 (Continued)

amount of energy at any instant but is not very efficient unless a large number of rucks are generated and propagated quickly. The analogy here is with a dislocation in a crystal that, after propagation across the slip plane, produces a unit-cell displacement of the crystal on one side of the slip plane with respect to crystal on the other side. The process requires relatively small amounts of energy and is efficient because of the very large numbers of dislocations that

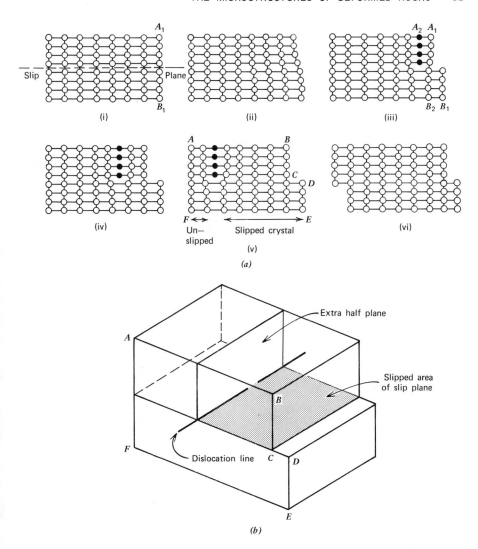

FIGURE 2.2 (a) Propagation of an edge dislocation through a crystal. In (i) there is slight distortion of the crystal lattice along the row of lattice points A_1, B_1. In (ii) this distortion is greater and in (iii) the distortion has been accommodated by what is equivalent to the introduction of an extra row of lattice points marked as black dots. This row originally would have been equivalent to a row of lattice points A_2, B_2. In (iv) and (v) this extra half plane has propagated through the crystal until in (vi) it has met the edge of the crystal producing a unit cell offset. [From Spry (1969).] (b) A block diagram representing the situation in (v) above. The face $ABCDEF$ is equivalent in both diagrams. The three-dimensional arrangement of the extra half plane, the dislocation line, and the slipped area are shown.

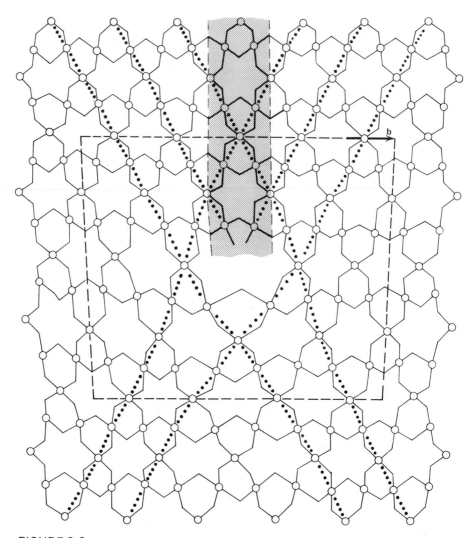

FIGURE 2.3   Sketch of the structure of a dislocation in quartz. The Burgers vector **b** is parallel to an **a** axis and the slip plane is (0001). The shaded region represents the extra half plane of material, one unit cell wide, that has been introduced to produce the edge dislocation. The rows of dots represent other possible half planes that could be introduced to produce the same structure. [From McLaren, Osborne, and Saunders (1971).]

may be present in a deforming crystal: the measure of dislocation density is the length of dislocation lines (measured in centimeters) present per cubic centimeter of crystal. In an undeformed natural crystal of quartz the density

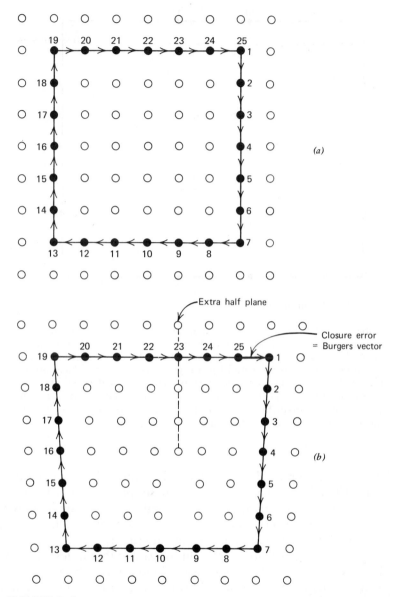

FIGURE 2.4   Definition of a Burgers vector. In (a) and (b) the crystal lattice is assumed to be cubic. (a) represents a closed clockwise traverse around a group of atoms numbered 1 through 25. In (b) this identical traverse has been carried out in a region of crystal where an extra half plane of material ends. There is now a closure error in the traverse and this is called the *Burgers vector*. The traverse is called a *Burgers circuit*. Sign conventions for Burgers vectors are discussed by Hirth and Lothe

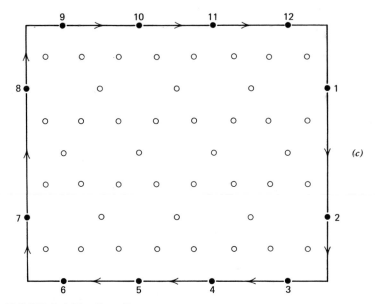

FIGURE 2.4 (Continued)

(1968, pp. 19–22). In *(c)* the silicon atoms from Figure 2.3 are marked out for a region of perfect crystal. A closed Burgers traverse is indicated by the line running from silicon atom 1 through silicon atom 12. In *(d)* the same traverse is indicated for a region around the dislocation of Figure 2.3. It can be seen that there is a closure error equal in magnitude to one unit cell dimensions, *a*, and indicated by the vector **b**, *(e)* Burgers traverses for the block shown in Figure 2.1e. In both cases the Burgers vector **b** is indicated.

of dislocations is generally about 10^3 cm of dislocation line per cubic centimeter. This is written 10^3 cm cm^{-3} or, even more simply, 10^3 cm^{-2}. Thus, dislocation densities are commonly quoted as a number per square centimeter. A heavily deformed quartz crystal might contain a dislocation density as high as 10^{12} cm^{-2} [see Hobbs, McLaren and Paterson (1972)].

One of the amazing features of the slip process is that it does not produce gross changes in the crystal structure even though it is able to produce gross changes in the shape of the crystal. This means that each dislocation leaves "good" crystal behind as it propagates within the slip plane. One way of producing this is illustrated in Figure 2.2, which represents one possible sequence of events at the atomic scale as a dislocation moves through the structure. This section through a dislocation can be represented by adding an extra plane of atoms above the slip plane. In minerals an extra layer one unit cell wide would need to be added above the slip plane as shown in Figure 2.3.

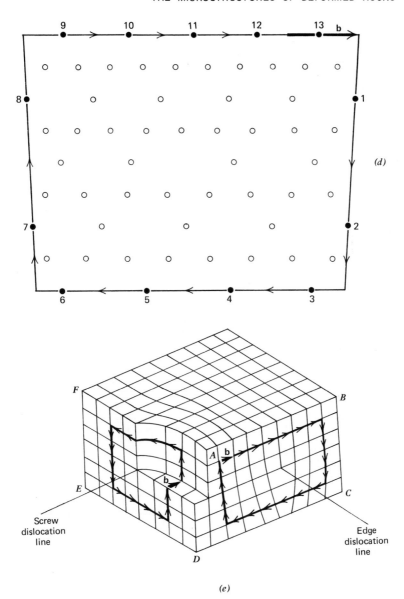

FIGURE 2.4 (Continued)

However, we will continue to refer to this as the *extra half plane*. The region at the end of the extra half plane, constituting the *core* of the dislocation, is ideal for the segregation of impurity atoms, which are able to satisfy the charge imbalance and the dilation field introduced by the dislocation. The structure

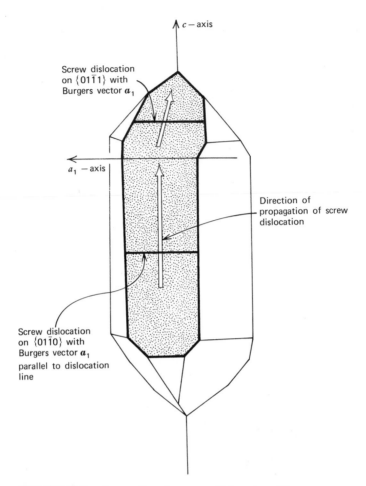

FIGURE 2.5   Cross slip of a screw dislocation. The two
stippled planes in the quartz structure indicate slip planes $(01\bar{1}0)$
and $(01\bar{1}1)$. A screw dislocation parallel to a_1 can move out of the
slip plane $(01\bar{1}0)$ and onto the slip plane $(01\bar{1}1)$. This motion is
called *cross slip*.

of dislocations, in reality, is probably more complicated than that illustrated in
Figures 2.2 and 2.3 in that the core region of a dislocation is probably several
unit cells wide.

Two geometrical elements are required in order to specify a dislocation
completely at a point in a crystal. One is the vector **l** tangent to the dislocation
line at the point, and the other is the vector **b**, which is in the slip direction and
has a magnitude equal to the smallest amount of slip in the direction that will

produce reregistry of crystal structure above and below the slip plane as the dislocation passes (Fig. 2.1 and 2.2). Vector **b** is called the *Burgers vector* and is defined in Figure 2.4. If **l** $\perp$ **b** the dislocation is called an *edge dislocation*. If **l** $\parallel$ **b** it is a *screw dislocation*. Otherwise, where **l** is oblique to **b**, the dislocation is *mixed*. Edge dislocations can only slip in the plane defined by **l** and **b**. Screw

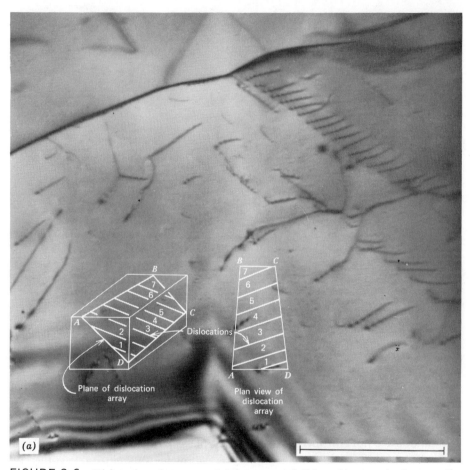

FIGURE 2.6   Dislocations in quartz. *(a)* An array of dislocations producing a wall. Each dislocation line in the array is parallel to an *a* axis. The inset diagrams indicate the geometrical features of the image. The array of straight dislocations lies in a plane that dips through the thin crystal slice examined in the electron microscope. Each dislocation plunges through the crystal slice intersecting the top and bottom of the slice. [From McLaren and Hobbs (1972), *Geophysical Monograph Series*, Vol. 16, pp. 55–66. Copyright © by American Geophysical Union.] Scale mark is 1 μm. *(b)* Dislocations on two slip planes in quartz. (Photographed by D. Morrison-Smith.) Scale mark is 1 μm.

FIGURE 2.6 (Continued)

dislocations are not confined to a single slip plane and may move in any plane parallel to **b.** The motion of a screw dislocation off one slip plane onto an inclined one is called *cross slip* (Fig. 2.5). For example, a highly probable mode of deformation in quartz is cross slip of screw $\langle \mathbf{c} + \mathbf{a} \rangle \{10\bar{1}0\}$ dislocations on to $\{10\bar{1}1\}$. Some examples of dislocations in silicates are shown in Figures 2.6 and 2.7.

Although dislocations have been extensively studied in metals using transmission electron microscopy since the 1950s, it was not until 1964 that the first images of dislocations were observed in silicates [McLaren and Phakey (1965)]. At that time the specimen preparation consisted of simply crushing

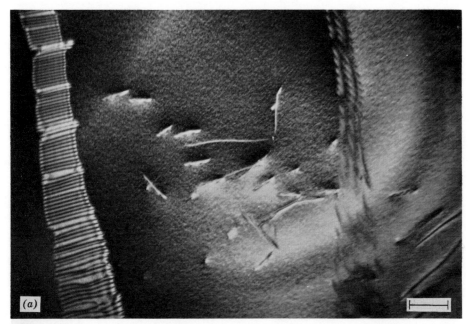

FIGURE 2.7   *(a)* Dislocations in naturally deformed peridotite, mylonite from Aneta Bay in New Zealand. [From Boland et al. (1971).] As in Figure 2.6*a* the straight dislocation lines represent dislocations plunging through the crystal slice and intersecting the top and bottom of the slice. Scale mark is 1 μm. *(b)* Dislocations in experimentally deformed plagioclase. The lamella regions with fringes parallel to their boundaries are albite twins. (Photograph by D. Marshall.) Scale mark is 1 μm.

the specimen and hoping that there would be thin areas (approximately 0.1 μm) on the edges of fragments where penetration by 100 kV electrons would be possible. Since about 1970 ion-thinning machines have become available [Gillespie, McLaren, and Boland (1971); Heuer et al. (1971)], and these are now in widespread use to prepare large areas of thinned material of known orientation.

On the atomic scale, a dislocation represents a linear region in which atoms are not quite in their proper crystallographic positions; that is, the atomic arrangement has been elastically distorted. The direction of slip in any particular slip plane is generally parallel to the closest-packed direction, since this tends to minimize the elastic distortion associated with the dislocation core. This distortion has associated with it a stress field that can exert forces on neighboring dislocations or other defects. The elastic distortion also means that there is strain energy associated with a dislocation so that the introduction of dislocations increased the free energy of a crystal. Dislocations are, there-

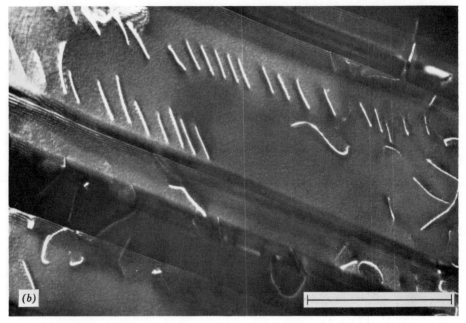

FIGURE 2.7 (Continued)

fore, unstable defects, and if a crystal with dislocations is heated, the dislocation density will decrease; for all practical purposes, the equilibrium concentration of dislocations in a crystal at a particular temperature is zero.

As the dislocation density increases in a crystal it requires an increasing
stress to move a dislocation through the "matrix" of dislocations surrounding
it. This is partly because of the forces exerted on the dislocations by others,
and partly because of the creation of extra segments of dislocation line that do
not lie in the slip plane and hence are difficult to move. These extra segments,
inclined to the slip plane of the dislocation, are called *jogs* (Fig. 2.8*a*). The
increased difficulty of movement is responsible for strain hardening (see Section 1.4.3) in crystals deformed at low temperatures.

Under conditions where the rate of solid-state diffusion can keep pace with
dislocation motion, that is, at high temperatures or at slow strain rates, edge
dislocations are no longer constrained to remain in their slip plane. By diffusion of material toward or away from the dislocation core, the length of the
extra half plane can be altered so that the dislocation *climbs* (Fig. 2.8*b*) from
one slip plane to the next. This process of climb means that dislocations can
circumvent obstacles to movement, so that it becomes easier to deform crystals
at high temperatures or at slow strain rates where diffusion processes are able
to become more prominent. Climb of edge dislocations and associated cross

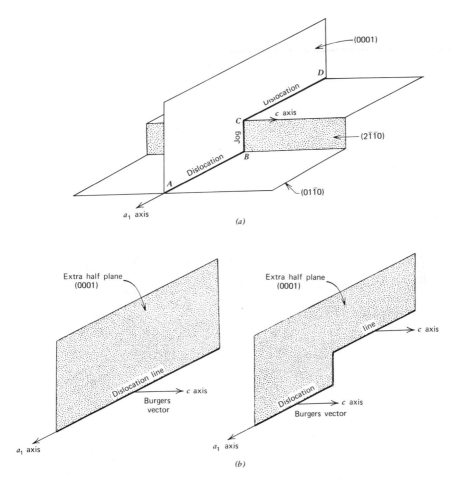

FIGURE 2.8 (a) Jog in a dislocation line. Dislocation line $ABCD$ has two
segments AB and CD that are parallel to the a_1 axis in quartz and lie in the slip
plane $(01\bar{1}0)$. These two segments are edge dislocations with extra half plane
parallel to (0001). The small segment BC of this dislocation line does not lie in
the slip plane and is known as a jog. It itself is an edge dislocation with Burgers
vector $\mathbf{c}$, extra half plane (0001) and a slip plane $(2\bar{1}\bar{1}0)$. At low temperatures,
slip on (0001) is relatively easy whereas slip on $(2\bar{1}\bar{1}0)$ is difficult. Thus, at low
temperatures, the jog would impede the motion of the two dislocation segments
AB and CD. (b) Climb of an edge dislocation that has Burgers vector $\mathbf{c}$ and is
parallel to an a_1 axis. The extra half plane is (0001). By changing the shape of the
extra half plane, that is, by extending it or shortening it at any one place, the
dislocation line moves out of one slip plane into a parallel, neighboring one. This
motion is called *climb*. (c) Distortion of a crystal by climb. By removal of the extra
half plane of atoms, marked as black dots, the crystal is shortened by one unit
cell.

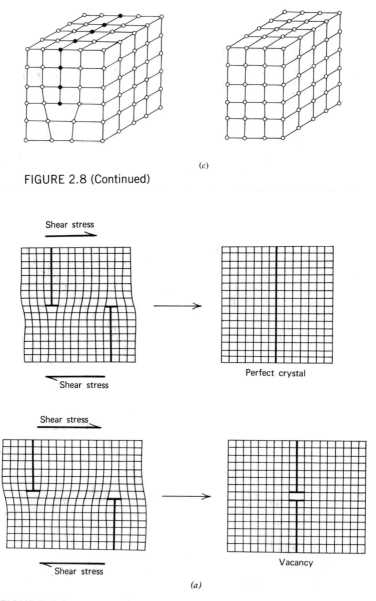

FIGURE 2.8 (Continued)

(c)

(a)

FIGURE 2.9   Interaction of dislocations. *(a)* Two edge dislocations
of opposite sign moving on the same slip plane will meet to form a
perfect crystal. They meet to produce a line of vacancies if they are on
adjacent slip planes. *(b)* A wall of edge dislocations aligned to produce
a simple tilt boundary. The extra half planes of the edge dislocations
are shown as solid lines. This is the structure illustrated in Fig. 2.7a.

slip of screw dislocations are probably important modes of deformation at the slow strain rates corresponding to geological deformations. When an edge dislocation climbs through an appreciable part of a crystal, the effective result is either an extension or a shortening normal to the extra half plane of the dislocation. Thus, under conditions where climb is important an extra mechanism of shape change, other than slip, is introduced [Groves and Kelly (1969); see also Fig. 2.8c].

Two edge dislocations moving in parallel slip planes with the same Burgers vector are said to be of *opposite sign* if one has the extra half plane above the slip plane and the other below. Two edge dislocations of opposite sign can annihilate each other, producing perfect structure if they meet on the same slip plane or producing a row of complicated point defects if they meet on neighboring slip planes (Fig. 2.9a). Annihilation of dislocations of opposite signs is one of the main ways of reducing the dislocation density in a strongly deformed crystal.

3. *Planar Defects.* Planar defects include structures such as subgrain boundaries, grain boundaries, deformation band boundaries, deformation lamellae, stacking faults, and twins. Each of these structures is considered briefly now.

Subgrain boundaries are relatively planar boundaries, within grains, across which the lattice misorientation is small, of the order of 1–5°. In the optical microscope they separate regions of slightly differing extinction position. Such boundaries are called *low angle boundaries.* Some are formed during deformation when dislocations group to form planar walls (Fig. 2.9b) that are

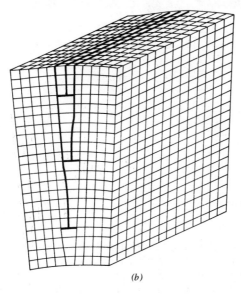

(b)

FIGURE 2.9 (Continued)

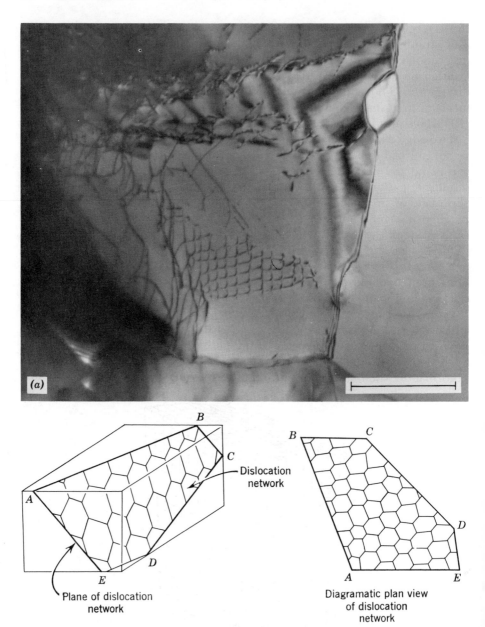

FIGURE 2.10 *(a)* Network structure, deformed quartzite. This network is within a grain containing other arrays of dislocations. Two bubbles are situated in a boundary that runs across the right-hand side of the photograph. The insets show the geometrical features of the image. [From McLaren and Hobbs (1972), *Geophysical Monograph Series*, Vol. 16, pp. 55–66. Copyright© by American Geophysical Union.] Scale mark is 1 μm. *(b)* Tangled dislocations in experimentally deformed synthetic quartz. A densely spaced

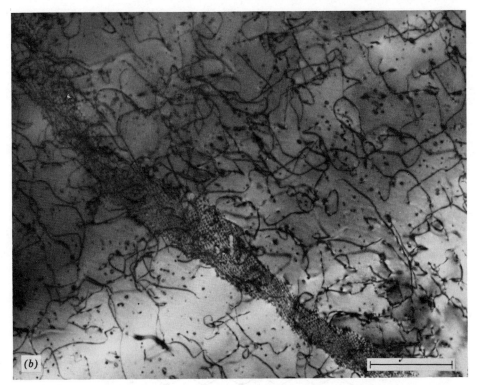

FIGURE 2.10 (Continued)

array of dislocations crosses the lower left-hand corner of the plate and this is a steeply dipping network structure. Notice the large number of isolated loops scattered over the background of the photograph. This indicates that a large amount of climb or recovery has taken place. (Photograph by D. Morrison-Smith.) Scale mark is 1 μm. (c) A screw dislocation on the basal plane of a hexagonal lattice. Solid circles represent lattice points in perfect structure whereas open circles represent lattice points defining the dislocation. Burgers vector is **b**. The shaded portion represents the layer one unit cell below. (d) Part of a screw dislocation network on the basal plane of a hexagonal lattice. There is no long-range strain associated with such a network but it does result in a slight rotation of the lattice above the plane of the network relative to that below the plane. This represents, in principle, the structure of the networks shown in Figure 2.10a and b.

able to move as a unit under an applied stress. Figure 2.7a shows examples of such walls in a peridotite mylonite. Other subgrain walls are formed after deformation ceases as the dislocation density attempts to return to a low value. During this stage, edge dislocations can group into low energy walls to form a particular type of low angle boundary called a *tilt boundary*. Such a boundary is

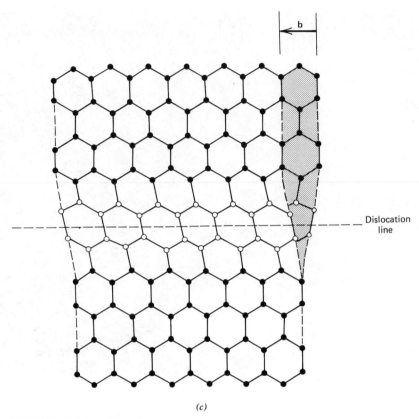

Dislocation line

(c)

FIGURE 2.10 (Continued)

shown diagrammatically in Figure 2.9*b* and an example from a quartzite is given in Figure 2.6*a*. Another type of low energy array forming a subgrain boundary is illustrated in Figure 2.10 where dislocations with three different Burgers vectors have combined to form a *network*. It is important to note that subgrain boundaries formed both during deformation and after deformation may appear identical in the optical microscope yet have entirely different dislocation structures [McLaren and Hobbs (1972)].

Grain boundaries separate grains of very different lattice orientations or of different composition. There is, of course, a complete gradation from subgrain boundaries to grain boundaries (see Fig. 2.9*c* to *f*). The precise nature of grain boundaries is still a subject of active research although it appears that, at the atomic scale, the boundary must be irregular with regions of fairly good lattice fit separated by regions of poor fit (Fig 2.9*c* to *f*). Grain boundaries are *high angle boundaries*. Figure 2.9*f* shows a symmetrical high angle grain boundary in two dimensions, the misorientation across the boundary being de-

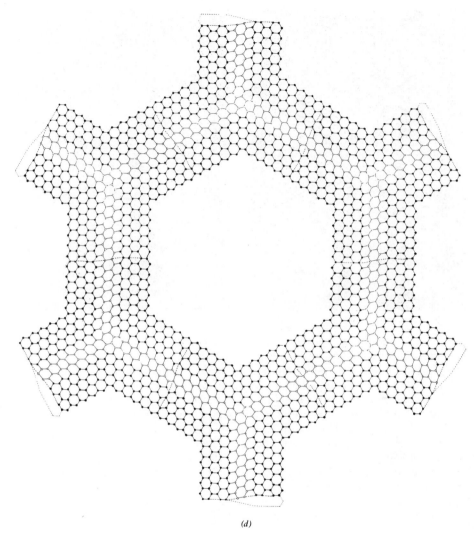

(d)

FIGURE 2.10 (Continued)

scribed by a rotation about a line perpendicular to the plane of the paper. In the most general two-dimensional situation the lattice on one side of the grain boundary is asymmetrically related to the lattice on the other side. In order to specify the most general grain boundary in three dimensions, five quantities are required: two to describe the orientation of the boundary with respect to one of the lattices and three to specify the relative orientation of the lattices either side of the boundary (i.e., successive rotations about three orthogonal axes, two in the plane of the boundary and one normal to the boundary). Such

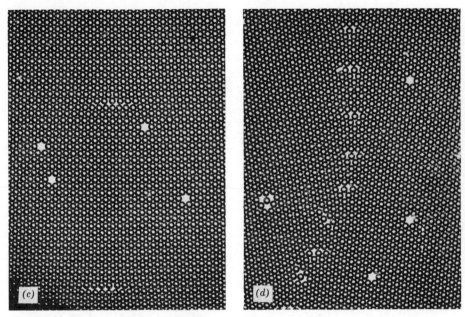

FIGURE 2.9 c to f Arrays of bubbles illustrating the structure of various grain boundaries. [From Lomer and Nye (1952), *Proc. Roy. Soc.*, Vol. 212.] (c) Symmetrical boundary with angle between lattices of 2°. The misorientation is best seen by viewing the array at a low angle. Several vacancies and other more complicated point defects are shown. (d) Boundary that changes direction. It is symmetrical in the top half of the figure and asymmetrical in the lower part. Angle between lattices is 10°. (e) High angle boundary. Angle between lattices is 20°. (f) High angle boundary. Angle between lattices is 25°. In each diagram of the series (c) through (f) the reader should lay tracing paper over the figures and construct the dislocation structure of the boundary in the manner of Figure 2.9b. In (c) and (d) the dislocation structure is readily established. In (e) only the symmetrical parts of the boundary are easily represented by dislocations. In (f) no dislocation structure can be readily established. The sequence, therefore, represents the changes in structure that accompany a change from low angle to high angle boundaries. See Lomer and Nye (1952) for a full discussion of this sequence.

a boundary appears to be composed of a complex array of defects comprising regions of fairly good and poor fit. More extensive discussions are given by McLean (1957) and Kelly and Groves (1970, Chapter 12); a summary of recent results is given by McLean (1973).

Deformation bands are planar regions within grains that have suffered a different kind of deformation to that in adjacent parts of the crystal. The deformation may differ only in the amount of slip that has occurred on a particular slip plane in which case the lattice is simply bent. A *kink* is a special

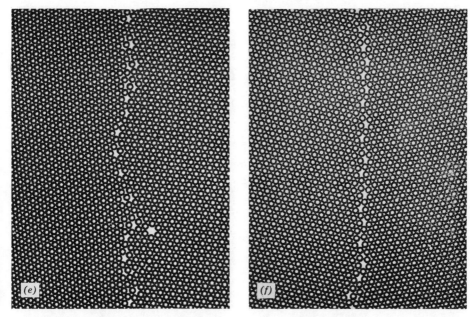

FIGURE 2.9 (Continued)

example of this in which the boundary between the adjacent regions is sharp (Fig. 2.11*a*). In more complicated deformation bands, different slip systems operate to different degrees in adjacent bands. An example is illustrated in Figure 2.11*b* [see Paterson and Turner (1970)]. Deformation band boundaries are commonly high angle boundaries.

Deformation lamellae are narrow (of the order of 0.5 to 5 μm), approximately planar features developed within single grains that have slightly different refractive index to the host grain (Fig. 2.12*a*) [Christie et al. (1964)]. They are common in quartzites deformed at low grades of metamorphism but have also been described in olivine [Boland et al. (1971)] and in plagioclase [Borg and Heard (1970)]. In some experimentally deformed quartz, deformation lamellae are parallel to slip planes [Christie et al. (1964)], and are the optical expression of the long-range stress field associated with a strained region [see Christie and Ardel (1974)]. In other examples, including the deformation lamellae in many naturally deformed quartzites, the structure of deformation lamellae is more complicated and is not associated with long-range strains (Fig. 2.12*b*). For a discussion see McLaren et al. (1970), McLaren and Hobbs (1972) and Christie and Ardel (1974).

Another term used in close conjunction with deformation lamellae is *Böhm lamellae*. These lamella features are planar arrangements of bubbles. They may or may not be associated with birefringence changes across the lamella.

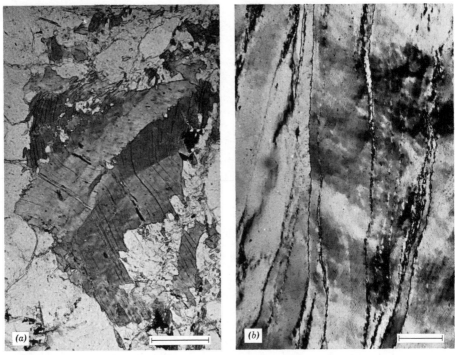

FIGURE 2.11   *(a)* Kink boundary in naturally deformed biotite. (Photograph courtesy of T. Bell.) Scale mark is 0.3mm *(b)* Deformation bands in naturally deformed quartz. Arunta Complex. Central Australia. Scale mark is 0.3 mm.

Their origin is poorly understood although some features resembling Böhm lamellae develop during the experimental annealing of deformed quartz bearing deformation lamellae [Hobbs (1968); Ardel et al. (1974)]. Also present in some deformed quartzites are more or less planar arrays of inclusions that cross many grains with the same spatial orientation independent of the crystallographic orientations of the grains. These planar features have been called *Tuttle lamellae* [see Tuttle (1949)].

Stacking faults are planar defects in a crystal structure across which there is a simple relative displacement of the crystal structure. These defects may be formed during the growth of the crystal or may arise during deformation. The vector that relates the structure either side of the defect is known as the *fault vector,* and the magnitude of this vector is always a fraction of a crystallographic repeat distance. The fault vector may be parallel or inclined to the stacking fault. Stacking faults are common in pyroxenes, plagioclases, kyanite, talc, sphalerite, and rutile. An example is shown in Figure 2.13*a* and the structure of a simple type of stacking fault is illustrated in Figure 2.13*b*.

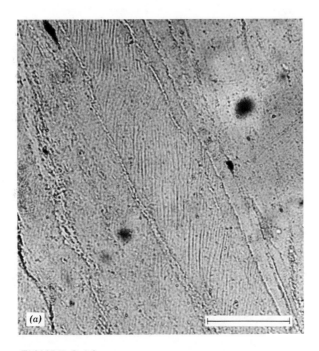

FIGURE 2.12   *(a)* Deformation lamellae within the
deformation bands of Figure 2.11*b*. Plane-polarized
light. Notice, that, in this example, the deformation
band boundaries have the same optical characteristics
as do the deformation lamellae. [From McLaren and
Hobbs (1972), *Geophysical Monograph Series*, Vol. 16,
pp. 55–66. Copyright © by American Geophysical
Union.] Scale mark is 0.3 mm. *(b)* Tangled
dislocations and dense arrays forming deformation
lamellae in experimentally deformed synthetic
quartz. The dense arrays forming deformation
lamellae are approximately parallel to $\{10\bar{1}0\}$.
(Photograph by D. Morrison-Smith.) Scale mark is
1 μm.

If a stacking fault ends within a grain, the fault boundary is a dislocation
that does not leave crystallographic registry behind as it moves. Such dislocations are called *partial dislocations* (Fig. 2.13*b*) and are common in minerals with
stacking faults. Stacking faults have no optical expression but are important
mechanically in that their presence influences the propagation of dislocations.
Some stacking faults act as nuclei for twinning.

FIGURE 2.12 (Continued)

Twin boundaries are planar boundaries in crystals across which the crystal structure is related in a particular symmetrical manner known as a *twin relationship*. The twinning operation involves an element of symmetry not present in the untwinned structure. The crystal structure either side of the boundary may be related by a simple reflection across some lattice plane (*reflection twin*) or by a rotation about an axis (*rotation twin*). An example of a reflection twin is the Brazil twin in quartz where the structure either side of the boundary is related by a reflection across (0001) [McLaren et al. (1967)]. An example of a rotation twin is the albite twin common in the plagioclases where the structure either side of the boundary is related by a rotation through 180° about the normal to (010).

Twins may be formed either during growth of the crystal or during deformation. Some of the deformation twins that have been observed in common minerals are given in Table 2.1. During deformation twinning the lattice points on one side of the twin boundary may be thought of as undergoing a homogeneous shear with respect to lattice points on the other side of the boundary. In addition to this homogeneous shear the atoms associated with each lattice point undergo minor displacements known as *shuffles* in order to preserve the crystal structure. An example is presented in Figure 2.14 for deformation twinning of calcite.

FIGURE 2.13   *(a)* Stacking faults (indicated by the fringe patterns) and partial dislocations (where the fringe patterns end) in experimentally deformed plagioclase. (Photograph by D. Marshall.) Scale mark is 1 μm. *(b)* A view normal to a stacking fault between the lines *AB* and *CD* on (0001) of a hexagonal close packed structure or $\{111\}$ of a face-centered cubic close-packed structure. *AB* and *CD* represent the lines of partial dislocations. [From W. G. Moffatt, C. W. Pearsall, and J. Wulff (1964), *The Structure and Properties of Materials,* Vol. I. Copyright © 1964 by John Wiley and Sons, Inc. Reprinted by permission of John Wiley and Sons, Inc.]

It is thought that deformation twinning takes place by a dislocation mechanism, but the dislocations involved (*twinning dislocations*) must commonly be partial dislocations since they do not produce lattice registry as they move through the crystal. Figure 2.15*a* shows twinning dislocations in a pyroxene, arranged along the boundary of a deformation twin.

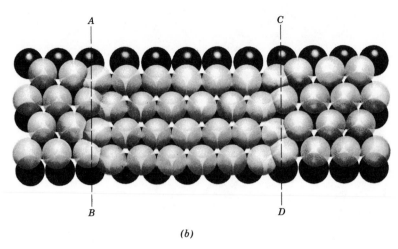

(b)

FIGURE 2.13 (Continued)

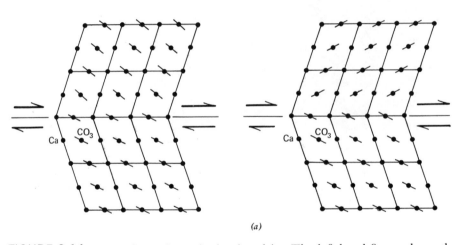

(a)

FIGURE 2.14 *(a)* Deformation twinning in calcite. The left-hand figure shows the result of a simple shear on the calcite structure. The solid dots represent Ca atoms and the lines with a dot in the middle represent CO_3 groups. The top half of this left-hand diagram has been sheared with respect to the bottom. Notice that although the calcium atoms and the carbon atoms are in their proper position in the sheared state, the plane of the CO_3 groups have not been rotated far enough by the simple shear to create a perfect $CaCO_3$ structure. In the right-hand part of the diagram a shuffle has been added to the simple shear to rotate each CO_3 group and thereby create a perfect calcite structure in the top half of the sheared structure. *(b)* Albite twinning in a peristerite, An_5. The thick lamellae are albite twins whereas the thin lamellae are peristerite unmixing lamellae. (Electron micrograph by D. Marshall.) Scale mark is 1 μm.

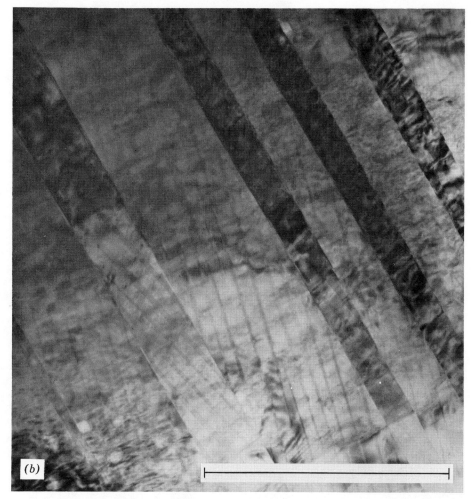

FIGURE 2.14 (Continued)

2.2.2 Principles of Microstructural Development

As a basis for the study of more complicated situations we first examine the sequence of events that takes place when a mineral is deformed at a relatively low temperature and is then heated at a higher temperature in the absence of a deforming stress. This is the type of microstructural sequence that a low-grade regional metamorphic rock might go through when it undergoes later contact metamorphism around a younger intrusion. Having established the changes that occur in this simple environment we can proceed to more com-

FIGURE 2.15   (a) Twin in experimentally deformed pyroxene. Notice the dislocations along the twin boundary. (Photograph J. Boland.) Scale mark is 1 μm. (b) Stacking faults outlined by the fringe patterns and partial dislocations where the fringe patterns end in a naturally deformed orthopyroxene. The alignment of the partial dislocations one above the other creates a polygonization structure. (Photograph A. McLaren and M. Etheridge.) Scale mark is 1 μm. (c) Subgrains in naturally deformed quartz grain. Mt. Isa, Australia. (Specimen courtesy C.J.L. Wilson.) Scale mark is 0.3 mm.

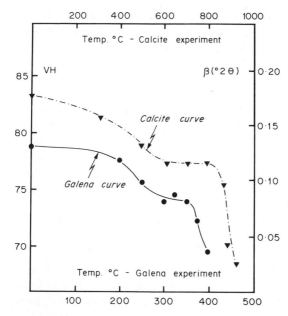

FIGURE 2.16   Experimental annealing of
galena and calcite. The calcite curve shows the
change in line broadening at various temperatures
whereas the galena curve shows the change in the
Vickers Hardness at various temperatures. [From
R. L. Stanton, *Ore Petrology*. Copyright © 1972 by
McGraw-Hill Book Co. Used with permission of
McGraw-Hill Book Co.]

plicated situations where deformation is taking place along with recrystallization. We develop the discussion mainly for a monomineralic rock but the
principles are valid for a polyphase rock as well.

When a crystal with a high density of dislocations is heated above the temperature at which it was deformed a number of microstructural changes take
place. These changes are all mechanisms of lowering the free energy of the
deformed crystal and the first step is to reduce the *stored energy of deformation*
by decreasing the dislocation density and by arranging dislocations into low
energy configurations. These changes may be divided into two overlapping
and competing processes called *recovery* and *primary recrystallization*. The second of these processes results in a polycrystalline aggregate forming where
once there was a single crystal so that the next step is to attempt to decrease
the free energy of the system further by increasing the average grain size of
this new aggregate. This step decreases the *grain boundary tension* [see Hirth
and Lothe (1968, pp. 637–638), for a discussion of this term]; this is often

loosely referred to as the *surface tension* by analogy with liquids. The processes that operate in this step are called *normal grain growth* and *secondary recrystallization*. These four processes are discussed below. It should be realized that all of these processes could be taking place at the same time in different parts of a single aggregate.

Recovery. Recovery is the name given to all those processes that attempt to return a deformed crystal to the undeformed state without the formation and migration of high angle boundaries. The introduction of a high density of dislocations into a crystal, together with high densities of point defects, results in drastic changes in a number of physical properties including the hardness, the width of X-ray diffraction peaks (this is called *X-ray line broadening*), the density and the electrical conductivity. During recovery these properties tend to return to values characteristic of the undeformed crystal and, at the same time, the stored energy of deformation decreases. Figures 2.16 illustrates some of these features for experimentally deformed marble and naturally deformed galena.

Most of these effects are associated with a general decrease in the dislocation density resulting from the edge components of dislocations climbing until they meet another dislocation of opposite sign and annihilating each other, or until they meet a grain boundary; the screw components cross slip to produce similar results. Since this process involves climb, recovery can only take place at a temperature where solid-state diffusion can proceed at a rate sufficient to produce an appreciable climb rate.

As the dislocations move they can group themselves into stable low energy configurations that usually take the form of low angle boundaries. This aspect of recovery may lead to an arrangement of polygonal subgrains within larger grains, the orientation difference between adjacent subgrains being only a few degrees at most. This is called *polygonization* (Fig. 2.15b) and some of the variations in extinction seen within grains of minerals, such as quartz and olivine in crossed polars, is an expression of the existence of these subgrains (Fig. 2.15c).

Primary Recrystallization. This is the process that removes the stored strain energy remaining after recovery. Primary recrystallization is characterized by the formation and growth to impingement of strain-free grains within the strained crystal. The strain-free grains always have a greatly different orientation to that of the strained crystal so that primary recrystallization involves the formation and migration of high angle boundaries. The nucleation stage for strain-free grains is the subject of considerable controversy; it had been thought that a nucleus formed during a thermally activated event when a group of atoms came together in such a way as to produce a submicroscopic volume of unstrained crystal structure within the highly strained imperfect host structure. If this region is large enough then the stored strain energy eliminated by growth of the nucleus into the strained host outweighs the increase in surface-free energy resulting from an increase in the surface area

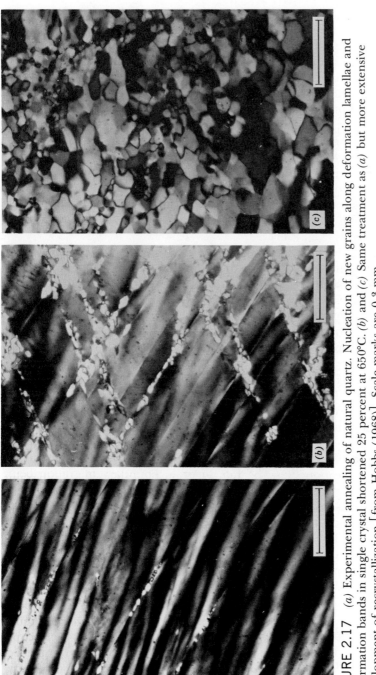

FIGURE 2.17   (a) Experimental annealing of natural quartz. Nucleation of new grains along deformation lamellae and deformation bands in single crystal shortened 25 percent at 650°C. (b) and (c) Same treatment as (a) but more extensive development of recrystallization [from Hobbs (1968)]. Scale marks are 0.3 mm.

of the nucleus. Orowan (1954) was the first to point out that such a nucleation process is energetically unlikely although calculations of the type used by Orowan fail to take into account small chemical changes that might be associated with such a nucleation process. Changes in free energy resulting from small chemical changes could supply the extra energy required to drive the classical nucleation process and could be very common in minerals; an example has been proposed in the case of recrystallization of mica [Etheridge and Hobbs (1974)]. In many situations, however, nuclei are strongly deformed portions of original grains which, during the initial stages of recovery, change their detailed dislocation structure, first becoming subgrains and then new grains by the development of high angle boundaries between themselves and the remainder of the original grains [Hu (1963)]. This is the beginning of primary recrystallization. In general the newly appearing grains form in regions of high strain such as along kink boundaries, other deformation bands, or grain boundaries (Fig. 2.17).

The high angle boundaries sweep through the deformed crystal and remove most of the dislocations remaining from the recovery stage. As they do so, the newly recrystallized grain grows. However, not all high angle boundaries move at the same rate, and it is found that certain special orientation differences between the new and old grain lead to high angle boundaries that move faster than others. The precise reason for this anisotropy of grain boundary migration has not been established, but it is clear that boundaries across which there is a maximum of coincidence of crystal structure move faster than others. Thus, in face-centered cubic metals those boundaries move fastest, for which a $\langle 111 \rangle$ direction is common from new to old grain and the two crystal structures differ by a rotation about $\langle 111 \rangle$ of 38°. This rotation produces the best fit between the new and old grains other than coincidence. Low angle grain boundaries are generally relatively immobile.

Grains formed during primary recrystallization grow toward each other from isolated nucleation sites, the rate of grain boundary migration being controlled by the orientation difference between the new and old grains and by differences in dislocation density within the old grain. Thus, when the new grains finally impinge, they have rather irregular shapes and a large range of sizes (Fig. 2.17). This irregularity and size range is soon modified by the adjustment of grain boundaries to reduce grain boundary tension.

Normal Grain Growth. Normal grain growth is a process that reduces the grain boundary tension of a polycrystalline aggregate. It may occur in an aggregate that has originated by primary recrystallization or it may occur in a strain-free, fine-grained aggregate (such as a flint) that is simply heated to a relatively high temperature.

If each grain in a rock was structureless then the grain boundary tension would be minimized by each grain adopting the shape that has the smallest surface area for a particular volume. If each grain was isolated in space this

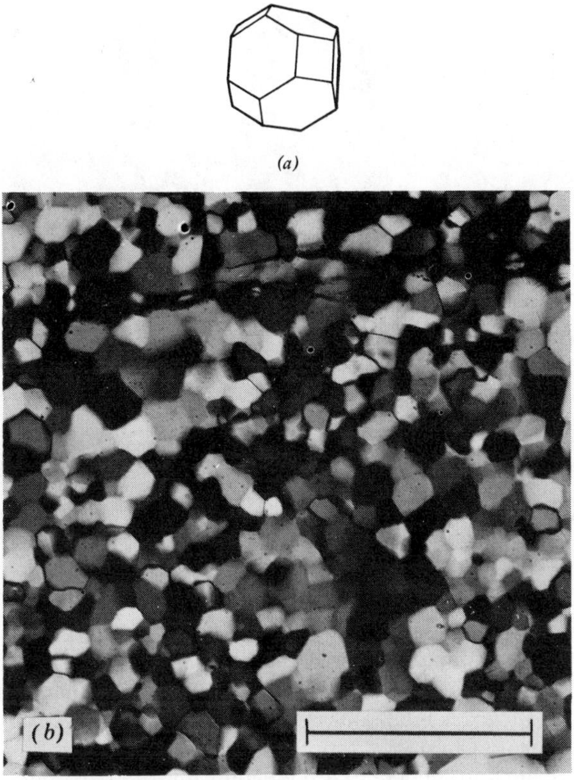

(a)

FIGURE 2.18   (a) Truncated octahedron. (b) The
shapes of grains that have grown together to produce
the minimum grain boundary tension configuration.
Experimentally deformed and annealed single crystal
of synthetic quartz [from Hobbs (1968)]. Scale mark
is 0.3 mm.

would be a sphere but if, as in a metamorphic rock, each grain fits together
with its neighbor to fill space with no voids, then the shape adopted is that
polyhedron that has the smallest surface area for a given volume. This is a
figure that approximates a truncated octahedron in shape [Smith (1948,
1964)]. This polyhedron has fourteen faces, eight of which are regular hexa-
gons and six of which are squares (Fig. 2.18a). In reality, the constraint that
three faces should include a dihedral angle of 120° means that the ideal shape
adopted by each grain is a slightly distorted truncated octahedron in which
each face is not plane but has a double curvature [Kelvin (1911, pp. 297–
309)].

FIGURE 2.19 *(a)* Variation in grain size related to volume density of small micas. The less the number of micas the larger the grain size. Mylonite from Arltunga Complex, Central Australia. Scale mark is 0.3 mm. *(b)* Quartz grain shape related to distribution of mica grains. Schist from Chester, Vermont, U.S.A. Scale mark is 0.3 mm. *(c)* Coarse grain size in quartzite, amphibolite grade metamorphism. The boundaries between quartz grains are largely independent of the distribution of micas. Mt. Isa, Australia. Scale mark is 0.3 mm.

These statements apply only to an aggregate of structureless grains [see Smith (1948)]. However, the crystalline nature of grains in a deformed rock means that the shape of a grain isolated in space that has the minimum surface tension is not a sphere but is a crystal having the common crystal forms developed. In an aggregate of quartz the ideal shapes of Figure 2.18*a* are common. In mica aggregates, on the other hand, there is always a strong tendency to modify the ideal shape of Figure 2.18 by developing the {001} form.

Once the ideal shape of Figure 2.18 is established the grain boundary tension may be further decreased by decreasing the number of interfaces in the aggregate. That is, the average grain size increases. There is no theoretical limit to the grain size that may be attained by this process, but in practice the average grain size becomes stationary, even if the temperature remains high, for one of two reasons. The first is called *orientation inhibition* where grain

FIGURE 2.19 (Continued)

growth ceases because a large number of grain boundaries become low angle boundaries due to an otherwise induced very strong crystallographic-preferred orientation in the aggregate. The second is called *inclusion inhibition* where a large number of grain boundaries become attached to minute second-phase particles such as micas so that grain growth virtually ceases. Inclusion inhibition appears to be very common in metamorphic rocks and is often responsible for grain-size variations in adjacent layers with varying mica contents (Fig. 2.19a). A preferred orientation of mica grains may also lead to a strong shape-preferred orientation of other growing grains (Fig. 2.19b). At relatively high temperatures the motion of grain boundaries may not be impeded by small second-phase particles, and microstructures, such as that illustrated in Figure 2.19c, develop where the small particles are not restricted to grain boundaries but are distributed within large grains. Such a microstructure is typical of rocks occurring at high grades of metamorphism. Examples of microstructures developed due to the adjustment of grain boundary tension are given by Vernon (1968, 1970).

Secondary Recrystallization.   Secondary recrystallization is a process that again tends to decrease the grain boundary tension of an aggregate. It occurs within an aggregate that has undergone normal grain growth and where growth has ceased due either to orientation or inclusion inhibition. If the aggregate is heated to a higher temperature some boundaries, which are slightly more mobile than others, either because they are high angle or because they do not contain a sufficiently high density of inhibiting inclusions, begin to move. This means that a few grains in the aggregate suddenly start to grow consuming the smaller grains around them.

(c)

FIGURE 2.19 (Continued)

2.2.3 Microstructures Developed in Rocks Undergoing Deformation

The sequence of events described in Section 2.2.2 is generally complicated in nature by deformation proceeding at a temperature such that recrystallization occurs at the same time. To some extent, then, the microstructures that develop during deformation are governed by the rate at which recrystallization occurs compared to the rate at which grains are internally deformed. Thus, the microstructures observed in deformed rocks are a function of the strain rate, the temperature at which deformation occurred and, for silicates, presumably of the (OH) content as well, since this appears to influence the rates of deformation and of recrystallization so drastically. The effects of hydrostatic pressure are not clear; there is some indication that the orientation of some intragrain structures such as deformation lamellae may be a function of hydrostatic pressure [Avé Lallemant and Carter (1971)] but, apart from the work of Avé Lallemant and Carter, little systematic investigation of the effect of hydrostatic pressure on microstructures has been made. The microstructures developed in rocks are also a function of the thermal history experienced by the rock since deformation ceased so that, for example, a rock strongly deformed at low temperatures may undergo recovery and primary recrystallization if maintained at moderate temperatures for long periods subsequent to deformation. One of the aims of the structural geologist is to unravel the relative influence of these various factors to arrive at a better understanding of the deformational history.

Thus, we expect that rocks deformed at relatively low temperatures or high strain rates will show widespread evidence of the deformation, the grains being flattened with deformation bands, deformation lamellae, and undulatory extinction common. The electron microscope may reveal high densities of tangled dislocations [McLaren and Hobbs (1972)], perhaps of the order of 10^{11} cm^{-2}, which is close to the saturation density that any crystal can contain.

FIGURE 2.20   (a) Serrated grain boundaries in quartzite that is just starting to undergo recrystallization around the grain boundaries. The grain boundaries are growing into regions of high dislocation density in surrounding grains. Quartzite from Arltunga Complex, Central Australia. Scale mark is 0.3 mm. (b) Recrystallization along grain boundaries in quartz mylonite. Wyangala Dam, New South Wales. Scale mark is 0.3 mm.

FIGURE 2.20 (Continued)

If deformation takes place at higher temperatures or lower strain rates so that solid-state diffusion becomes important on the scale of a grain, dislocations have the opportunity to climb under the influence of the stress causing deformation [Hirth and Lothe, 1968 (pp. 78–82)]. If diffusion is important at this scale, then grain boundaries may move or newly recrystallized regions may grow. The precise microstructure that forms is the result of a number of competing processes and no definite statements can be made at present. If the initial grain size is small so that the total grain-boundary area of grains is large, then normal grain growth may be the only microstructural change [Green et al. (1970)]. Then, even though the rock is strongly deformed, there may be little obvious sign of the deformation in the microstructure. On the other hand, the deformation may proceed under conditions where high densities of dislocations are built up locally in subgrain or deformation band boundaries [Tullis et al. (1973)]. Preexisting grain boundaries may migrate more quickly along these regions producing the serrated grain boundaries characteristic of some rocks (Fig. 2.20a). Under other conditions, regions of high strain, such as grain boundaries, may recrystallize during the deformation to produce an aggregate of small, strain-free grains surrounding obviously deformed grains [Fig. 2.20b. See also Tullis et al. (1973)].

The types of microstructures that are able to form in deformed rocks are

FIGURE 2.21   (a) Flattened grains of quartz mylonite. Arltunga Complex, Central
Australia. Scale mark is 0.3 mm. (b) Deformation bands in quartz mylonite, Wyangala
Dam, New South Wales. One large quartz grain has developed a regular pattern of
subgrains whereas a large neighboring grain has now been subdivided into a large
number of ribbon-shaped deformation bands. Scale mark is 0.3 mm. (c) Experimental
recrystallization of naturally deformed galena. New grains nucleate in deformation
band boundaries and then grow preferentially along these boundaries to form an
aggregate of elongate grains [from Stanton (1970)]. Scale marks are 0.2 mm. (d)
Elongate quartz grains controlled in their shape by the distribution of elongate
sillimanite needles. Sillimanite gneiss, Broken Hill, New South Wales. Scale mark is 0.3
mm.

quite varied and there is ample scope at the present time to systematically
describe and relate them to the type of deformation and grade of metamorphism the rock has experienced. Certainly the observation that the grains in a
deformed rock show no obvious signs of deformation is not sufficient to show
that the rock has recrystallized *after* the deformation ceased [see Green et al.
(1970)].

Similarly, the presence of flat grains of minerals such as quartz, calcite or
olivine in a deformed rock is not sufficient to indicate that the grains have
been flattened. Such flat grains may originate in one of four ways:

1. The grains may have been flattened during the deformation by any of a
    number of processes including slip, diffusion, and solution (Fig. 2.21a).

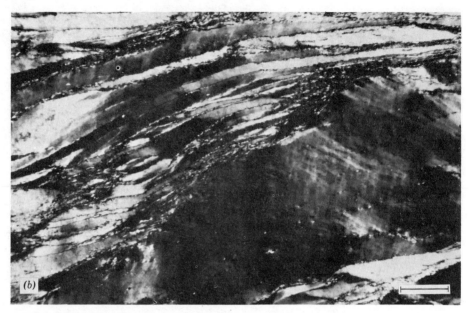

FIGURE 2.21 (Continued)

2. The "grains" may be elongate deformation bands within grains (Fig. 2.21*b*). Where this microstructure is well developed it may be very difficult to distinguish from (1).

3. The microstructure may develop during recrystallization by new grains growing preferentially along planar regions of strong deformation such as deformation lamellae (Fig. 2.21*c*).

4. The microstructure may develop during normal grain growth due to preferential inhibition of grain boundary migration by a preferred orientation of micas (Fig. 2.21*d*).

2.3 CRYSTALLOGRAPHIC PREFERRED ORIENTATIONS IN DEFORMED ROCKS

When a rock is deformed it commonly develops a preferred orientation of the crystallographic directions in the constituent minerals. It is normal practice to represent the pattern of preferred orientation of a particular crystallographic direction by a density map showing the manner in which the distribution of that direction varies in space. Such maps are generally in the form of an equal area projection and are known as *pole figures* or *fabric diagrams* (see Appendix A). Examples of common patterns of preferred orientation are illustrated in Figure 2.22. Until recently it has been usual to construct these patterns using a

FIGURE 2.21 (Continued)

universal stage on an optical microscope, but X-ray techniques are becoming more widely used [Baker et al. (1969)]. Although many deformed rocks display strong crystallographic preferred orientations, there are many that display random patterns of preferred orientation. There has been a widespread opinion that such random patterns indicate that the rock has recrystallized in a nondeviatoric stress field subsequent to deformation. This may be true for some rocks but, for instance, there is also an indication that there exists a distinct deformation environment, characteristic of medium grades of metamorphism, where random quartz patterns develop during deformation [Hopwood (1968); Wilson (1973)].

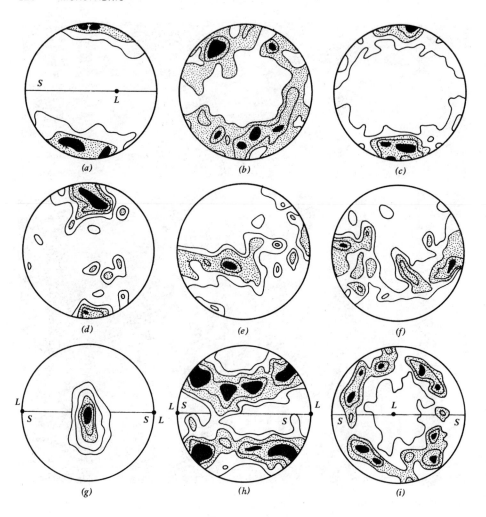

FIGURE 2.22 Common patterns of crystallographic preferred orientations in naturally deformed rocks: (a), (b), and (c) Common patterns of preferred orientations of c axes in calcite-rich rocks. (d), (e), and (f) A common pattern of preferred orientation in olivine-rich rocks. (g) to (o) Common patterns of preferred orientation of c axes in quartz-rich rocks.

(a) 240 c axes in deformed, large calcite grains. Contours are 1, 3, and 6 percent per 1 percent area. Vennatal, Tyrol. [After Felkel (1929).] S is foliation, L is lineation. (b) 263 c axes, calcite. Contours are 1, 2, and 3 percent per 1 percent area. [After Sander, (1950), D99.] (c) 209 c axes, calcite. Contours are 1, 3, 5, and 7 percent per 1 percent area. Undeformed Yule marble. [After Griggs (1940).] (d) X, (e) Y, and (f) Z from 120 grains of olivine. Contours are 1, 2, 4, and 6 percent per 1 percent area. Ilmenspitz, Italy. [After Andreatta (1934).] (g) 138 c axes of quartz. Contours are 0.5, 4, 8, 12, and

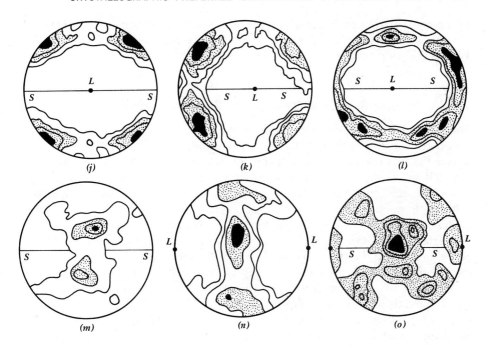

FIGURE 2.22 (Continued)

16 percent per 1 percent area. Slickenside in Melibokus granite. [After Sander (1930).] *(h)* 284 *c* axes of quartz. Contours are 0.5, 1, 2, and 3 percent per 1 percent area. Maximum concentration 10 percent per 1 percent area. Hartsmanndorf. [After Sander (1930).] *(i) c* axes of quartz. Contours are 1, 3, 4, and 5 percent per 1 percent area. Quartzite, Brome County, Quebec. [After Fairbairn (1939).] *(j)* 300 *c* axes of quartz. Contours are 1, 3, 5, and 8 percent per 1 percent area. Mylonite, Moine Thrust. [After Christie (1963, D2).] *(k)* 200 *c* axes of quartz. Contoured according to Kamb (1959a). Contours are 2, 4, 6, 8, and 10 σ. Newly recrystallized grains in mylonite, Central Australia. [After Bell and Etheridge (1974).] *(l)* 200 *c* axes of quartz. Contours are 1, 2, 3, and 4 percent per 1 percent area. Mylonite, Moine Thrust. [After Christie (1963, D9).] *(m) c* axes of quartz. Contours are 1, 5, 7, and 10 percent per 1 percent area. From shear zone, Teshima granite, Japan. [After Hara et al. (1973).] *(n) c* axes of quartz. Contours are 5, 10, 20, and 40 percent per 1 percent area. Mugle gneiss, Upper Styria. [After Schmidt (1932).] *(o)* 115 *c* axes of quartz. Contours are 1, 2, 3, and 4 percent per 1 percent area. Granulite, Saxony. [After Sander (1930).]

There appear to be two main mechanisms by which a crystallographic preferred orientation may develop in deformed rocks. First, at low temperatures or high strain rates where recrystallization does not occur, a preferred orientation may develop due to rotation of inequant grains or to slip within grains and accompanying grain rotations. Second, under conditions where recrystallization is widespread, a preferred orientation may be associated with the recrystallization process.

2.3.1 Preferred Orientation Developed by Slip and Rotation

Inequant grains embedded in a deforming matrix develop a preferred orientation related to the strain that the body undergoes. The general situation, where the inequant grains are capable of deforming and the matrix has complicated mechanical properties, has so far received little mention in geology [however, see Gay (1968b)] but some progress has been made for two different simplified situations. One of these considers the case of rigid rods and plates embedded in a deforming, perfectly viscous fluid [Gay (1968b)]. The preferred orientation here develops due to the torques exerted by the deforming fluid on the rigid objects. The results of this analysis indicate that any prolate ellipsoid will rotate until its long axis is parallel to the axis of maximum extension if the deformation is by pure shear. This is a steady-state orientation. The axis of an oblate ellipsoid will rotate towards the axis of maximum shortening. The other treatment is of plates that behave passively in the deforming matrix. The word "passively" is meant to indicate that there is no mechanical interaction between the matrix and the plates. March (1932) considered the case of an initially uniform distribution of plates deformed by general strain, and Owens (1973) has extended the argument to include any initial distribution of plates. An example is considered in Section 5.2.2. These analyses indicate that during a given strain the plates rotate exactly as the radii of an initial sphere would rotate if the sphere were to undergo the same strain. Tullis (1971) has elegantly demonstrated that mica flakes can rotate during deformation to produce a preferred orientation of (001) normal to the principal axis of shortening. The degee of preferred orientation at least for moderate strains varies with the amount of strain in a manner that is predicted by the March analysis. On the other hand, Etheridge et al. (1974), in very similar experiments to those of Tullis (1971), have shown that crystal growth contributes to preferred orientation development of micas. This presumably arises through anisotropic supply of material to the growing grain. Rotation of mica flakes during deformation is presently a popular mechanism for producing a preferred orientation of platy minerals in slates (see Section 5.2.2).

For some time now, geologists have noted that crystallographic preferred orientations develop in the grains in deformed quartz-rich* rocks without any

---

*Because there is much recent work on quartz, emphasis is placed on this mineral here. There is a host of work on calcite and reference should be made to Turner and Weiss (1963, pp. 335–355).

signs of recrystallization, the preferred orientation development being associated with distortion of the initial grain shapes [Trener (1906); Schmidt (1925); Sander (1930, 1932, 1948); Sahama (1936); Fairbairn (1937); Hietanen (1938); and Phillips (1945)]. Recently, preferred orientations have been produced experimentally without recrystallization in deformed quartzites [Tullis et al. 1973)]. Schmidt in 1925 seems to have been the first to postulate that these preferred orientations developed during slip; he proposed that slip took place in directions of closest packing of Si atoms in the quartz structure. These slip directions include the **c** axis, the **a** axes and the [**c** + **a**] axes. Slip planes included (0001), $\{01\bar{1}0\}$, and $\{01\bar{1}1\}$. However, the strains envisaged by Schmidt in order to predict preferred orientations were unrealistic in that all of them involved only simple shear parallel to the foliation. Sahama (1936), Fairbairn (1937), and Hietanen (1938) diagrammatically present some of the common preferred orientations known for quartz and discuss their development in terms of these slip systems. Although there was no experimental justification for these slip systems at that time, it now appears that many such systems are physically realistic [Christie and Green (1964)]. Furthermore, the theoretical basis for the development of preferred orientation by slip and consequent rotation is now better established [e.g., Taylor (1938); Calnan and Clews (1950, 1951); Bishop (1953, 1954); and Bishop and Hill (1951)].

The reason for the development of preferred orientation as the result of plastic deformation by slip can be directly related to the mechanics of dislocation glide. In each of the constituent crystalline grains of the plastically deforming rock mass the slip process results in a deformation of the individual crystalline grains *but* the crystallographic structure remains more or less undistorted. This is an important point because it means that the crystal axes can be used as an internal reference system for each grain during the deformation.

One slip system operating (see Figure 2.23 described below) results in a progressive simple shearing taking place (with respect to crystal axes). If many slip systems operate simultaneously the resulting deformation may still be approximated by a progressive simple shearing. Such a deformational history (with respect to crystal axes) is rotational, that is, if we consider any deformational increment it may be considered as the combination of a strain and a rigid body rotation. A more precise way of looking at the situation is to examine it kinematically. The deformation described above leads to the crystalline material of each grain developing a *vorticity** with respect to crystallographic axes. (At any instant, the axes of the incremental strain ellipsoid have an angular velocity with respect to the crystal axes.)

---

*Vorticity is related to the curl of the velocity field of the deforming continuum. If infinitesimal paddle wheels were dispersed throughout the continuum in suitable orientations, their rotation would reflect the vorticity at each point.

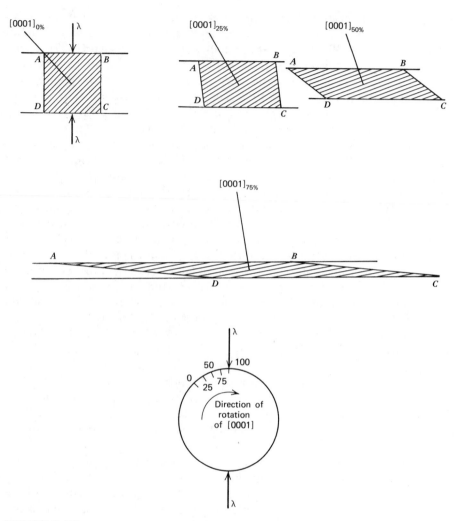

FIGURE 2.23   The principle involved in the development of preferred orientations by crystallographic slip. See text for discussion.

Continuing to discuss the situation kinematically, consider the imposed deformation with respect to an external frame of reference. At any instant the *vorticity of the material in any crystalline grain* (with respect to this external reference frame) *must be that imposed vorticity.* However, the vorticity of the deforming material with respect to crystal axes need *not* be the same as this imposed vorticity with respect to external coordinates and, in general, is different. The geometric constraints of the situation require that the crystallographic axes of the individual grains rotate with respect to these external axes at a compensating rate to ensure that the above condition (in italics) is maintained. These lattice rotations of the grains in a plastically deforming rock mass usually conform to some sort of pattern, and a strong pattern of preferred crystallographic orientation can develop as a result.

The principle governing the development of a preferred orientation by slip is illustrated in Figure 2.23. This shows a cross section $ABCD$ through a single isolated grain, which is to be shortened homogeneously parallel to the direction λ as though it was held between the jaws of a vise. We suppose that there is just one slip plane parallel to the diagonal BD and one slip direction in this plane, parallel to BD. The faces AB and DC are held constant in orientation during deformation. If slip is the only mechanism of deformation then the important geometrical features of the deformation are that

1. The dimension of the grain measured parallel to the slip direction cannot change during deformation.
2. The projection of the distance between two points measured normal to the slip plane cannot change during deformation.

Thus, the sequence of shapes adopted by the grain as it deforms is that shown in Figure 2.23. In this particular example, the crystallographic slip plane remains parallel to the short diagonal of the resulting parallelogram, but rotates progressively relative to the direction of shortening such that the normal to the slip plane rotates towards the axis of shortening. If the slip plane was (0001) in quartz, for instance, large amounts of shortening would orient the c axis near to the shortening axis. If all grains in a quartzite, initially with random c axes, were free to behave in this manner, then the result of higher and higher strains will be a progressively higher concentration of c axes near to the axis of shortening.

However, the grains in a rock are not free to behave in this manner. If a rock is to behave ductilely during a deformation, then the grain boundaries must remain in contact, otherwise cracks will develop and the rock will ultimately fail in a brittle manner. One way for the rock to undergo a general strain, maintaining continuity across grain boundaries, is to require that strain be homogeneously distributed throughout its constituent crystalline grains. In order for these grains to undergo a general strain by intracrystalline slip, five independent slip systems must generally operate within each grain. This is the von Mises condition (see Paterson, 1969).

Two slip systems are said to be *independent* if a shape change produced by shear on one system cannot be produced by shear on the other system. This applies to infinitesimal deformations. For example, shear parallel to the a axis on the basal plane of quartz leads to a shape change identical to that produced by an equal amount of shear parallel to the c axis on the $(2\bar{1}10)$ plane of quartz (See Fig. 2.24). In fact for the 12 slip systems in the two symmetry sets of mechanisms $\langle \mathbf{a} \rangle \{0001\}$ and $\langle \mathbf{c} \rangle \{2\bar{1}\bar{1}0\}$ not more than 2 slip systems can be chosen at any one time that are linearly independent. If a third system is chosen, then any shape change produced by the 3 systems together can be duplicated by 2 systems operating together (as long as they are linearly independent). These matters are discussed in detail by Paterson (1969).

The requirement of 5 independent slip systems to produce a general deformation arises because the slip process is a constant volume deformation. In order for an increment of strain to be a constant volume deformation, the relationship $\epsilon_{11} + \epsilon_{22} + \epsilon_{33} = 0$ holds [see Jaeger (1969, p. 47 and Section 1.3.1)]. This means that, of the six independent components of strain defined by the array (1.16), only five are independent for an infinitesimal, constant

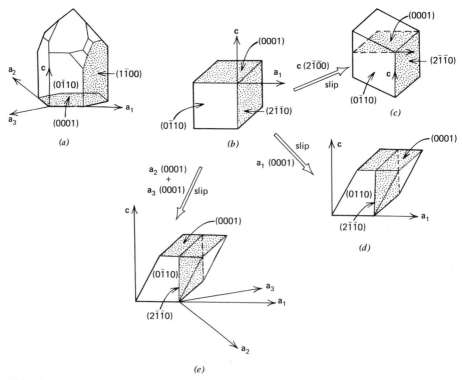

FIGURE 2.24   Independent slip systems. (a) The crystallography of α quartz. (b) A cube of quartz with faces (0001), (2$\bar{1}\bar{1}$0), and (0$\bar{1}$10). (c) The same cube distorted by **c** (2$\bar{1}\bar{1}$0) slip. The cube is distorted to form a parallelepiped with faces (2$\bar{1}\bar{1}$0), (0$\bar{1}$10), and another plane that is irrational. (d) The cube of (b) distorted by $\mathbf{a}_1$ (0001) slip. The cube becomes a parallelepiped with faces (0001), (0$\bar{1}$10), and an irrational face. (e) The cube of (b) deformed by simultaneous slip parallel to $\mathbf{a}_3$ (0001) and $\mathbf{a}_2$ (0001). This parallelepiped is identical to the one produced in (d).

volume, deformation. In order to produce such a strain, then, five independent simple shears are required; this is what gives rise to the requirement of 5 independent slip systems to produce a general deformation of a crystal.

The operation of these slip systems produces a distortion of each grain together with a rotation of crystallographic axes relative to the strain axes. At high enough strains a developing pattern of preferred crystallographic orientation may become recognizable. This pattern continues to develop as strain increases although its nature may change.

The requirement of 5 independent slip systems is modified if cross slip or climb become important modes of deformation: in general the number of systems required to produce an arbitrary strain is decreased [see Groves and Kelly (1969)]. For deformations at higher temperatures where cross slip, climb, and some degree of grain boundary adjustment occurs, preferred orientation development under some circumstances could be caused by grain boundary sliding if there is shape asymmetry related to crystallography. If this mechanism is not important then, unless there is some degree of crystallographic slip taking place, no preferred orientation can develop. If all the deformation were by climb, for instance, then no preferred orientation develops [see Groves and Kelly (1969)].

The advances made in the theory of preferred orientation have been considerable in the last few years. Some early theories such as those of Schmid and Boas (1935) led to models such as that proposed by Calnan and Clews (1950). These workers considered that grains stayed together only by virtue of a high degree of heterogeneous deformation at the grain boundaries. Such models are capable of predicting some patterns of preferred crystallographic orientation. They are, however, incapable of quantitative application to the problem and are restricted to rather simple qualitative predictions for simple deformational histories. These matters are discussed by Lister (1974).

In 1938 Taylor developed a theory that took into account the difficulty in determining the stress state in a grain in an aggregate and, in a somewhat simplified manner, the compatibility of strain that must exist between grains in order for the aggregate to behave ductilely. He simply postulated that the strain in every grain should be the same, that is, the strain throughout the aggregate should be homogeneous. There has been much criticism of the Taylor model over this point but Taylor has supplied what seems to be adequate answers to these criticisms (Taylor, 1955). Taylor calculated for each grain that combination of five independent slip systems that would produce the imposed strain and at the same time do the minimum amount of work. Having determined those slip systems which would satisfy these constraints, he was able to calculate the lattice rotations which would result during the imposed increment of strain. In this way he calculated the grain rotations which would lead to the gradual development of a pattern of preferred orientation. Taylor's approach has been criticized on three main grounds:

1. The approach assumes homogeneous strain whereas it is commonly observed that the strain varies considerably from one grain to another within a deformed aggregate.

2. In 1940 Barrett and Leverson performed a series of experiments designed to check Taylor's predictions. They found that Taylor was in agreement with about one-third of their observed rotations, disagreed in another one-third, and in the remainder no definite statement could be made. This test is commonly cited when attempting to show that the Taylor approach is invalid.

3. It is commonly argued that five independent slip systems do not operate in every grain but, at the moment, such evidence is inconclusive. It is suggested that two or three active simultaneously appear to be more the normal situation.

These criticisms are discussed briefly below:

1. Taylor made the assumption of homogeneous strain because at that time this was the only way in which he could proceed efficiently with the calculations. After all, Taylor (1938) only had the help of "Mallock's equation-solving machine." The development of large, fast computers has made the consideration of more complicated situations possible. Wonsiewicz and Chin (1970) have shown the Taylor theory can be applied to inhomogeneous strains and Lister (1974) presents a discussion of how general inhomogeneous strains could be treated using the Taylor model. The indications are that the patterns of preferred orientation that develop during the inhomogeneous straining of an aggregate would be much the same although perhaps weaker than those that develop by applying the strict Taylor model.

2. In 1954 the calculations performed by Taylor were repeated by Bishop [see Bishop and Hill (1951); Bishop (1953); Bishop (1954)]. These authors adopted a different approach to preferred orientation development although it is now clear that the approach is mathematically related to that of Taylor [see Chin and Mammel (1969)]. Bishop found that Taylor had made a considerable number of mistakes in his calculations [see Taylor (1955)] and that, in fact, when the calculations were performed correctly the theory produced results that were very closely in agreement with the observations of Barrett and Leverson. A large number of subsequent tests have been made [Hosford (1965, 1966); Chin et al. (1966); Mayer and Backofen (1968); Dillamore et al (1968); Chin et al. (1969); Kallend and Davies (1972)], and it is now well established that the Taylor model is capable of predicting observed patterns of preferred orientation in metals. This is true not only for axially symmetric strains but for general strains as well. The fact that Taylor was wrong in his calculations seems to have been neglected by many sub-

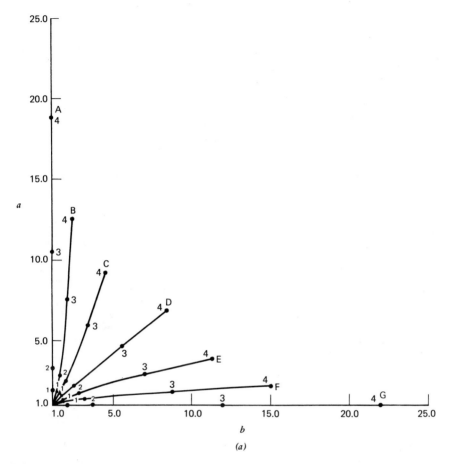

FIGURE 2.25   (*a*) Flinn diagram showing strain paths used to calculate progressive development of quartz *c*-axis orientations by crystal slip accompanied by grain rotation. Each point such as B4, for example, refers to a pattern of preferred orientation shown in Figure 2.25 (*b*), (*c*), and 2.26. (*b*) Progressive development of patterns of preferred orientation of *c* axes in quartz. For principle used in calculations see text. Density diagram printed by computer as multiples of a uniform distribution. These diagrams calculated on the basis that prism slip does not occur, basal slip is relatively easy, and the ratio of critical resolved shear stresses for negative to positive rhomb slip is greater than 1.3 so that negative rhomb slip does not operate; the strain history is coaxial. Labels such as A4 refer to strain state and strain path shown in Figure 2.25*a*. *X* is direction of maximum principal extension. *S* is principal plane of flattening. *L* is also the principal axis of extension. Maximum concentrations are A1: 8; A2: 8; A3: 8; A4: 16; D1: 8; D2: 8; D3: 8; D4: 16; G1: 8; G2: 64; G3: 128; G4: 400 times uniform concentration. (*c*) End products of deformation along strain paths intermediate to those shown in Figure 2.25*b*. Labels and conditions are the same as in Figure 2.25*b*. Maximum concentrations are B4: 16; C4: 16; E4: 16; F4: 64 times uniform concentration. [See Lister and Hobbs (1974) for details.]

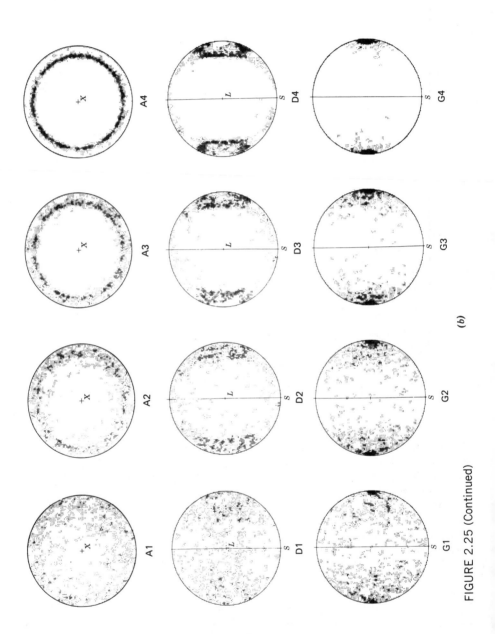

FIGURE 2.25 (Continued)

(b)

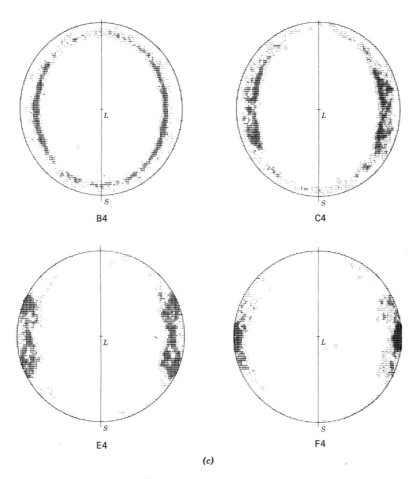

B4

C4

E4

F4

(c)

FIGURE 2.25 (Continued)

sequent critics of the Taylor approach and, for instance, it is not referred to in Barrett and Massalski (1966).

3. Although five independent slip systems are demanded in the strict Taylor approach, when the calculations are actually performed (Lister, 1974) it is found that the major part of the strain in each grain is accommodated by slip on three, or at the most four systems. Although the other systems are operative, the shears parallel to these systems are often very small. This observation, combined with the relaxation of the five independent slip system requirement if climb or cross slip is operative, means that one does not expect to see evidence of five slip systems in most grains. Perhaps two or three would be more the average situation.

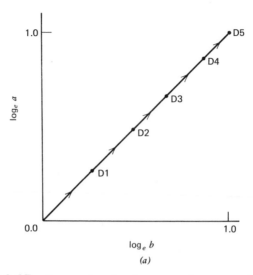

FIGURE 2.27 Progressive development of pattern of
preferred orientation of c axes of quartz in a deformation history
involving progressive simple shear together with progressive
extension parallel to the axis of shear. The deformation history is,
therefore, similar to that involved in the rolling of dough. The
strain path is D1 to D5 inclusive, shown in Figure 2.27a.
Calculations assume the same conditions for slip as is shown in
Figure 2.25. In Figure 2.27b, S_1 is the shearing plane and S_2 is the
principal plane of flattening. X, Y, and Z are principal axes of
strain with X the principal axis of extension and Z the principal
axis of shortening. The sense of shear is shown by the arrows in
figures D4 and D5.
Maximum concentrations are D1:8; D2:8; D3:8; D4:16; D5:16
times uniform concentration. [See Lister and Hobbs (1974) for
details.]

The results of a number of fabric simulations [Lister (1974); Lister and
Hobbs (1974)] for quartz involving coaxial strain histories are presented in
Figure 2.25; in these calculations it is assumed that $\langle \mathbf{a} \rangle \{0001\}$ slip is relatively
easy. It can be seen that the type of preferred orientation varies from a point
maximum of c axes parallel to the shortening axis for a strain history involving
only pure flattening through to a 75° small circle of c axes about the extension
axis for a strain history involving only pure extension. For a strain history
involving only plane strain the preferred orientation pattern consists of a 20°
to 30° small circle about the axis of principal shortening. Results for non-

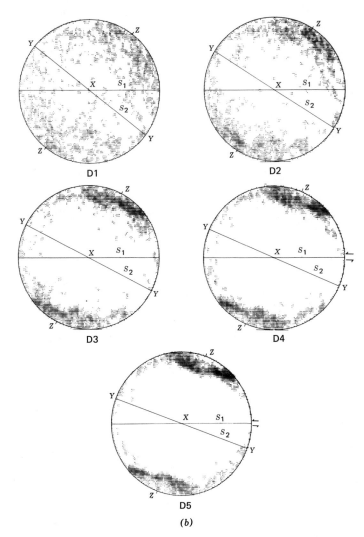

FIGURE 2.27 (Continued)

coaxial strain histories are given in Figure 2.27. A number of points arise immediately from these diagrams.

1. The type of deformation history whether it be coaxial or noncoaxial is immediately obvious in that noncoaxial deformation histories produce diagrams asymmetric with respect to the principal axes of finite strain in accord with symmetry principles [see Paterson and (Weiss, 1961)].

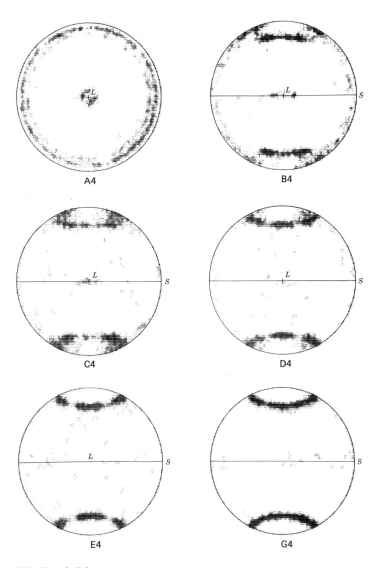

FIGURE 2.26   Preferred orientation patterns developed for *c*
axes of quartz by crystal slip and grain rotation. The diagrams
calculated on the basis that the critical resolved shear stresses for
positive and negative rhomb slip are about equal, basal slip is
relatively easy, and prism slip does not occur; the strain history is
coaxial. Label as in Figure 2.25. Maximum concentrations are A4:
31; B4: 40; C4: 45; D4: 79; E4: 91; G4: 61 times uniform
concentration. [See Lister and Hobbs (1974) for details.] Compare
these results with those of Hara (1971, Fig. 18).

2. The type of deformation history, whether it be uniform shortening, uniform extension, simple shear, or somewhere between these classes of history, is also clear from the nature of the pattern of preferred orientation.

3. The amount of strain is indicated by the degree of preferred orientation since this steadily increases as the amount of strain increases.

4. Something regarding the physical conditions of deformation can be determined since the relative ease with which slip systems operate is a function of temperature, strain rate, and perhaps OH content (see Section 2.2.1). Thus, for instance, the fabrics simulated assuming easy $\langle \mathbf{a} \rangle$ $\{0001\}$ slip for seven coaxial deformation paths (see Figure 2.25) will not be the same as fabrics simulated assuming easier rhomb $\langle \mathbf{c} + \mathbf{a} \rangle$ $\{10\bar{1}1\}$ and $\langle \mathbf{c} + \mathbf{a} \rangle$ $\{01\bar{1}1\}$ slip than in this case. At a critical ratio of yield stresses on the slip systems an entirely different set of patterns develops for the same imposed deformation paths. For example, for axially symmetric shortening, rather than a point maximum of c axes developing about Z, it is found that a strong 25° girdle appears (Fig. 2.26). Fabric transitions are discussed by Lister (1974) and Lister and Paterson (1974) in great detail for quartzite masses.

In summary, one can ultimately expect to gain the following types of information from patterns of preferred orientation developed by crystallographic slip and rotation:

1. The type of strain history, whether it be progressive shortening, extension, plane strain, or somewhere in between.
2. The strain history, whether it be coaxial or noncoaxial.
3. The amount of strain.
4. The physical conditions of deformation.

Some natural examples of preferred orientations developed by slip and a correlation with the type of strain is indicated in Figure 2.28.

2.3.2 Preferred Orientations Developed by Recrystallization

If a mineral recrystallizes in the absence of a nonhydrostatic stress, and after deformation has ceased, there is commonly a crystallographic relationship between the new strain-free grains and the old deformed grain within which the new grains grow. Thus in calcite the new and old grains tend to have an a axis in common and are related by rotation of approximately 30° about this common axis [Griggs et al. (1960a,b); Ferreira and Turner (1964)]. A similar relationship has been reported for quartz (Hobbs, 1968). In olivine the two grains are inclined at 20 to 40° [see Carter (1971)], and in dolomite a weak tendency for a 30° rotation about an a axis has been reported (Neumann,

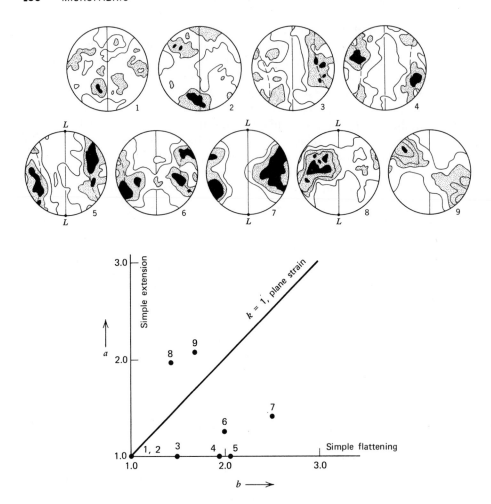

FIGURE 2.28   Natural examples of preferred orientations of c axes produced by crystallographic slip and rotation in deformed quartzites. The state of strain in each case has been established by measuring the shapes of distorted detrital quartz grains [after the method of Dunnet (1969)]. The states of strain in each case are shown in the Flinn diagram. Each projection has been contoured according to Kamb (1959a). First contour is 2σ, contour interval is 2σ. The number of grains for each diagram is 1: 320; 2: 208; 3: 320; 4: 208; 5: 320; 6: 320; 7: 56; 8: 256; 9: 104 [from Marjoribanks (1974)]. The line across the center of each projection is the foliation. L is a lineation where developed. A small circle has been sketched in for diagrams 3, 4, 5, and 6.

1969). This crystallographic relationship may be related to the anisotropic grain boundary migration mentioned in Section 2.2.2, although no investigation of grain boundary structure has been made for these minerals to deter-

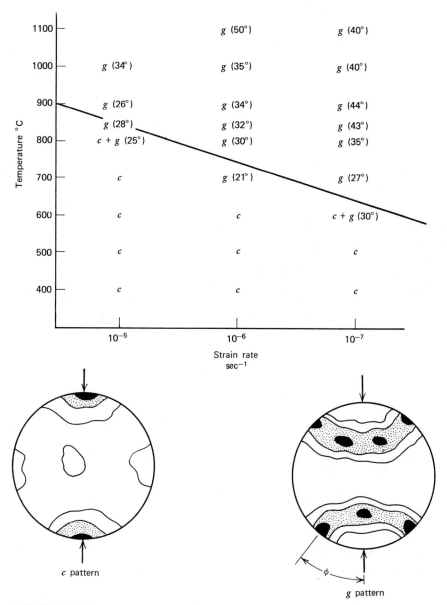

FIGURE 2.29   Experimental results produced by deformation of quartzites. The graph shows a plot of the type of preferred orientation produced at various strain rates and temperatures. Letter c stands for the c pattern shown in the stereogram below and consists of a point maximum of c axes of quartz close to the axis of shortening. The g stands for the g pattern: a small circle of c axes of opening angle ϕ symmetrical about the shortening axis. The number after g in the graph is an estimate of the angle ϕ [after Tullis et al. (1973)].

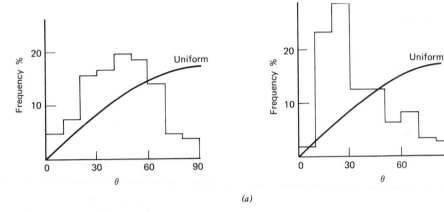

FIGURE 2.30   Preferred orientations produced where there is a host relationship between the newly recrystallized grains and the older host grain. *(a)* Histograms of the angle θ between the c axes in the newly recrystallized grain and that in host grain. That on the left is after Wilson (1973). That on the right is after Bell and Etheridge (1974). *(b)* The preferred orientation of c axes in old and new grains for the left-hand histogram in *(a)*. *(c)* The preferred orientation of c axes for the rock corresponding to the right-hand histogram of *(a)*. *(b)* Both diagrams are 150 grains. Both are contoured according to Kamb (1959a). Contour interval is 0, 2, 4, and 6 σ [after Wilson (1973)]. *(c)* Both diagrams are 300 grains. Both are contoured according to Kamb (1959a). Left: contour interval is 2, 7, 12, 17, and 22 σ. Right: 2, 5.5, 9, 12.5, and 16 σ [after Bell and Etheridge (1974)].

mine if the reported orientations correspond to grain boundaries of high lattice coincidence [see also Bell and Etheridge (1974)].

The effect of such a crystallographic relationship in metals is to alter any preexisting preferred orientation [Beck and Hu (1966)]. Nothing has been done experimentally so far on mineral aggregates except for Green (1967) where a flint initially deformed at low temperatures was heated at a higher temperature so that recrystallization took place. No deformation accompanied the recrystallization. During this process an exceptionally strong point maximum developed from a small circle girdle of c axes formed during the deformation. However, the mechanism by which this was achieved is unknown.

The situation is not clear for minerals that recrystallize during deformation. Thus, Figure 2.29 shows in a diagrammatic form the preferred orientations produced experimentally by Tullis et al. (1973). This diagram, which includes a plot of temperature against the strain rate, shows two fields, one in which recrystallization took place during deformation and one where there was no recrystallization. The diagram shows that the same patterns of preferred

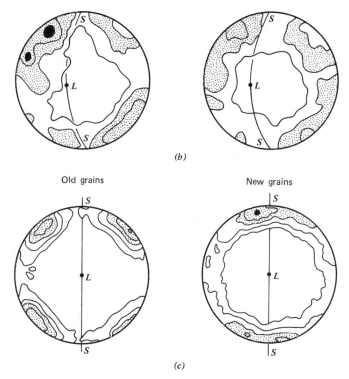

(b)

Old grains   New grains

(c)

FIGURE 2.30 (Continued)

orientation were developed under some conditions with or without recrystallization. Thus, it does appear that, at least in some situations, there is no difference between the patterns of preferred orientation that develop due to slip and those that develop during recrystallization. However, the situation has not yet been fully explored experimentally.

It has been widely proposed in the geological literature that the stress responsible for deformation has an influence on the pattern of preferred orientation that develops and a considerable amount of literature exists [e.g. see Kamb (1961) and Paterson (1973)]. Certainly in experiments a symmetrical relationship between the shortening direction and the preferred orientation pattern commonly exists (Fig. 2.29) but the mechanism by which this develops has never been established. There are some indications in experiments that a crystallographic control of new grains by old grains exists [Ferreira and Turner (1964); Hobbs (1968)]. However, in many cases a stress control over preferred orientations can be proposed equally well. One mechanism for the production of a preferred orientation during deformation is the formation of subgrains early in the deformation which steadily rotate relative to their neighbors during continued deformation as more and more dislocations are added to the subgrain boundaries [Hobbs (1968); Bell and Etheridge (1974)]. Ultimately the misorientation increases to the stage where the subgrain is identified as a new grain and grain boundary migration becomes important. This appears to be the mechanism in some experimentally deformed quartz [Hobbs (1968)]. However, Ardel et al. (1974) who examined the experimentally deformed material of Tullis et al. (1973) using transmission electron microscopy have shown that, at least in these experiments, there is no subgrain development and recrystallization during deformation takes place through the nucleation of discrete strain-free grains which ultimately grow to produce an aggregate of polygonal grains. However, whether this gives rise to the preferred orientations that are observed, or whether that preferred orientation is somehow developed solely by means of a slip process, is not understood. A crystallographic control over newly recrystallized grains has been reported from natural examples in some quartzites [Ransom (1971); Wilson (1973) and examples are presented in Figure 2.30 (See page 138). Commonly, in quartzites, although a crystallographic control may be strong, there is not a marked change in the pattern of preferred orientation between new and old grains (Fig. 2.30*b*). In many quartzites, however, such a crystallographic control does not exist and some other mechanism, controlled perhaps by the stress or the strain, must be operative. That marked changes in preferred orientation do sometimes accompany recrystallization in nature is shown in Figure 2.30*c*.

The influence of nonhydrostatic stress on the preferred orientation that develops during recrystallization has been reviewed recently by Paterson (1973). At the present time, although such an influence cannot be denied, it is difficult to design an experiment that will test whether such an influence exists or not. There is an urgent need for better theoretical and experimental investigation of this problem.

3

PRIMARY STRUCTURES

3.1 INTRODUCTION

This chapter discusses the features of rocks that are present before the onset of deformation. Such structures are called *primary structures* because they are original features of sedimentary or igneous rocks, resulting from deposition or emplacement. Structures reflecting subsequent deformation or metamorphism, which are the subject of most of this book, are *secondary structures*.

Interest in primary structures is twofold; first, the ultimate goal is to understand the total history of a deformed rock and not just its deformational history. Furthermore the processes of deposition and deformation are not necessarily isolated in time. Thus, the history of, for example, a mountain chain, comprises an intermingling of the processes of deposition and deformation in such a way that the type of sediment being deposited can often provide considerable information on the tectonic environment. For example, a 5-km thick sequence of shallow water sediments must have been deposited in a gradually subsiding basin, otherwise the lowest members of the sequence would show deep water features; in this context primary structures are very significant. In addition to this however, primary structures are important for the interpretation of the geometrical, as opposed to historical, aspects of the deformation. It is this aspect of the problem with which we are principally concerned here.

3.2 **PRIMARY STRUCTURES AS MARKERS**

Primary structures play an important role in the interpretation of the structure of deformed areas. Thus, for example, if there were no bedding in sedimentary rocks there would be far less evidence of deformation in deformed rocks. This point becomes important in areas of deformed igneous rocks, such as a deformed ophiolite sequence. In such rocks there may be abundant microstructural evidence indicating that the rock is deformed, but because the predeformational shape of the various units is unknown, and their contacts were probably initially irregular, it is generally not possible to interpret the structure in the sense of being able to predict, with much confidence, the configuration of a given unit or contact beyond the surface of observation. Far more confidence can be attached to the interpretation of a deformed, bedded sequence because it can be assumed that the bedding was initially more or less planar. Thus, in Figure 3.1a without further evidence we

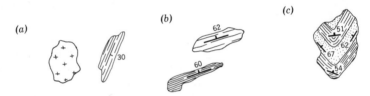

FIGURE 3.1   Interpretation of outcrops; see text for discussion.

can only guess at the orientation of the contact between the igneous and sedimentary rocks; some limitations on its orientation in outcrop are imposed by the position of the outcrops, but the only constraint on the angle of dip is that it cannot be parallel to the topographic surface. In Figure 3.1b, on the other hand, the contact can reasonably be interpreted as being parallel to the bedding orientation observed in the outcrops. The only limitation of this interpretation is that the contact, though locally planar, can be expected to be curviplanar on a larger scale. However, the evidence presented does not indicate whether a change in orientation occurs a few meters or a few hundred meters below the surface. Finally, in Figure 3.1c, if measurement in outcrop indicates that the fold is cylindrical, then projection of surface data down the plunge will generally give a very reliable interpretation of the three-dimensional structure, within the area of the map. But again, if one of the folded units is an intrusive sill, that unit cannot be interpreted with the same confidence as the sedimentary units, because it is known that sills generally jump from horizon to horizon. Similarly some sediments, such as fluvial deposits, are impersistent so that their shape also cannot be reliably interpreted.

   The point, then, is that in order to make a reasonable interpretation of the postdepositional history, and in order to be able to evaluate the probability of an interpretation being correct, it is necessary to have as good a knowledge as possible of the nature of the starting material.

As a marker, bedding is perhaps the most useful primary structure because it is very common and its predeformational nature is usually predictable. Also common and of considerable use, is layering of other origins, for example, layering of igneous origin or the layering common in basement rocks, the origin of which is generally obscure. This layering is used in the same way as bedding but is not as reliable since its predeformational configuration is not as predictable.

Less common, but of considerable value in areas where they occur, are primary features that can be used to analyze the strain of the deformed rock (see Section 1.3). These include pebbles [e.g., Flinn (1956); Ramsay (1965); Hossack (1968); Gay (1969); Ramsay and Sturt (1970)], ooids [e.g., Cloos (1947, 1971); Badoux (1970)], and fossils [e.g., Wettstein (1886); Breddin (1957, 1964); Badoux (1963); Nickelsen (1966)] in sedimentary rocks; and vesicles, lapilli, and deformed crystals [e.g., Stauffer (1967); Oertel (1970, 1971); with discussions by Helm and Siddans (1971) and Mukhopadhyay (1972); Shelley (1971)] in rocks of igneous origin.

3.3 PRIMARY STRUCTURES AS EVIDENCE OF YOUNGING

As well as acting as markers, certain primary structures can also provide very valuable additional information. These structures indicate the direction in which the surrounding rocks get younger or, as it is generally expressed, the *younging* direction of the sequence.

Thus, for example, it is possible, given the right structures, to tell which was the top and which was the bottom of a sedimentary layer or a basalt flow at the time that they were laid down, and this indicates the direction in which the rocks get younger. This can be very valuable information in the interpretation of an area; it may lead to recognition of folds that are otherwise difficult to recognize as in Figure 3.2a or may lead to the recognition of a complex structure in what is otherwise an apparently simple situation, as in Figure 3.2b.

Some of the more reliable and commonly occurring structures, from the point of view of younging criteria, are discussed here; the interested reader however, is referred to a much more exhaustive treatment by Shrock (1948, pp. 1–418) and to the work of Potter and Pettijohn (1963), Pettijohn and Potter (1964), and Allen (1970) on sedimentary structures.

CROSS BEDDING

Cross bedding is defined by Pettijohn et al. (1972, p. 108) as ". . . . . a structure confined to a single sedimentation unit and characterized by internal bedding or lamination, called foreset bedding, inclined to the principal surface of accumulation." It has been treated at some length by these writers and also by such writers as Allen (1968) and Blatt et al. (1972). The terms *current bedding* and *cross lamination* are commonly used for the same structure.

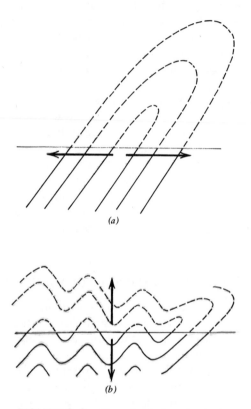

FIGURE 3.2   Hypothetical geological
sections. Horizontal line represents ground
surface and the arrows represent directions
of younging. See text for discussion
(Section 3.3).

This structure is found in a variety of water and wind-deposited clastic sediments and also in some igneous rocks [e.g., Nesbitt et al., (1970)] and can often be used to determine the direction of younging of the sequence. As deposited it may be a symmetrical structure (as in Fig. 3.3a and b) in that angles between the foreset lamination and the bed boundaries are the same at the top and at the bottom. This type of cross bedding is of no value as a younging criterion. However, cross-laminated units commonly suffer erosion

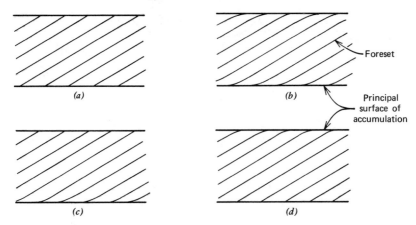

FIGURE 3.3   Various types of cross bedding as seen in the plane perpendicular to the line of intersection of the foreset beds and the principal surface of accumulation. See text for discussion.

prior to the deposition of the next bed, so that if the structure was originally sigmoidal (as in Fig. 3.3*b*) the resulting structure is asymmetrical as shown in Figure 3.3*c*. It is this asymmetry that is used to determine the direction of younging and, in general, it is very reliable. However, cross bedding is sometimes deposited as an asymmetrical structure [e.g., Hills (1972, p. 12)] with an apparently truncated surface at the base (Fig. 3.3*d*). If subsequent erosion removes the upper parts of such foreset beds the structure becomes symmetrical and, therefore, of no consequence as far as younging is concerned. If however, there is no erosion before deposition of the next bed, interpretation of the structure in the normal way will give the wrong direction of younging. Fortunately, investigation of undeformed sediments indicates that this variety of cross bedding is very rare and the method can therefore be used with reasonable confidence.

Cross bedding may survive quite intense deformation and metamorphism [see e.g., Tobisch, (1965)] so that it is a very useful younging criterion in deformed metamorphic rocks. When working in such rocks, however, care must be taken to ensure that the structure observed is of sedimentary, and not tectonic, origin (see Section 5.4.3).

RIPPLE MARKS

Ripple is the name given to a group of wavelike depositional structures that may form in water or in air. Structures of this type vary in amplitude from a few millimeters to mega ripples, such as sand dunes, which have an amplitude measurable in meters or tens of meters. They can be divided into two groups:

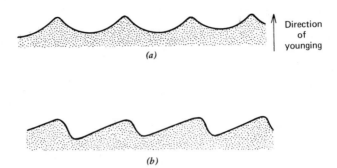

FIGURE 3.4   Ripple marks: (*a*) Oscillation type; (*b*)
Current type.

oscillation ripples and *current ripples.* Cross lamination, as described above, is
intimately related to the development and migration of current ripples [see
Allen (1968, p. 96)] so that the younging of many ripple-marked sequences
can be determined from the cross lamination. However, cross lamination is
not present in all ripples and, even when present, it is not always visible in the
field, especially in the deformed rock. Nevertheless, it is often still possible to
determine younging and here we are concerned with the criteria that may be
used when cross lamination is not visible.

Oscillation ripples, in profile, are commonly seen to comprise angular
ridges separated by arcuate troughs (Fig. 3.4*a*). This difference between the
shape of the ridge and that of the trough, often makes it possible to tell the
direction of younging of an oscillation ripple-marked sediment.

It is sometimes more difficult to determine the direction of younging from
current ripples. They are asymmetrical in profile but both the ridges and the
troughs have the same shape (Fig. 3.4*b*), so that when the structure is inverted
their appearance is unchanged. Thus the direction of younging cannot be
determined from the shape of the structure alone. However, in many cases,
heavy minerals or organic matter accumulates in the ripple troughs so that the
latter can be distinguished from the crests of the ridges, and the direction of
younging can therefore be determined. Similarly, grains larger than the bulk
of the sediment, but of the same composition, also accumulate in the troughs
of water-formed ripples, thus making identification of the troughs possible.
However, this fact must be used with care because if the sediment is wind
deposited the coarse material tends to accumulate at the crest of the ridges
[see Shrock (1948, p. 100)].

GRADED BEDDING

In many clastic rocks there is a systematic variation in grain size within a bed,
such that the sediment at one side of the bed is coarse and becomes progres-

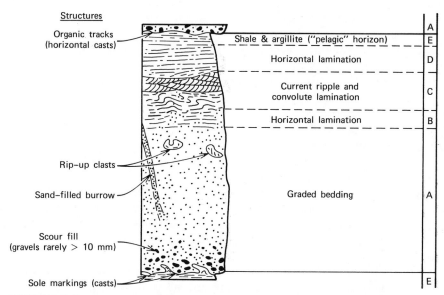

Structures

Organic tracks
(horizontal casts)

Shale & argillite ("pelagic" horizon)   E

Horizontal lamination   D

Current ripple and
convolute lamination   C

Horizontal lamination   B

Rip-up clasts

Sand-filled burrow

Graded bedding   A

Scour fill
(gravels rarely > 10 mm)

Sole markings (casts)   E

A

FIGURE 3.5   Generalized form of a turbidite bed showing the various lithological units and the structures associated with them. The letters in the column at the right-hand side are the letters ascribed to the various units by Bouma (1959). [From Stanley (1963), with permission of the Society of Economic Paleontologists and Mineralogists.]

sively finer toward the other side. Such a bed is said to be *graded* and generally, although not invariably [see Twenhofel (1936); Selley (1970, pp. 68, 77)] , the coarser material is at the base or oldest side of the bed. *Graded bedding* is common in all kinds of sediments including varved clays, detrital limestones, tuffs (particularly those that have been deposited in water), and sandstones deposited in a variety of environments. It is also found in some igneous intrusions [e.g., Irvine (1965)]. However, it is perhaps best developed in turbidite deposits, that is, sedimentary deposits of very variable composition, grain size, and mineralogical maturity, that are believed to have been deposited from a turbidity current [e.g., see Kuenen (1953)]; examples are the flysch of the European Alps and the greywacke sequences found in many ancient orogenic belts.

A turbidite sequence can be divided into a number of beds, each of which is graded, fining upward, and each of which can ideally be divided [Bouma (1959)] into five intervals, A-E (Fig. 3.5). The basal interval A is coarse and generally graded. It typically lies on a fine-grained pelite (the E interval of the underlying turbidite unit) and the interface between the two is therefore sharp and well defined, although it is commonly irregular. The A interval is massive and is overlain by the planar-laminated B interval. This is followed by the C interval, which is characterized by small-scale current ripples, then by another planar-laminated interval D, of finer, silty grain size, and finally the planar-laminated, pelitic* E interval. In a given outcrop not all intervals are normally found in every turbidite bed, but each bed generally starts with an A interval and there is usually sufficient evidence to determine the younging direction of every individual bed. This particular form of graded bedding is very reliable as a means of determining the direction of younging. It is particularly useful in orogenic belts, for two reasons. First it is a fairly common, primary structure in such regions and can survive intense deformation and high-grade metamorphism; it is still recognizable, for instance, at sillimanite grade, in multiply deformed rocks near Cooma, Australia. Second, recognition of the structure is not solely dependent on grain-size variation; this is important because metamorphic processes might easily give rise to a grain-size variation resembling that seen in graded bedding but it is highly improbable that they would give rise to a structure resembling a complete turbidite bed with its five distinct intervals.

Shrock (1948, p. 420) has pointed out that metamorphic processes can give rise to a reversed grading due to coarser grain size of metamorphic minerals such as garnet or andalusite in the more pelitic parts of a graded bed. While we agree with this observation we would caution against the use of the distribution of metamorphic minerals as a younging criterion because there can be no certainty that the distribution reflects an original sedimentary structure, unless there is additional supporting evidence. For example, Shrock (1948, p. 422) quotes a case of reversed grading from Broken Hill, Australia, but there is now reason to believe [see Ransom (1968)] that the bedding referred to in this example is, in fact, a layering of metamorphic origin and that the "grading," therefore, is of metamorphic origin also.

There are many factors governing the variation, in grain size, of an aggregate of metamorphic minerals and they are not all readily related to the microstructure and composition of the original sediments. Thus the fact that a layer in a metamorphic rock shows a systematic variation in grain size is not sufficient to prove that the layer is bedding and that the grain size is directly or indirectly related to sedimentary grading. If on the other hand, however, the various intervals of a turbidite bed can be recognized, the layering is demonstrably bedding and the younging can be determined.

SOLE MARKING

The name *sole marking*, is given to certain irregularities in the interface between a pelite and the coarser material (conglomerate, sandstone, or limestone) stratigraphically overlying it. The structure is referred to as sole mark

---

*A pelite, strictly speaking, is a clastic sediment of clay grain size and therefore generally of aluminous composition. The term is commonly used in hard rock geology, in a loose sense, to describe metasediments that are believed to have originally had such a grain size and composition.

FIGURE 3.6   Sole marks, viewed from below, on the underside of an A horizon in Normanskill turbidite. Relief has been accentuated by load casting. [New Baltimore Service Area, New York State Thruway (Interstate 87), U.S.A. Photograph by David Graham.]

because it is generally observed on the original lower surface of the sandstone after the pelite has disintegrated and fallen away. However, it develops in the pelite and exists in the sandstone as a cast. The irregularities therefore, represent the relief on the mud surface immediately prior to deposition of the sand and they are formed in various ways. Some are due to current erosion, others are due to material being dragged, bounced, or rolled across the surface and others are animal tracks [see Dżułyński and Walton (1965, Chapter 3)].

These structures are commonly preserved in deformed rocks (see Fig. 3.6) and recognition of them, on the underside of the sandstone, gives the direction of younging.

LOAD CASTS

The name *load cast* is given to structures that are believed to have developed in soft sediment by deformation under the influence of gravity. The structure is generally found at the interface between a sandstone or fine-grained conglomerate and pelite or coal that is stratigraphically below. It is believed to form by the sand or gravel sinking into the underlying sediment which, due to its high water and/or gas content, is very ductile and less dense than the overlying sand. The detailed morphology of the resulting structures varies

FIGURE 3.7   Sandstone balls in Hunsrückschiefer from the Moselle Valley, Germany. These structures have probably been flattened perpendicular to bedding, either by sedimentary compaction or by later tectonic deformation.

considerably and some varieties have specific names. Thus *sandstone balls* (Fig. 3.7) or *pillow structures* are one example of a structure believed to form by load casting [Allen (1970), p. 86)] and *flame structures* are another [e.g., Kelling and Walton (1957, p. 487); Potter and Pettijohn (1963, p. 148)]. Sole marks originate in a different way (see above), but they are very commonly modified by load casting; the effect is principally one of accentuation of the original relief.

All of these load cast structures have one property in common: invariably the coarser bed of the pair forming the deformed interface is the younger of the two.

DESSICATION CRACKS

Dessication cracks are a fairly common feature of sediments that have been deposited on land. They are very commonly associated with the ephemeral lakes of arid regions. They form when the water that has deposited the sediment drains away or evaporates, leaving the sediment to dry out subaerially. Examination of this structure in present-day sediments reveals that the cracks that develop during drying have a polygonal form in plan and that, in section, the individual polygons of sediment become turned up at the edges, so that they have a concave upper surface. This form is commonly preserved in the rock and the upward concavity indicates the direction of younging.

3.4 RECOGNITION AND NATURE OF BEDDING IN DEFORMED ROCKS

A lengthy discussion of bedded sediments is outside the scope of this book, but we stress that people working in areas of deformed sediments need a good knowledge of the sedimentary aspects of the rocks that they are concerned with.

Two points concerning bedding are selected for discussion because of their bearing on structural problems and because they are not normally discussed in sedimentary texts. The first concerns the recognition of bedding and the second the sedimentary layer silicate fabric of deformed rocks.

RECOGNITION OF BEDDING

It is not uncommon for workers in areas of deformed rocks to assume that the layering that they observe is bedding. However, this is not always a safe assumption, for layering is known to develop in other ways. Even at low grades of metamorphism good layering can develop by metamorphic differentiation, a process by which material becomes redistributed throughout the rock so as to define a new compositional layering (see Sections 5.2.6 and 5.3.5). Lenticular layering can also develop mechanically by flattening of grains or aggregates of grains that existed before the deformation. This is commonly observed in shear zones [e.g., Ramsay and Graham (1970); Vernon and Ransom (1971)] and could conceivably be an important mechanism for the development of layering, on a regional scale, in basement rocks. In light of modern ideas on plate tectonics (see Section 10.2) the occurrence of large areas that have undergone very large shear strains would seem very probable. For example, such a situation would seem compatible with the present interpretation of Benioff zones (see Section 10.2.2) and it seems equally probable that, in such zones, strain will be high enough to produce a regular layering from any initial configuration of rock units. Thus to argue, for example, as to whether a given amphibolite was originally a dolomitic limestone or a basalt flow may be pointless. If the area has suffered such large shear strain, it could equally well have been a dolerite plug or dyke. Similarly, banded ironstones, that are a common feature of Archaean shields, are generally interpreted as chemical sediments. While this may be very reasonable, their present layered disposition should not be taken, by itself, as compelling evidence of their sedimentary origin.

Now if layering can develop in deformed rocks by more than one process it is obviously unreasonable to make the automatic assumption that a given layering is, in fact, bedding, since the significance of the layering is entirely different if it developed by another process. Thus, workers in deformed areas should endeavor to *prove* that the layering is bedding and not assume it. This is generally best done by demonstrating the presence of other sedimentary structures associated with the layering in a way that is consistent with a sedimentary origin for the layering itself. This, however, must be done with care since some sedimentary structures have morphological counterparts of secondary origin in deformed rocks. For example, a structure closely resembling cross bedding can result from tight folding (see Section 5.4.3), mullion structure (see Section 6.2.6) may superficially resemble sedimentary flame structures, differentiation can produce layering (see Sections 5.2.6 and 5.3.5) that closely resembles sandstone dykes or structures that have been described as such [Moench (1966), Braddock (1970), and Clark (1970)], metamorphism can produce grain-size distributions related to composition that closely resemble graded bedding, folding can produce structures resembling ripples [Potter and Pettijohn (1963, p. 64)], and so forth. Obviously a good knowledge of sedimentary structures, as well as of tectonic structures,

0.5 mm

FIGURE 3.8   Photograph showing preferred orientation of muscovite and biotite parallel to bedding, in unconsolidated sediment from the Sohm Abyssal Plain, North Atlantic. (Specimen provided by J. R. Conolly. Crossed nicols; photograph by David Graham.)

is essential in order to determine the origin of layering. Furthermore it is important to remember that even if the layering can be shown to be bedding the possibility remains that it has been transposed (see Section 5.4).

SEDIMENTARY LAYER SILICATE FABRIC

Strong preferred orientation of layer silicates is known to be a characteristic of shales [O'Brien (1970); Moon (1972)] and may be common in argillaceous, sedimentary rocks in general. It is also known to occur in coarser sediments (e.g., Fig. 3.8) and is an obvious consequence of the shape of the mineral grains. However, failure by structural geologists to recognize the existence of this sedimentary fabric has led to the common, but erroneous, belief that all layer-silicate preferred-orientation fabrics in deformed rocks are of deformational origin. Furthermore in areas where the earliest recognizable generation of folds have an axial plane crenulation cleavage the folds are commonly labelled as second generation and earlier folds are sought, simply because crenulation cleavage is believed to develop only in rocks that have an earlier preferred orientation of layer silicates, and it is assumed that the earlier fabric must be of deformational origin. There is no foundation for this argument. The preferred orientation of layer silicates in sediments can be as good and even better than that found in many slates, so that there is no reason for arguing that all crenulation cleavages must be second generation or later structures.

It should never be assumed, therefore, that a preferred orientation of layer silicates is of deformational origin; the point must be demonstrated. In doing so, the possibility that the layer silicates may adjust to metamorphic conditions and increase in size while more or less maintaining their original, sedimentary orientation must be kept in mind. Maxwell and Hower (1967) have traced the layer silicates in a sequence of shaly sedimentary rocks through progressive diagenesis and metamorphism to greenschist facies. They describe the higher grade rocks as phyllites, presumably indicating that the sedimentary preferred orientation has been preserved during development of the metamorphic rocks (they do, however, point out that there is a decrease in the degree of preferred orientation). Thus it is not sufficient to demonstrate that the layer silicates are of metamorphic origin to prove that their preferred orientation is of deformational origin.

3.5   UNCONFORMITIES

Unconformities are major primary structures and they play an important role in the investigation of deformed rocks. They represent breaks in the stratigraphic sequence, that is, they record periods of time that are not represented

(a)

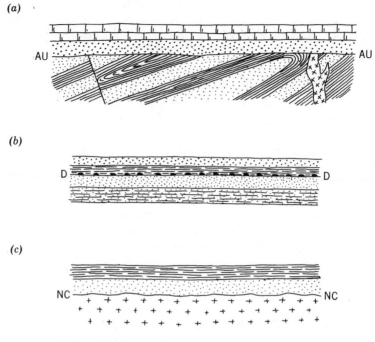

(b)

(c)

FIGURE 3.9   Unconformities: (*a*) Angular unconformity,
AU–AU. Younger beds overlie an older folded sequence and
truncate bedding and earlier faults and igneous rocks. (*b*)
Disconformity D–D. Younger and older beds lie parallel but
there is a sudden marked change in age across the
disconformity. (*c*) Nonconformity, NC-NC. Younger sedimentary
rocks overlie earlier plutonic igneous or massive metamorphic
rocks.

in the stratigraphic column, at the place in question, by rock units. This break
may be due to a period of nondeposition or to a period during which previously deposited rocks have been eroded. Its significance in tectonic studies is
that it records a fundamental change in the environment, from one of deposition to one of nondeposition or erosion, and in the case of large unconformities, this change itself generally represents an important tectonic event.

Various terms have been used to describe different types of unconformities.
Following Krumbein and Sloss (1963, pp. 304, 305) we use unconformity as
the general term and angular unconformity, disconformity, and nonconformity as more specific terms. An *angular unconformity* is an erosional surface
and separates an older sequence of rocks, in which bedding is inclined to the
unconformity, from a younger sequence in which bedding is essentially parallel to the unconformity (Fig. 3.9*a*). A *disconformity* separates two sequences,

both of which are bedded parallel to the unconformity, and may represent a period of erosion or simply a period of nondeposition (Fig. 3.9b). A *nonconformity* is an erosional surface that separates plutonic igneous rocks or massive metamorphic rocks from a younger sequence (Fig. 3.9c).

Criteria for the recognition of unconformities have been discussed by Krumbein (1942) and Shrock (1948, p. 46) and are summarized as follows:

1. Truncation of earlier structures such as bedding, folds, faults or dykes by a sedimentary sequence or lava (see Fig. 3.9a).

2. Juxtaposition of two rock types that could not have formed, in a continuous time sequence, in such close proximity. For example, unmetamorphosed sediments sitting directly on a sequence of basement gneisses or on a coarse-grained granite that lacks a chilled contact cannot represent a continuous stratigraphic sequence. In either example the contact must be an erosional surface or a fault plane.

3. Existence of old weathering surfaces or nondepositional surfaces within a sequence. Krumbein (1942) has considered recognition of such surfaces in some detail and discusses such features as buried soil profiles and manganese nodules which may have developed on the sea floor during a period of nondeposition. Many of the criteria he lists are indicative only of minor disconformity and are likely to be very difficult to recognize in metamorphic terrains. Others, however, such as existence of pebbles of one rock unit in an overlying unit, are more readily applied to deformed metamorphic rocks.

4. Marked difference in the age of adjacent members of a stratigraphic sequence as evidenced by dating based on fossils or on isotopic methods.

5. Existence of a more complicated history on one side of a given surface than on the other. For instance, one side may show evidence of two generations of folding or metamorphism whereas only the second generation can be recognized on the other side. This, of course, overlaps with (1) and (2) above.

Many of these criteria indicate that there is a discontinuity in the sequence but do not prove the nature of the discontinuity. In many cases it may be difficult to distinguish between an unconformity and a large-scale thrust simply on the basis of the criteria listed here. This difficulty may be compounded by the fact that unconformities commonly become fault planes during subsequent deformation simply because they separate rocks of very different mechanical properties.

Like other surfaces, unconformities may become complexly folded and overturned [e.g., Forman (1971)]. In areas where this is so, they may constitute a valuable piece of younging evidence, of value in interpreting the geometrical aspects of the structure as well as in throwing light on other aspects of the stratigraphic and structural history.

3.6 SEDIMENTARY VERSUS TECTONIC STRUCTURES

It is not always easy or even possible, in deformed metamorphic rocks, to distinguish between sedimentary and tectonic structures. The problem of distinguishing between bedding and layering of metamorphic origin has already been discussed (Section 3.4, see also Section 5.5). A similar problem exists in distinguishing between some sedimentary and tectonic folds.

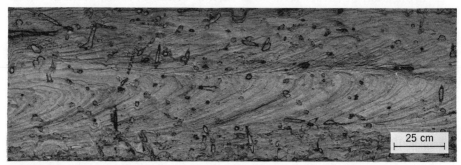

FIGURE 3.10   Foldlike structure in unconsolidated, Oligocene sands near Louvain, Belgium. (From lacquer peel provided by J.W.A. Bodenhausen. Photograph by W. C. Laurijssen.)

Folds are believed to form in unconsolidated sediments in several ways. They may form by the sliding of a superficial layer down an inclined surface so that it piles up as a folded body at the foot of the slope. Such *slump structures* are commonly found in the foresets of cross-bedded sequences [e.g., see Conybeare and Crook (1968, plate 36A)]. Another fold structure, referred to as *convolute lamination,* forms when laminated sediments, near to the depositional surface, deform in response to currents flowing over the sediment [see Sanders (1960)]. Similarly, the structure illustrated in Figure 3.10 is believed to have formed by deformation of cross bedding in preexisting, but still unconsolidated sediments, in response to stress induced in the sediment by the body of water flowing over it [see McKee et al. (1962a and b); Allen and Banks (1972)].

Yet other folds (e.g., Fig. 3.11) are believed to form — by flow in shallowly buried, unconsolidated sediment — in response to instabilities resulting from variations in the density of the sediment. The density is a function, not only of the composition of the solid particles, but also of the water and gas (from decomposition of organic material) content [e.g., Monroe (1969)]. Some writers have suggested that the deformed sediment was initially thixotropic* and

---

*A thixotropic material is a material that rapidly loses strength when disturbed and that regains its strength if allowed to stand undisturbed for a given period of time [see Terzaghi and Peck (1948); Grim (1962)].

that deformation occurred when the gelled state was destroyed by an earthquake [Williams (1960); Boswell (1961)] or by an increase in the superincumbent load [Sugden (1950)].

When these structures occur in otherwise undeformed sedimentary rocks they are not likely to be mistaken for tectonic folds. When they are preserved in deformed metamorphic rocks, however, recognition can be much more difficult and failure to recognize their sedimentary origin can lead to an unduly complicated interpretation of the deformation history. For example, interpretation of folds like that illustrated in Figure 3.10 as a tectonic structure would probably lead to the belief that the sequence was isoclinally folded and that the bedding was a transposed surface (see Section 5.4). Obviously it is important to recognize the true origin of such structures and this, in general, cannot be done purely on the basis of the fold morphology. The geometry of the structure indicates that the layering has been deformed but not, as a rule, its physical state at the time of deformation. In the example cited above (Fig. 3.10), we know that the fold formed in soft sediment because the sediment is still unlithified. However, even after lithification, recognition of the undeformed crab burrows would still indicate that the structure formed in soft sediment, but this conclusion could not be drawn from the shape of the fold.

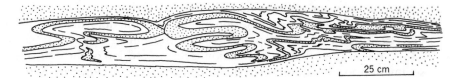

FIGURE 3.11   Convolute lamination in Bunga Beds, South Coast Region, NSW Australia. These rocks are unmetamorphosed and tectonic folds are only very weakly developed. [From Williams et al. (1969), Copyright © 1969, The University of Chicago. Reproduced with permission of the University of Chicago Press.]

Several writers have described criteria for distinguishing between soft sediment or penecontemporaneous structures (i.e., those approximately contemporaneous with deposition and predating lithification) and hard rock or tectonic folds (i.e., folds postdating lithification). These various criteria are now discussed briefly.

1. The existence, in the deformed material, of the undeformed burrows, of animals that lived in soft sediment (see Fig. 3.10) is good evidence that the fold formed in soft sediment.
2. The bending of metamorphic foliations or mineral grains around fold closures is good evidence that the fold formed in a hard rock. For exam-

ple, a fold that deforms a metamorphic foliation that formed parallel to the axial surface of an earlier fold or deforms sillimanite porphyroblasts, can hardly have developed in soft sediment.

3. Rettger (1935, p. 292) points out that deformed fossils are indicative of hard rock folding. Of course, the deformation of the fossils may be something that has been imposed after folding but if a systematic relationship can be shown between the strain of the fossils and their position in the folds, then this would seem to be a very good criterion.

4. Rettger (1935, p. 291) points out that experimentally produced, soft sediment folds are characterized by "normal and reverse structures . . . in the same bed within short distances." In other words, he obtained folds that were close to one another in the same bed, and were overturned in opposite directions as in Figure 3.11. This situation is not uncommon in penecontemporaneous deformation structures [e.g., see Fig. 3.11 and also ten Haaf (1959, p. 49)], but it is also found in conjugate, tectonic folds. However, no examples are known to the authors of opposite facing, tight to isoclinal folds, sandwiched between undeformed beds (as in Fig. 3.11), in tectonically deformed rocks, except where the folds can be shown to belong to more than one distinct generation.

5. Truncation of some of the folded beds by the overlying undeformed bed is cited by some writers as a criterion of soft sediment deformation [Rettger (1935); Nevin (1949, pp. 201–204); Potter and Pettijohn (1963, p. 143)]. It is argued that the truncation is due to erosion of the deformed sediment prior to deposition of the undeformed layer. However, Rettger (1935), in his experimental work, showed that the truncation could occur on either side of the deformed bed and could, in fact, be a product of deformation rather than erosion. Such truncation is known in hard rock structures, especially where the deformation has occurred near the brittle ductile transition [e.g., see Williams (1971, Fig. 5e-h)].

6. It is claimed that penecontemporaneous structures are generally unrelated to the tectonic deformation of an area and should, therefore, differ in appearance and orientation from hard rock folds in the same area [Rettger (1935); Nevin (1949, pp. 201–204); Potter and Pettijohn (1963, p. 143)]. However, Potter and Pettijohn (1963, p. 153–154) point out that some soft sediment deformation may occur in response to tectonic deformation, and measurements made by Kühn Velten (1955) support the hypothesis that some soft sediment folds form by slumping, on the flanks of large tectonic structures, perpendicular to the tectonic fold axes. Such structures will have a similar orientation to the large tectonic folds which may also involve lithified material. However, even where this is not true there remains the possibility of interpreting both types of folds as products of hard rock deformation, produced during two

different periods of deformation. Thus, this would seem to be a very difficult criterion to apply.

7. The absence of joints and veins that can be related to the folds is also said to indicate that the latter formed in soft sediment [Rettger (1935); Nevin (1949, p. 201–204); Potter and Pettijohn (1963, p. 143)]. Certainly it is unusual for hard rock folds to lack, at least, the *ac joints* (see Section 7.2.3) but the absence of veins is less remarkable and not a good criterion. The presence of joints, however, does not prove that the folds are hard rock structures; soft rock folding may be followed much later, after lithification, by the development of joints that owe their orientation to the preexisting mechanical anisotropy of the folded rock.

8. Axial plane foliation is cited by a number of writers [e.g., Rettger (1935); Nevin (1949); Potter and Pettijohn (1963)], as diagnostic of hard rock deformation but recent observations reveal the presence of cleavage, defined by preferred mineral orientation, in penecontemporaneous fold structures [Williams et al. (1969); Moore and Geigle (1972)].

9. Deformation of soft sediment is said to be restricted to a single bed [Potter and Pettijohn (1963, p. 143)] or a group of adjacent beds. This statement, however, applies equally well to some hard rock structures (see Fig. 4.20*b*).

10. Soft sediment folds are said to be chaotic [Rettger (1935); Nevin (1949, pp. 201–204); Potter and Pettijohn (1963, p. 143)] but this is not a good criterion because it is also true of some hard rock structures (e.g., see Fig. 4.22*a*).

Of the criteria listed above only the first three are definitive if used singly, but if several of the other criteria are taken together, and account is taken of the general geological setting of the structures, it may be possible to make a good case for calling a given set of folds, either soft sediment or hard rock structures. However, this division into two distinct groups is probably artificial since it would seem more reasonable to assume that folds form in rocks representing all stages of lithificaton. Furthermore, if folds are forming in soft sediment and in response to tectonic folding, as suggested by Kühn Velten (1955), it would seem difficult and misleading to draw a line between the two.

4

FOLDS

4.1 INTRODUCTION

In many rocks, what were once approximately planar surfaces have been deformed to produce curved or nonplanar surfaces. These new structures are called folds.

Folds are perhaps the most common, obvious manifestation of ductile deformation of rocks. They form under very varied conditions of stress, hydrostatic pressure, pore pressure, and temperature as evidenced by their presence in soft sediments (penecontemporaneous structures, see Section 3.6), sedimentary rocks, the full spectrum of metamorphic rocks, and even as primary flow structures in some igneous rocks. Their presence certainly indicates some form of ductile deformation, although it is well to remember that the absence of folds does not indicate the absence of a pervasive deformation, as has been very well demonstrated by Sylvester and Christie (1968).

This chapter deals first with the description (Sections 4.2–4.5) and classification of folds (Section 4.6) and finally with a brief consideration of present ideas concerning the development of folds in sedimentary and metamorphic rocks (Sections 4.7 and 4.8).

4.2 DESCRIPTION OF SINGLE FOLDS

The word fold (or folds) can be used to describe the configuration of any nonplanar surface resulting from deformation. Since there is considerable variation in fold morphology an extensive vocabulary has developed to describe these structures; the list of terms presented here is by no means complete but is a working collection of those most commonly used.

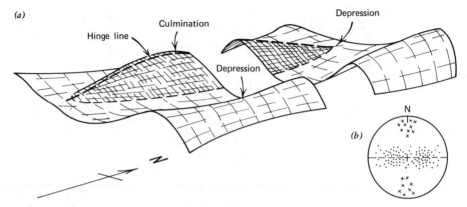

FIGURE 4.1   (*a*) Folded surface. Medium broken line and heavy shading indicate the extent of a single fold. Heavy dashed line represents the hinge line of the fold. (*b*) Equal area projection representing distribution of poles to the surface (dots) and the distribution of segments of fold hingelines (crosses).

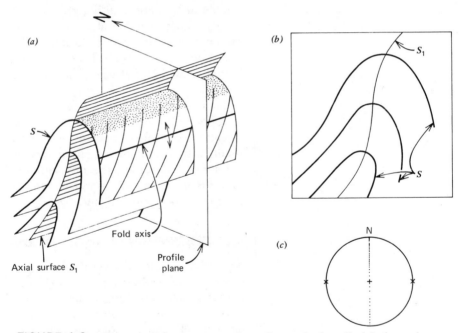

FIGURE 4.2   Diagrammatic representation of a perfectly cylindrical, east/west trending, horizontal fold. (*a*) Sketch of fold showing folded surfaces (*S*), axial surface (*S*₁), and profile plane. On one *S* surface the hinge and limb areas are indicated by stippling and hatching respectively. (*b*) Fold profile. (*c*) Equal area projection showing the distribution of poles to *S* and *S*₁ (dots) and the orientation of the fold axis (crosses).

Figure 4.1 represents a typically folded surface and the extent of a single fold is outlined by the dashed line. In this fold, as in most natural folds, the radius of curvature of the folded surface varies from point to point, and an area with a small radius of curvature, known as the *fold closure* or *hinge,* is flanked by two areas of larger radius of curvature known as the *limbs* (see also Fig. 4.2). The heavy broken line in Figure 4.1, drawn along the hinge of the fold, is the locus of points of minimum radius of curvature and is known as the *hinge line.* It is possible, although geologically unusual, for a folded surface to have a form that can be defined in terms of segments of a circular based cylinder (Fig. 4.3). For such a surface the preceding definitions breakdown and the fold limbs are redefined as the area around the line of inflection, and the hingeline is redefined as the line midway between adjacent limbs.

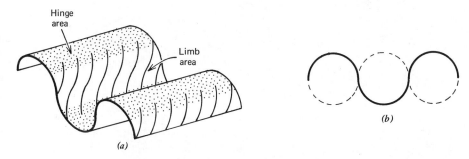

FIGURE 4.3   (*a*) Sketch of an idealized, cylindrical fold in which the profile (*b*) is composed purely of segments of a circle.

It is possible for a fold to be confined to a single surface as depicted in Figure 4.1 (e.g., a folded fault plane) but, in general, it includes a whole family of surfaces such as a number of adjacent bedding planes. Such a fold is illustrated diagrammatically in Figure 4.2. The surface (S_1 in Fig. 4.2) joining the hingelines in adjacent folded surfaces (S in Fig. 4.2) is referred to as the *axial surface* or, if it is planar, the *axial plane* (unfortunately, this term is commonly used loosely to describe the axial surface even when it is not planar). A fold that has a planar axial surface is described as a *plane fold* otherwise it is a *nonplane fold.* The structure represented in Figure 4.2 is a special type of fold in that it can be generated by moving a straight line parallel to itself in such a way as to outline the surface. Because it has such a generator the fold is said to be *cylindrical* and the generator is referred to as the *fold axis.* Poles (see Appendix A) to the folded surface of a cylindrical fold plot in stereographic projection on a great circle (Fig. 4.2*c*) and deviation from this ideal plot is a measure of the deviation from perfect cylindrical form (compare Figs. 4.1*b* and 4.2*c*). Since natural folds are rarely, if ever, perfectly cylindrical the fold

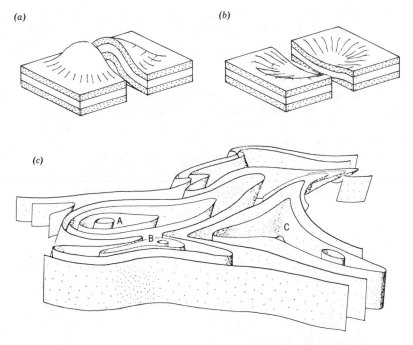

FIGURE 4.4   Domes and basins. Diagrammatic block diagram of a dome
(*a*) and a basin (*b*). (*c*) Complex, tight dome and basin structures in the Glen
Cannich area, Scotland. A and B are domes and C is a basin. (From
Tobisch, 1966, with permission of the Geological Society of America.)

axis is generally defined statistically and is the pole to the great circle that best
fits the poles to the folded surface (e.g., the dashed line in Fig. 4.1*b*). The
profile of a fold is the section drawn perpendicular to the fold axis (Fig. 4.2*a*
and *b*); this contrasts with a *geological section* which is normally drawn in a
vertical plane.

   In general, folds deviate markedly from the ideal geometry of Figure 4.2
and resemble more closely the structures represented in Figure 4.1. Most such
noncylindrical folds, can be subdivided into segments that approximate cylindrical geometry but there are others that cannot be subdivided this way; for
example, the *dome* and *basin* structures depicted in Figure 4.4*a* and *b*. Figure
4.4*c* however, shows a series of very tight dome and basin structures that are
again divisible, in part, into cylindrical segments. This is true also of the
specimen depicted in Figure 4.5. Other terms used to describe noncylindrical
folds include *doubly plunging,* which means that the plunge of the fold reverses
along its length (e.g., the heavily shaded fold in Fig. 4.1) and *conical fold,* which
describes a noncylindrically folded surface that has the approximate geometry

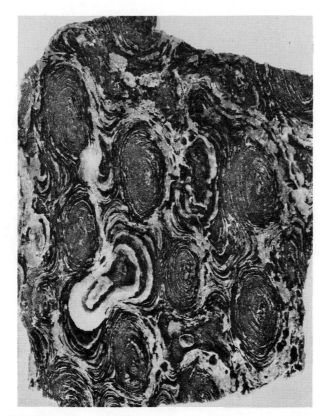

FIGURE 4.5   "Outcrop" of a system of asymmetrical domes
and basins on a flat surface. The structures are very tight and
resemble long, narrow test tubes except for their elliptical
cross section. Alternating layers comprise quartzite and
marble. Specimen from Rum Jungle, Australia.

of a cone [e.g., Stauffer (1964)]. It is important to remember that although
folds are commonly drawn as plane cylindrical structures, and although they
may appear so in outcrops where only a small portion of their total extent is
exposed, they are commonly nonplane and almost invariably noncylindrical.

A fold, with limbs that converge upward, (e.g., Fig. 4.6*a* and *b*) is referred to
as an *antiform* and one with the limbs that converge downward is a *synform*
(e.g., Fig. 4.6*c* and *d*). *Anticline* and *syncline* are terms with stratigraphic significance and refer, respectively, to folds in which the oldest strata are in the
core of the fold and to folds in which the youngest strata are in the core. Thus,
in Figure 4.6, (*a*) is an antiformal anticline known simply as an anticline, (*b*) is
an *antiformal syncline*, (*c*) is a synformal syncline, and (*d*) is a *synformal anticline*.

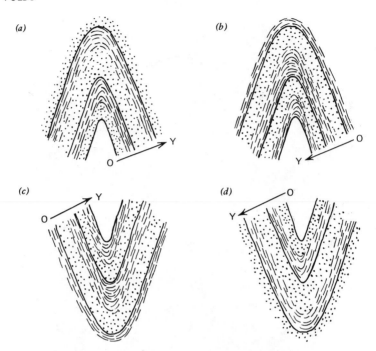

FIGURE 4.6 Naming of folds on basis of shape and direction of younging; the latter is indicated by diagrammatic grading of folded layers and by arrows pointing from oldest (O) to youngest beds (Y). (*a*) Anticline. (*b*) Antiformal syncline. (*c*) Syncline. (*d*) Synformal anticline.

Antiformal synclines and synformal anticlines may be referred to as *downward facing folds* since the stratigraphy is completely inverted in passing along the axial surface or, conversely, anticlines and synclines can be referred to as *upward facing folds* [Shackleton (1958)]. A common situation in which downward facing folds are formed is during refolding of the stratigraphically inverted limb of a recumbent fold (see Fig. 8.10).

Folds that are neither antiformal nor synformal are said to be *neutral folds*. They comprise vertically plunging folds, folds with horizontal axial surfaces, and folds that plunge parallel to the dips of their axial surface (see Fig. 4.25*e*, *f*, and *g*).

4.3 DESCRIPTION OF FOLD SYSTEMS

A group of folds spatially and genetically related is referred to as a fold system. Figure 4.7 indicates what is meant by the *amplitude* and *wavelength* of a fold within such a system. Since most fold systems are not strictly periodic and are rarely symmetrical, these terms are generally used very loosely or other parameters are defined [e.g., see Fleuty (1964)]. Where folds of one size occur on the limbs or in the closure of larger folds they are referred to as *parasitic folds* with respect to the large structures. They may also be referred to as *second-order folds* while the large folds are referred to as *first order*.

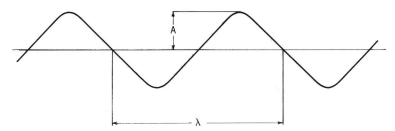

FIGURE 4.7   Fold amplitude (*A*) and wavelength (λ). Strictly speaking these terms should only be used to describe folds that are symmetrical and periodic as illustrated.

It is often convenient to consider the general or overall orientation of a tightly folded surface. This is given by the orientation of the *enveloping surface,* a surface that is tangential to the hinges of some or all of the folds in a given surface. Which folds are touched by the enveloping surface depends on the scale at which the structure is being considered. Thus in Figure 4.8 on the scale of the area represented by the diagram, enveloping surface A gives the overall orientation of the folded surface but in the smaller areas I and II, enveloping surface B gives the general orientation. If the axial planes of a group of folds are perpendicular to the enveloping surface drawn tangential to those folds then the folds are said to be *symmetrical;* if not, the folds are *asymmetrical.* Thus, the first-order folds in Figure 4.8 are symmetrical whereas the second-order folds are asymmetrical and their asymmetry varies systematically across the axial surfaces of the first-order folds. This systematic variation is referred to as vergence* and the folds in limb I as viewed in the diagram are said to have *dextral* or *clockwise vergence* or are described as *Z folds,* whereas those in limb II have *sinistral* or *anticlockwise vergence* or are *S folds.* This property, of course, depends on the direction in which the observer is looking along the fold axis. Thus, if Figure 4.8 could be observed from behind the folds, looking out of the picture, the sinistral folds would become dextral and vice versa (this point is demonstrated in Fig. 4.9). The property of vergence is put to considerable use in the mapping of some areas [e.g., Wood (1963)] and is discussed further in Section 8.4.

*This term has been used differently, with genetic connotations, by German writers [see Fleuty (1964, p. 476)].

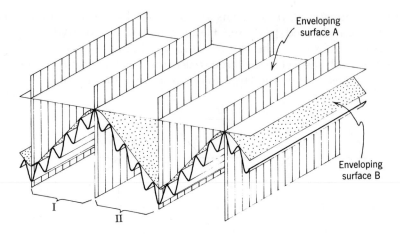

FIGURE 4.8   System of folds with enveloping surface A drawn tangential to large fold hinges and enveloping surface B drawn tangential to parasitic folds.

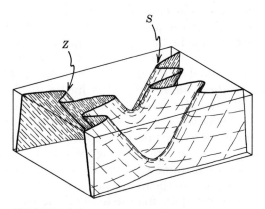

FIGURE 4.9   Diagrammatic representation of fold system demonstrating dependence of sense of vergence on the direction of observation along the fold axis. If observer looks down the plunge of these folds, the sense of vergence of a given fold changes as the direction of plunge and, therefore, as the direction of observation changes; see, for example, the fold marked *Z* and *S* at two points of outcrop.

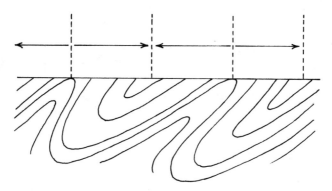

FIGURE 4.10   Overturned folds; one limb is
stratigraphically overturned as is indicated by the arrows
representing the direction of younging.

There are many areas in which the axial surfaces of folds are approximately
vertical but elsewhere other orientations are predominant and the sequence in
one limb may be inverted relative to the sequence in the other limb as depicted
in Figure 4.10. Thus, if the folded surface is bedding, in some limbs it will
young in the opposite direction to which it dips. Such folds in which one limb
is stratigraphically inverted are said to be *overturned*.

   In folds possessing a vertical axial surface the hinge line passes respectively
through the highest and lowest point of each antiform and synform. This is
not generally true for overturned folds as can be seen in Figure 4.11, which

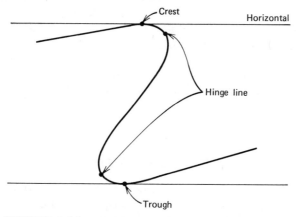

FIGURE 4.11   Fold profile showing how the crest
and trough generally differ from the hinge lines
when the axial surface is dipping.

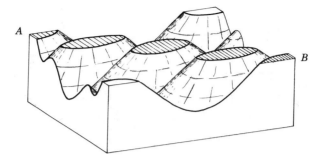

FIGURE 4.12 A system of *en échelon* folds along the line *A-B*.

represents a vertical section through a fold, cut perpendicular to the trend of its hinge line. The highest and lowest points with respect to a horizontal datum line do not coincide with either hinge line and, in such folds, the highest point is called the *crest* and the lowest point the *trough*. Lines joining corresponding crest and trough points in successive sections are *crest lines* or *trough lines*. In noncylindrical folds such lines are neither horizontal nor rectilinear but vary in height along their length. This can be seen in Figure 4.1 where crest and trough lines happen to be coincident with hingelines. The high points in both crest and trough lines are referred to as *culminations* and the low points are *depressions* (Fig. 4.1). In some noncylindrically folded surfaces, doubly plunging folds with steeply dipping axial planes are arranged spatially such that culminations and depressions in successive folds lie along lines that make an acute angle with the axial surfaces as shown in Figure 4.12. Such folds are said to be arranged *en échelon*. Note, however, that this term describes the geometry of the folded surface and is independent of the relationship of the structure to the horizontal and vertical. Thus, if the structure represented in Figure 4.12 is rotated until its initially horizontal enveloping surface is vertical, the folds will still be arranged *en échelon* but for their particular shape in section it is no longer possible to use crest and trough lines. This is because the folds are now neutral.

4.4 DESCRIPTION OF FOLDS AS SEEN IN PROFILE

Much of the description of the morphology of folds is concerned with the appearance of the profile. An accurate description of this aspect of the fold is very useful but often difficult to communicate because of considerable variation in the usage of some of the terms and because of loose usage of others. For example, an *isoclinal fold* is generally defined as a fold with parallel limbs, but it is not uncommon to find the term being used to describe folds with

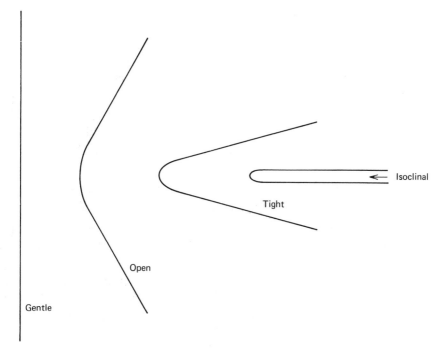

FIGURE 4.13   Terms used to describe folds on the basis of interlimb angle.

limbs inclined to one another by as much as 30°. For this reason we stress the value of well-prepared sketches and photographs; sketches are particularly valuable if outlines are traced from photographs and details then added in the field.

An important point in describing fold profiles is to indicate whether the structures are open or tight. The most precise way of doing this would be to measure the angle subtended between the two limbs (the *dihedral angle*) but, in general, the value of the result does not warrant the work necessary to obtain it. It is more practical to divide the spectrum of dihedral angles into a number of groups that, to a reasonable approximation, can be recognized in the field. Figure 4.13 presents such a scheme.

Fold closures may be described as *rounded* (Fig. 4.14a) or *angular* (Fig. 4.14b). Folds with planar limbs and very angular hinges (Fig. 4.14c) are re-ferred to here as *kinks* although some writers use the term in a more restricted and often genetic sense. If the limbs of a kink are of equal length so that the folds are symmetrical (as defined above, Section 4.3) the more specific name *chevron fold* can be used to describe them. Where kinks are markedly asym-metrical as in Figure 4.15, the long narrow zone defined by the short limb may

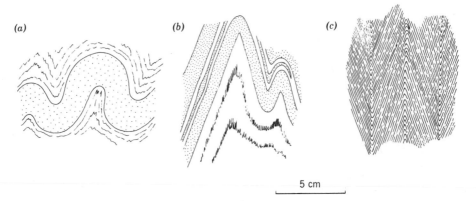

(a)   *(b)*   *(c)*

5 cm

FIGURE 4.14   Fold profiles in rocks from the Chester Dome area, Vermont, U.S.A.. *(a) Rounded fold* in garnetifferous layer in biotite schist. *(b) Angular fold* in quartz feldspar layers in biotite muscovite schist. *(c) Kink folds* of chevron type in biotite schist.

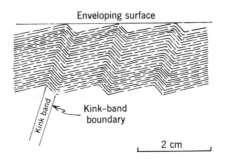

Enveloping surface

Kink band

Kink-band boundary

2 cm

FIGURE 4.15   Asymmetric kinks in phyllite near Bermagui, N.S.W., Australia.

be referred to as a *kink band* and the axial plane traces may be referred to as *kink band boundaries*. Very commonly the angle between the limb and the axial plane of a kink is approximately the same for both limbs, and some writers would therefore describe the fold as symmetrical. However, we prefer to define symmetrical and asymmetrical folds in terms of the relationship of the axial surface to the enveloping surface as in Section 4.3.

In areas of asymmetrical kinks, axial planes can occur in two orientations such that the kink bands may intersect one another (e.g., Fig. 4.16*a*). Such structures are referred to as *conjugate kinks* or *conjugate kink bands*. The more general term *conjugate fold* is used to describe any pair of apparently related folds that have axial surfaces that are inclined to one another at a high angle

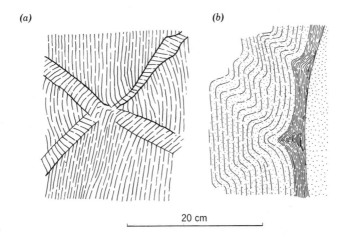

(a) *(b)*

20 cm

FIGURE 4.16 (*a*) Conjugate kinks in slates from near
Eifel, Germany (Rheinisches Schiefergebirge). (From
photograph by H.J. Zwart.) (*b*) Conjugage fold in
metasediments from near Bermagui, N.S.W., Australia.

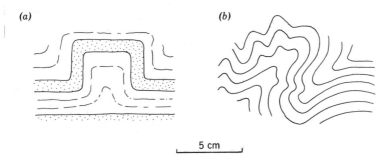

(a) *(b)*

5 cm

FIGURE 4.17 (*a*) "Boxfold" in thin sandstone interbedded with
shales. (*b*) Polyclinal folds in layered cherts. Both are drawings of
outcrops near Bermagui, N.S.W., Australia.

as shown in Figure 4.16*b*. Such folds generally terminate in an angular fold, of
yet another orientation, at the point where their axial surfaces meet; this point
is generally in an incompetent layer (Fig. 4.16*b*). Figure 4.17*a* shows a special
case of conjugate folds. This structure which describes three sides of a rectangle is called a *box fold*. In some structures axial surfaces may have several
orientations (Fig 4.17*b*) instead of only two and the folds are then said to be
polyclinal.

Folds in which the thickness of folded layers, measured normal to the bed,

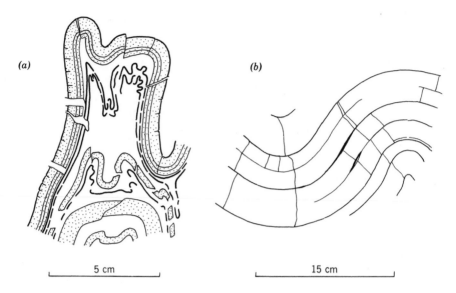

(a)   *(b)*

5 cm   15 cm

FIGURE 4.18   (*a*) Parallel folds in shale (stippled) and sphalerite-rich layers (heavy black lines) in ore from Mt. Isa, Australia. Unornamented layers are rich in galena. (*b*) Concentric folds in quartzite near Mt. Isa, Australia.

is constant are said to be *parallel* (Fig. 4.18*a*). *Concentric folds* are a special case of the latter in which the folded surfaces, as seen in profile, define circular arcs (Fig. 4.18*b*). Folds in which the dimension of the layers, measured parallel to the axial surface, is constant (Fig. 4.19) are said to be *similar*. Approximations to these two morphologies are found in rocks, but the majority of natural folds lies somewhere between the two. Furthermore, as illustrated in Figure 4.18*a* some layers may approximate one of the ideal morphologies while other layers do not. Similar folds tend to be persistent in profile that is, the folded outline of a given layer is repeated by all other layers visible in the outcrop. Such folds, whether similar or not, are said to be *harmonic* (Fig. 4.20*a*) and folds that are impersistent on the scale of observation are said to be *disharmonic* (Fig. 4.20*b* see also Fig. 4.18*a*). The latter are particularly common in areas of parallel folds because they are a necessary consequence of parallel fold geometry [see de Sitter (1964, pp. 182–187 and p. 213); Busk (1929, pp. 9 and 10)].

In areas of intense folding it is not uncommon to find isolated, tight fold closures in rocks that are otherwise not obviously folded; such structures are referred to as *intrafolial folds* or, if dismembered as in Figure 4.21, as *rootless intrafolial folds* (see also Figs. 5.30–5.33).

Convolute and *ptygmatic* folds are characteristic of high-grade rocks. The former have markedly curviplanar axial surfaces and are generally dishar-

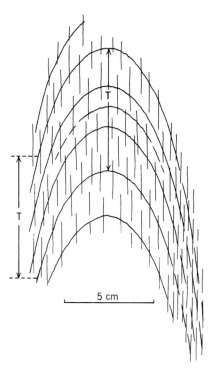

FIGURE 4.19   Similar fold in
graphitic slate from Rum Jungle,
Australia.

monic (Fig 4.22*a*). Ptygmatic folds (Fig. 4.22*b*) occur mostly in migmatites.
They generally involve a single layer, typically a pegmatite vein, in a schistose
or gneissic matrix. The folds have a lobate appearance and tend to be polyclinal. They are commonly concentric.

4.5 **ORIENTATION OF FOLDS**

The orientation of a fold is specified by the orientation of its axial surface and
the orientation of its hinge line. If the fold is plane and cylindrical (or closely
approximates these conditions) only one value is required for each; if not,
more values are required and a map or block diagram is required to show how
orientation varies with position. We concern ourselves here with plane, cylindrical folds.

    The orientation of the axial plane is expressed as a *dip and strike* or, in a
more compact form, as a *dip* and *direction of dip*. As for any surface the *strike is*

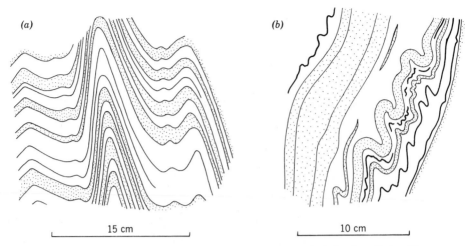

FIGURE 4.20 (*a*) Harmonic folds in micaceous quartzite Plattjen, Swiss Alps. Sketched from photograph taken by C.J.L. Wilson. (*b*) Disharmonic folds in "lead ore," Mt. Isa, Australia. Shale layers are represented by heavy black lines and stippling, remaining layers are composed largely of galena.

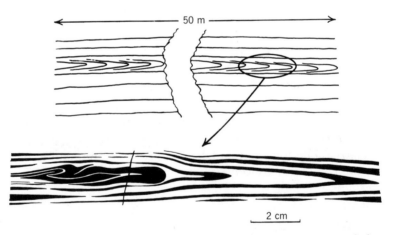

FIGURE 4.21 Rootless intrafolial folds in silt layers in meta shales, Mt. Isa Mine, Australia. Outcrop is continuous for approximately 50 meters.

the trend of the horizontal line contained in the surface; there is only one such line in any planar surface unless the surface is horizontal, in which case horizontal is a full description of its orientation. *The angle of dip is the angle between the surface and a horizontal plane;* consequently, it is measured in the vertical plane perpendicular to the strike (Fig. 4.23). *The direction of dip of a surface is the trend of*

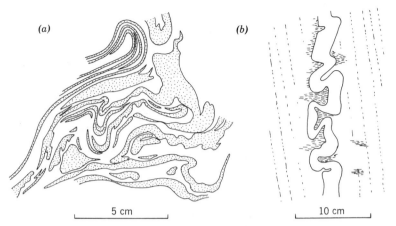

(a) (b)

5 cm   10 cm

FIGURE 4.22   (*a*) Convolute folds in diopside biotite gneiss from
Garwood Valley, Antarctica. Diopside rich layers are stippled. (*b*)
Ptygmatic folds in a pegmatite vein in quartz, feldspar, biotite gneiss.
Potosi gneiss quarry, Broken Hill, Australia. Ornamentation indicates
distribution and orientation of biotite.

the line perpendicular to the strike of the surface looking down the dip. Direction can
be represented by compass points, in terms of quadrants or by means of a 360°
circle. Thus the following are different ways of expressing the same trend:
NW, N 45° W, W 45° N, and 315°, whereas strike N 45° W dip 30° SW and dip
30° toward 225° are examples of two of the various ways of expressing the
same dip.

The orientation of the hinge line or, since the fold is cylindrical, its axis, is
expressed by its *plunge and direction of plunge. The plunge of a line is the angle
between the line and a horizontal plane. The direction of plunge of a line is the trend of
its projection on the horizontal plane measured looking down the plunge* (Fig. 4.24).
The orientation of a line can also be expressed by means of its *pitch*, which can
be defined as *the angle between the line and the strike of a plane containing the line.*
Obviously for pitch to be meaningful the orientation of the plane must be
known. This convention is most useful where the "line" is a lineation of some
description, traversing a surface that also has to be measured. Fold hinges do
not commonly come into this category, and their orientation is therefore not
commonly expressed in this manner. However, there are exceptions, for
example, microfolds in a schistosity, in an outcrop in which the enveloping
surface is planar. This situation is represented in Figure 4.24, where the pitch
is 72° N and the full specification of the orientation is schistosity (more precisely the enveloping surface) dips 55° toward 270° and the folds pitch 72° N.
The N here specifies only that the pitch is measured from the north end of the
strike line, the pitch could equally well be expressed as 108° S but, by convention, we use the acute angle.

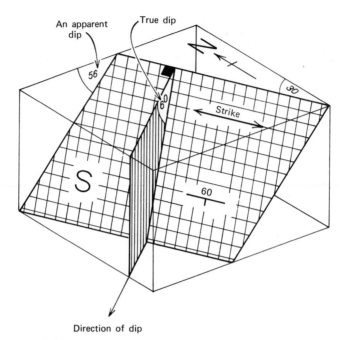

FIGURE 4.23   Dip and strike: orientation of
cross-hatched plane can be expressed as follows: strike 330°
dip 60° W or dip 60° toward 240°. For further explanation
see text.

In the past, pitch and plunge have been used synonomously and this has
lead to confusion. Clark and McIntyre discussed the problem in 1951 and
suggested the usage adopted here. Their suggestion has been followed by
most writers but pitch is still used in place of plunge by some economic
geologists.

4.6   CLASSIFICATION OF FOLDS

One of the most useful classifications of folds is based on geographic orientation of axial surface and hingeline (Fig. 4.25 and Table 4.1).

Another means of classification is based on the appearance of folds in
profile. Ramsay (1967) divides folds this way by means of *dip isogon* patterns: a
dip isogon being a line joining points of equal dip (Fig. 4.26). The isogons are
plotted on profiles, and the patterns that they define are the basis of classification. Variation in layer thickness around the fold could equally well be used as
the basis of classification, the significant point being that both parameters
provide information concerning the strain involved in the development of the

TABLE 4.1

		Orientation of Hinge Line		
		Horizontal	Plunging	Vertical
Orientation of axial surface	Vertical	Horizontal normal	Plunging normal	Vertical
	Dipping	Horizontal inclined	Plunging inclined (strike of axial plane oblique to trend of fold axis)	
			Reclined (strike of axial plane perpendicular to trend of fold axis)	
	Horizontal	Recumbent		

Classification of approximately plane cylindrical folds by orientation; after Turner and Weiss (1963).

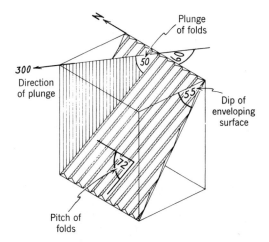

FIGURE 4.24 *Plunge and pitch.* The orientation of the small kinks can be specified by their plunge (50° toward 300°). Alternatively they can be treated as a lineation lying in a planar surface (their enveloping surface) and they are then said to pitch 72° N in a surface dipping 55° toward 270°. For further explanation see text.

fold. The value of this classification is that it divides folds morphologically into groups that are also significant in terms of strains [Hobbs (1971)]. The classification is illustrated in Figure 4.26.

A number of writers have classified folds on the basis of the shape of the

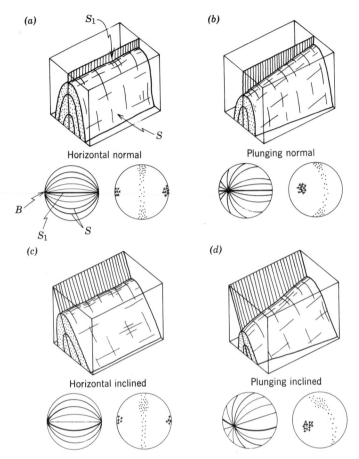

FIGURE 4.25   Classification of folds based on orientation of hinge line and axial surface (see Table 4.1); after Turner and Weiss (1963). Diagrammatic projections below each block diagram show orientation data; left-hand projections show cyclic representation of several S planes and the axial plane (heavy line) for ideally cylindrical folds. The right-hand diagrams are a more realistic representation of poles to S ($\bot S$) and of the orientation of the fold axes measured at different points on a fold that is statistically cylindrical.

fold hinge line and the shape of the axial surface. This type of classification is outlined by Turner and Weiss (1963) and Fleuty (1964).

4.7   THE DISTRIBUTION OF STRAIN IN FOLDS

The history of deformation at any point in a fold is likely to have been fairly complicated due to continuous changes in the states of stress and of strain [see

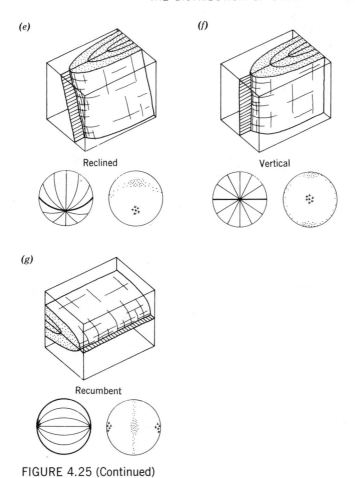

FIGURE 4.25 (Continued)

Fig. 1.20 and the figures in Chapple (1968c) and Dieterich (1970)]. The fabrics developed in the folded layers presumably reflect this complicated *strain history*. Such fabrics include the relationships of porphyroblasts to the adjacent foliation [see Spry (1969)], the microstructure of "pressure shadows" [see Elliott (1972); Durney and Ramsay (1973)], and crystallographic preferred orientations (see Section 2.3.1). However, the shape of the folded surface as seen now helps place constraints on the strain that the layer has undergone. It is important to note, however, that the shape of the folded layer by itself does not enable the strain to be established at each point. Other types of information are required in addition [see Hobbs (1971, pp. 361–368) for a discussion]. Figure 4.27 shows three folds of identical shapes but with three different distributions of strain.

Our knowledge of strain variation throughout folded rocks rests on the results of strain analyses such as those of Cloos (1947, 1971) and of Wood (1974) and on the strain distributions observed in various modeling experiments [for instance, Biot et al. (1961); Ramberg (1963a); Ghosh (1966, 1968);

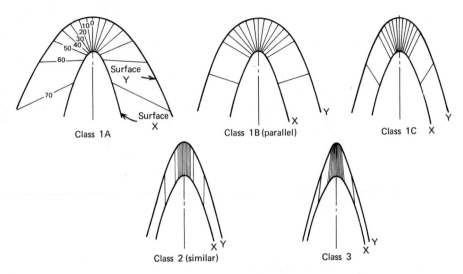

FIGURE 4.26 Classification of folds based on dip isogons; Dip isogons have been drawn at 10° intervals from the lower to upper surfaces X and Y. [From J. G. Ramsay (1967), *Folding and Fracturing of Rocks*. Copyright © 1967, McGraw-Hill Book Co. Used with permission of McGraw-Hill Book Co.]

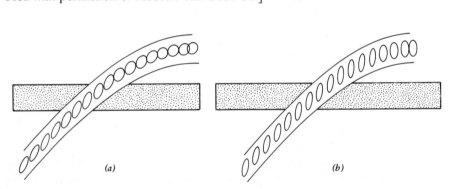

FIGURE 4.27 Folds of identical shape but with different distributions of strain. The stippled bar in each case is the undeformed shape of the bar. Each fold is a similar fold. (*a*) 10 percent shortening. (*b*) 30 percent shortening. (*c*) 50 percent shortening [after Hobbs (1971)].

Singh (1968); and Hudleston (1973a)]. In addition a number of theoretical treatments of strain distributions in folded layers exist such as those of Chapple (1968c) and Hudleston and Stephansson (1973); these are treated in more detail in Section 4.8.

This section discusses the states of strain that might exist in folded layers by examining the shape changes that are involved in producing simple kinds of folds.

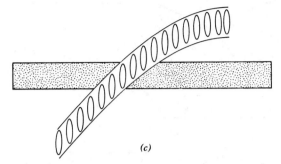

(c)

FIGURE 4.27 (Continued)

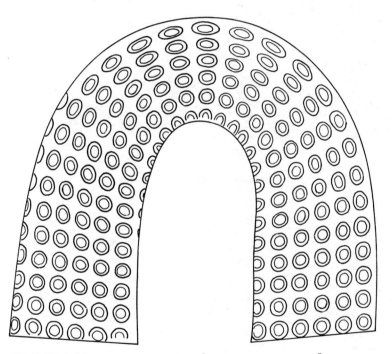

FIGURE 4.28   Bending of a bar. [From Sander (1948).]

4.7.1 Classical Models of Fold Formation

Three quite simplified models for the development of folds have been discussed for many years now in the geological literature. The first of these involves folding of a layer so that a parallel or concentric fold is produced but so that there is a surface of no strain toward the central part of the layer. This surface is commonly referred to as the *neutral surface* (Fig. 4.29*b*). This is the

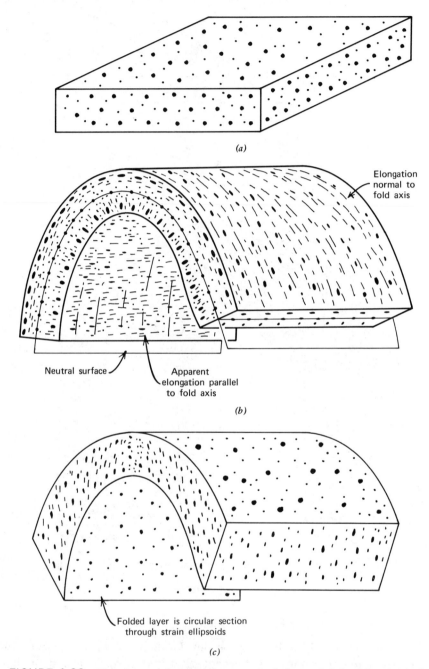

FIGURE 4.29   Simple models of folding. (*a*) Undeformed slab with spherical markers. (*b*) Slab deformed with a neutral surface. (*c*) Slab deformed by flexural shear. (*d*) Slab deformed by slip on planes oblique to the plane of the slab.

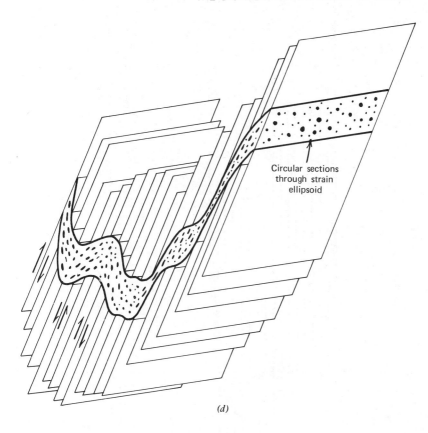

(d)

kind of fold that might develop in a bar that is end loaded and deformed by buckling [see Bayly (1971)] although, in general, buckling can be expected to produce more complicated folds (e.g., see Fig. 4.28 where the hinge is substantially thinned).

The second model (Fig. 4.29c) again produces concentric or parallel folds and involves shear on surfaces parallel to the layer as the layer folds. This process is known as *flexural slip* if the surfaces of shear are descrete with a finite thickness, or *flexural flow* if deformation takes place at the granular scale with no descrete surfaces of slip [see Donath and Parker (1964) and Fig. 4.29c)]. A fundamental feature of this model is that there is no distortion in the plane of the folded layer.

The third model involves shearing on closely spaced planes oblique to the layer being folded. This process produces ideally similar folds (Fig. 4.29d) and is called *shear or slip folding*.

Using the classification of Ramsay (1967), presented in Figure 4.26, the first and second models produce class IB folds whereas the third produces class II folds. Neither of these two types of folds are particularly common in nature

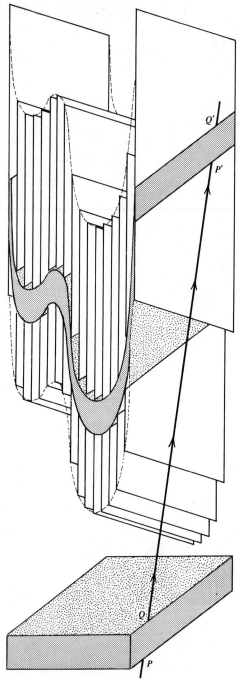

FIGURE 4.30 Ideally similar fold formed by slip on planes oblique to the folded layer. The direction of slip is not necessarily normal to the fold axis. In this diagram the direction of slip is parallel to *PQ* so that *P* is displaced to *P'* and *Q* to *Q'*.

(except for some kink folds which presumably are produced by flexural slip folding), but still the models are worth considering in some detail here because they represent simplified versions of the more complicated behavior that must exist in nature.

The important features of the deformation involved in the first model are summarized below:

1. Since the deformation involves only bending about the fold axis, there is ideally no extension parallel to the fold axis and the strain throughout the fold is plane. The fold axis is parallel to the intermediate principal axis of strain at each point in the fold.

2. There is a surface toward the middle of each layer (the neutral surface) that maintains its initial area; at all points within it there has been no strain.

3. The layer undergoing deformation maintains its initial thickness measured normal to the layer at each point. This is the reason such folds are commonly called parallel folds. However, the layer has been extended on the outer arc of each fold and shortened on the inner arc. The distribution of strain in such a fold is therefore of the form shown in Figures 4.29b and 4.33a. The neutral surface toward the middle of the layer is the surface that marks the changeover from shortening to extension within the layer. The surfaces on the outer side of the neutral surface have had their areas increased whereas surfaces on the inner side of the neutral layer have had their areas decreased. The strain increases with distance measured normal to the neutral plane. Notice that the $\lambda_1 \lambda_2$ planes of the strain ellipsoids fan across each fold, the fan diverging away from the fold hinge on the outer side of the neutral plane, and converging toward the fold hinge on the inner side.

4. An initially straight lineation, lying in the surfaces to be folded and inclined at θ to the fold axis before folding, becomes curved during the folding. Since the area of the neutral surface is not distorted during folding the angle between the lineation and the fold axis in this surface remains constant and equal to θ. For surfaces on the outer side of the neutral surface this angle is increased but it is not constant unless the extension within the surface is everywhere the same. For surfaces on the inner side of the neutral surface this angle is decreased but, again, it is not constant unless the shortening within the surface is everywhere the same (see Fig. 4.32b). The lineation in the neutral surface lies on a cone with the semiapical angle θ and with the fold axis as the axis of the cone. On a stereographic projection this lineation plots as a small circle. Lineations on surfaces other than the neutral surface plot as more complicated curves [see Hobbs (1971, pp. 350–354), for a discussion of the geometry of some of these redistribution patterns. See also Weiss (1959a) and Ramsay (1960, 1967, pp. 463–486)].

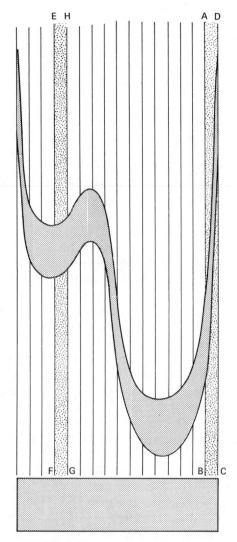

FIGURE 4.31 Profile of the ideally similar fold formed by the mechanism illustrated in Figure 4.30. The thickness measured parallel to the shear planes is constant and equal to the original thickness of the slab measured in the same direction. The volume of folded material in the thin lamella ABCD on the limb of a fold is the same as that in the thin lamella EFGH on the hinge of a fold.

The important features of the deformation involved in the formation of flexural slip or flexural flow folds are outlined below:

1. Since the deformation involves bending about the fold axis and shearing on surfaces in directions normal to the fold axis, the strain is plane at all points in the fold and the fold axis is parallel to the λ_2 axis of the strain ellipsoid at each point.

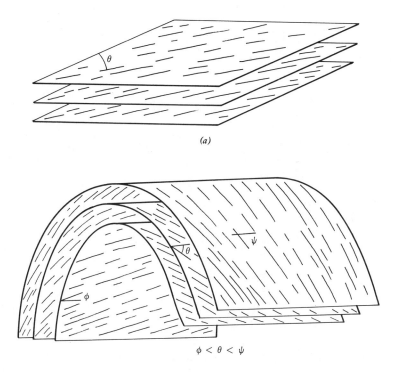

(a)

$\phi < \theta < \psi$

FIGURE 4.32   Redistribution of an initially straight lineation by the three simple models of folding illustrated in Figure 4.29. (a) Initial angle between lineation and fold axis is θ. (b) Folding so that a neutral surface exists. In the neutral surface the angle between the fold axis and the lineation remains at θ. For layers that have been extended this is increased; for layers that have been shortened this angle is decreased. Redistribution of the lineation is shown in the stereographic projection. (c) Folding by flexural shear. The angle between the fold axis and the lineation remains equal to θ. Redistribution of lineation is shown in the stereographic projection. (d) Folding by shear on planes oblique to the layer. Lineation is distorted so that it lies in a plane defined by the initial orientation of the lineation and the slip direction. (e) Kinematic reference axes a, b, and c.

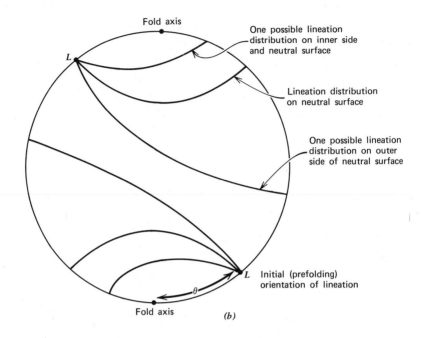

Fold axis

One possible lineation
distribution on inner side
and neutral surface

Lineation distribution
on neutral surface

One possible lineation
distribution on outer
side of neutral surface

L Initial (prefolding)
orientation of lineation

Fold axis   *(b)*

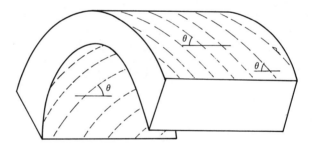

FIGURE 4.32 (Continued)

2. The layer maintains its initial thickness measured normal to the layer at each point so the fold is concentric or parallel. However there is no neutral surface.
3. The folded surface is parallel to a circular section of the strain ellipsoid at each point. The $\lambda_1 \lambda_2$ planes of the strain ellipsoids define a divergent fan across the fold (see Fig. 4.29c).

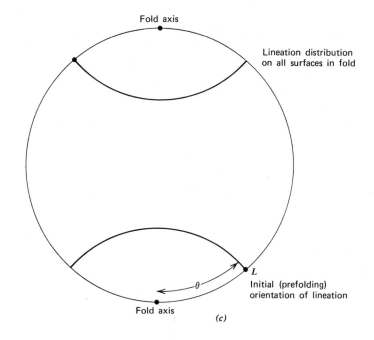

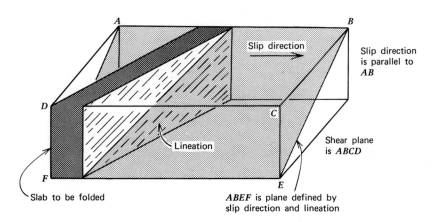

FIGURE 4.32 (Continued)

4. Since there is no distortion within the planes of the folded layers (these are circular sections of the strain ellipsoid at each point), an initially straight lineation lying within the surfaces to be folded, and inclined at θ to the fold axis before folding, remains at this angle to the fold axis at all

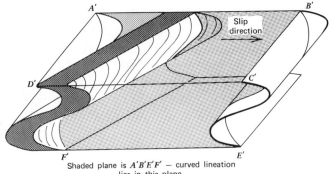

Shaded plane is $A'B'E'F'$ — curved lineation
lies in this plane

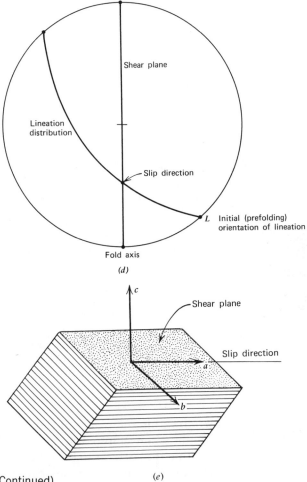

(d)

(e)

FIGURE 4.32 (Continued)

points in the fold. The lineation therefore lies on a cone of semiapical angle θ and with the fold axis as the axis of this cone. On a stereographic projection the lineation plots as a small circle (See Fig. 4.32c).

The important features of the deformation involved in the formation of shear or slip folds are outlined below:

1. Since the deformation is by simple shear, the shearing planes are circular sections of the strain ellipsoid at each point in the fold and the strain is a plane strain everywhere (Fig. 4.29d).
2. Notice, however, that there is no reason for the direction of shearing to be normal to the fold axis that develops. The only constraint on this is that the shearing direction is not parallel to the layer. For any other angle between the direction of shearing and the layer a fold develops although it reaches its maximum amplitude for a given amount of shear when the shearing direction is normal to the fold axis (Fig. 4.30). The strain is plane at each point with the λ_2 axis everywhere in the shearing planes and normal to the direction of shearing.
3. Since the shearing plane is a circular section of the strain ellipsoids, the layer maintains constant thickness when measured parallel to the axial plane in profile (Fig. 4.31). This results in greatly thickened hinges and thin limbs. Notice, however, that there is no need to postulate flowage from limbs to hinges to explain this. The volume of material between the two shearing surfaces AB and CD on a limb in Figure 4.31 is the same as the volume between EF and GH near a hinge.
4. For a layer initially normal to the shear surfaces, the shears are of opposite sense either side of the hinge line of the fold. Hence the distribution of strain is as shown in Figure 4.34. At all places in the fold the principal plane fan must be divergent.
5. There is no neutral layer within these folds. The strain is the same at all points within a layer along a particular shearing surface.
6. An initially straight lineation is distorted during folding such that each point on the lineation is displaced parallel to the slip direction. The lineation is therefore distorted so that it lies in a plane defined by the initial orientation of the lineation and the slip direction [see Fig. 4.32d; also Weiss (1959a); Ramsay (1960, 1967, pp. 469–480)].

The shear or slip fold model has been very popular in the geological literature, especially between 1945 and 1960 when it formed the basis for considerable controversy regarding the interpretation of lineations [see Anderson (1948); Kvale (1953); Lindstrom (1958a,b)].

At that time a system of coordinate axes was in widespread use, based on a terminology developed by Sander (1930). These coordinate axes described the movements involved in a progressive simple shear and were termed *kinematic axes*. These axes are illustrated in Figure 4.32e: the *a* axis is parallel to

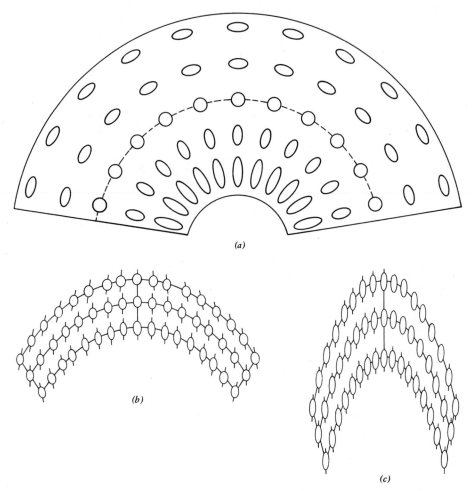

FIGURE 4.33   (*a*) Distribution of strain in a slab folded so as to have a neutral surface. (*b*) A fold of the type produced in (*a*) with a homogeneous additional shortening of 20 percent. (*c*) A fold of the type produced in (*a*) with a homogeneous additional shortening of 50 percent [from Hobbs (1971)].

the direction of slip or shearing within the shear planes; the *b* axis is normal to *a* in the shear plane and *c* is normal to both *a* and *b*. The shear plane is therefore the *ab* plane. The *ac* plane was commonly referred to as the *movement plane*. The direction *a* was commonly referred to as the *direction of tectonic transport* and considerable argument raged as to whether lineation in rocks is parallel to the *a* or the *b* direction. Since such argument was based on the oversimplified assumption that the deformation history in rocks amounted to a progressive simple shear, the controversy was rather pointless. The terminology is in little use now except when the movements involved in a progressive simple shear are being described.

4.7.2 More Realistic Models of Fold Formation

The problem with the classical models of fold formation outlined in Section 4.7.1 is that they lead to only class IB and class II folds. These types of folds are rare in nature, the common ones being class IC and class III [see Hudleston (1973b)]. Theoretical and model studies have indicated that the reason for this is that first, the folding layer does not maintain its original thickness throughout the deformation and second, we are normally involved with a multilayered sequence in which the individual layers possess different mechanical properties and thicknesses. This means that, in general, only one or two layers begin to fold initially and it is these layers that control the deformation from then on. Other weaker layers are more or less constrained to behave in the manner which the stronger layers determine. These two points are discussed below:

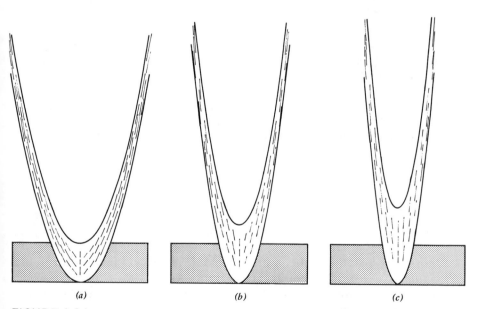

(a)   (b)   (c)

FIGURE 4.34   (a) Distribution of strain in a slab folded to produce a similar fold. No homogeneous additional strain. (b) The fold of (a) with a homogeneous additional shortening of 25 percent. (c) The fold of (a) with a homogeneous additional shortening of 50 percent. The stippled area in each case is the undeformed cross section of the slab [from Hobbs (1971)].

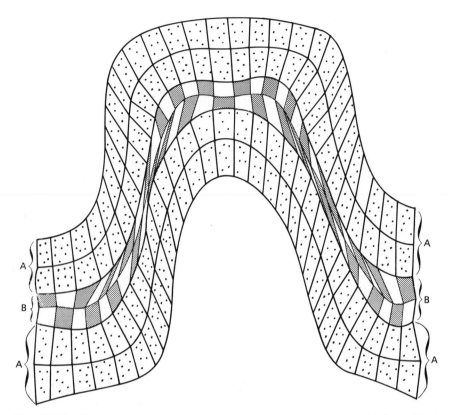

FIGURE 4.35 Experimentally produced fold with markers that initially defined rectangles that were approximately squares. Layer A is composed of NaCl and mica. Layer B is composed of KCl and mica. The material in layers A is about 10 times stronger than that in layer B. Within layer B, each black-and-white pair taken together across the layer represents an initial square in the undeformed state.

1. *Layer shortening before folding and homogeneous shortening during folding.*
Figure 1.10 is an example of a layered specimen undergoing buckling to produce a fold. It will be noticed that in the first 20 percent or so of shortening parallel to the layers no fold develops and instead the individual layers simply shorten at the same time as they increase their thickness. At around 20 or 30 percent shortening a fold nucleates and from then on begins to grow.

This is an example of the more general situation that accompanies the buckling of layered materials. The amount of layer shortening that occurs before folding begins appears to be a function of the strain rate and of the relative mechanical properties of the layers undergoing buckling [see Dieterich (1970); Sherwin and Chapple (1968)].

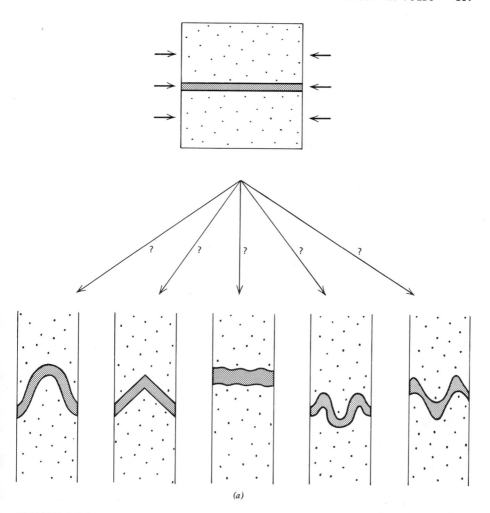

FIGURE 4.36   Some of the problems that are involved with the development of folds. (a) Single layer embedded in material with different mechanical properties and loaded parallel to the layer. Does the layer fold to produce a rounded or angular shape? Does it simply thicken up with little folding? Does the layer maintain initial thickness as it folds or do the limbs thin and the hinges thicken? (b) Multilayered situation. Do the layers buckle to produce folds with axial planes normal to the axis of shortening or do kink planes oblique to the axis of shortening develop?

In the example illustrated in Figure 1.10 the specimen continues to shorten after the fold nucleates. This progressive shortening may be thought of as composed of two parts: one part is directly associated with the bending of the layers to form folds, the other part consists of an *additional strain* at each point with a component of shortening approximately normal to the axial plane and

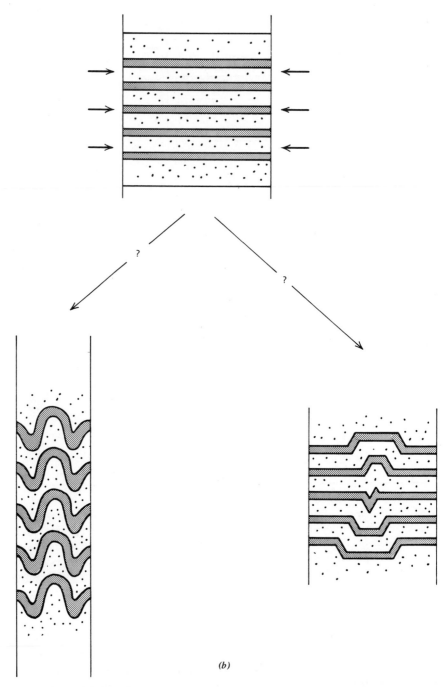

(b)

FIGURE 4.36 (Continued)

a component of extension in the axial plane and normal to the fold axis. To date there has been no theoretical analysis of this additional strain although computer simulations of folding [see Dieterich (1970) and Fig. 4.40] indicate that, for high viscosity contrasts between the folding layer and surrounding medium, this additional strain is small (Fig. 4.40a). For low viscosity contrasts the additional strain is an important contribution to the overall deformation (Fig. 4.40c).

It is to be expected that the distribution of additional strains throughout the folding body will depend on the manner in which the mechanical properties of the layers change with the progressive deformation. If there is little relative change then the additional strains are likely to be homogeneous. If, on the other hand, strain hardening is important in high strain areas, or there are relative changes in mechanical properties associated with the development of anisotropy, such as foliation or crystallographic preferred orientations, then it is unlikely that the additional strain field will be homogeneous. As indicated, there is no theoretical treatment to date but many workers [e.g., Ramsay (1962a); Hudleston and Stephansson (1973); Milnes (1971)] have assumed the additional strains to be homogeneous.

Some of the distributions of strain that develop because of layer shortening and the superposition of additional strains are illustrated in Figure 4.40. An example of the influence of a homogeneous additional strain on the strain distribution in a layer deformed with a neutral surface is shown in Figure 4.33. Another example of the effect of a homogeneous additional strain superposed on a shear fold is given in Figure 4.34. The general effect is to decrease the angle between the $\lambda_1\lambda_2$ plane at each point and the axial plane. Folds that nucleate as class IB folds develop into class IC folds with an additional homogeneous strain (see Fig. 4.33c and Ramsay 1962a) although under some circumstances folds very close to class II may form [see Hudleston and Stephansson (1973)]. Folds that are initially class II remain class II under the influence of an additional homogeneous strain (see Fig. 4.34c).

2. *Modifications introduced by multi layers.*

Folds initiated in multilayered sequences are generally controlled in their distribution and wavelength by the more competent members of the sequence. Although there is mechanical interaction between the competent and incompetent layers, the incompetent members conform to the shape changes that are largely prescribed by the other stronger layers. Thus, in Figure 4.35, two competent layers marked A are shown deforming by buckling but with additional homogeneous shortening so that two large class IC folds are produced. If the incompetent material in between, marked B, conforms to the shape changes prescribed by the other layers it must be deformed to produce a class III fold. If the strain is continuous across the boundaries between layers A and B, that is, there has been no slip at the interfaces, then the B layers have been shortened on the outer arc and extended on the inner arc. This may be envisaged more clearly by considering the shape of a grid, initially square, after folding has ceased (see Fig. 4.35). As an exercise the reader should sketch in the rough shape of the strain ellipsoid for each of the distorted squares in Figure 4.35. Notice that the $\lambda_1\lambda_2$ planes of the strain ellipsoids delineate a convergent fan for the A layers but that both convergent and divergent orientations are present within the B layer.

4.8 DEVELOPMENT OF FOLDS

The problem to be considered in this section is illustrated in Figure 4.36 and takes the two aspects shown in *a* and *b* of that figure. Figure 4.36*a* shows a thin layer embedded in material with different mechanical properties and compressed parallel to the layer. The questions to be answered are: *What governs the behavior of this body that might lead to the development of folds in the layer?* For instance, if a buckling* instability develops, what are the mechanical properties that control this? What governs its wavelength? On the other hand, is the amplification of local irregularities all that is important? If so, what governs the wavelength that is likely to grow? Will the folds that grow be rounded or sharp in profile? Will layer thickness be maintained during folding or will the limbs thin and the hinges thicken? What decides if a fold will develop or not? Under what conditions will the layer simply thicken up with little or no folding?

The same questions are posed for Figure 4.36*b*, except in this case the material to be deformed is a multilayered specimen consisting of layers of different thicknesses and mechanical properties. Another important question may be added in this case or for any situation where a strong anisotropy exists due to the presence of a foliation: Under what conditions will folds form with axial planes normal to the axis of shortening and what other conditions govern the formation of kink planes oblique to the axis of shortening? [See also Johnson (1970), Figure 8.20.]

Needless to say, the precise answers to most of these questions are presently unknown but a large amount of work exists, mainly based on materials with simple mechanical properties that can guide thinking with respect to these questions.

Biot (1957) published one of the first treatments of this problem and was closely followed by Ramberg in 1959. The early work treated the problem as an elastic one, that is, all of the layers were assumed to behave elastically. However Biot (1957, 1961, 1965) annunciated a correspondence principle between viscoelastic and elastic materials and on the basis of this principle treated some classes of viscoelastic materials. Even so, most of the work published to date has treated the various layers as though they were perfectly

*See Bayly (1971) for a discussion on this term.

viscous materials. This is the approach adopted in theoretical treatments by Ramberg (1959, 1961, 1963b, 1964) who has approached the whole problem as essentially one in fluid dynamics. The assumption of linear relationships between stress and strain or between stress and strain rate is simply to make the whole problem tractable. There are numerous attempts to demonstrate that rocks should behave as ideally viscous fluids under geological conditions, but such demonstrations usually amount to rationalizations of why simple mathematics has been used. After all, the use of linear relationships leads to differential equations that are difficult enough to handle; the use of nonlinear relationships between stress and strain rate in general leads to insurmountable problems mathematically. What is needed most of all now are better mathematical tools for handling the problems involved. Nevertheless the results arising from work done to date are extremely useful and they are outlined below. For a more detailed review of work done to date see Hudleston (1973a and b) and Hudleston and Stephansson (1973). Chapple (1968a and b) has discussed some of the basic mathematical relationships. In what follows the assumptions made are that, first of all, the folds are so small that gravity has not been an important factor in their development. Second, compression has been parallel to the layer to start with, and third, the deformation has only involved a plane strain. The situation where gravity is important has been considered by Biot (1959, 1965) and by Ramberg (1967, 1968). Treagus (1973) has considered the case where the imposed compressive stress is oblique to the layer.

1. Nucleation of Folds. Both Biot and Ramberg have discussed the two aspects of the problem illustrated in Figure 4.36. One of the most important results of their work is the concept of a *dominant wavelength*. This was first indicated by Biot in 1957. The mathematical treatments suppose that there are small low-amplitude sinusoidal irregularities in the layers about to be folded. These might be present in the layers prior to the imposition of compressive stress or may be local instabilities that develop as the compressive stress is applied. The theory indicates that, although all of these irregularities might start to grow, as the deformation proceeds one particular wavelength out of all those present will be selected and amplified to appear as a dominant wavelength in the folds that ultimately form.

For the situation illustrated in Figure 4.36a, if the layer is assumed to have a viscosity η_1 and is embedded in a less viscous medium of viscosity η_2 then both Biot and Ramberg derive the following equation for the dominant wavelength w_d:

$$w_d = 2\pi t \sqrt[3]{\eta_1/6\eta_2} \tag{4.1}$$

In this equation t is the thickness of the layer. Equation 4.1 has been experimentally verified by Biot et al. (1961).

Biot (1961) has indicated that if the contrast in viscosity between layers is small then folding is unlikely to develop and, instead, most of the deformation will consist solely of layer shortening. Some degree of layer shortening is likely to be present for most viscosity contrasts and the question arises as to what difference such layer shortening will make to the dominant wavelengths predicted by Equation 4.1. Sherwin and Chapple (1968) have taken layer shortening into account and show that the dominant wavelength changes with the amount of strain. Hudleston (1973a) has rewritten the Sherwin-Chapple equation for the dominant wavelength in the following form so that it can be easily compared with the Biot-Ramberg equation (4.1):

$$w_d = 2\pi t \sqrt[3]{\frac{\eta_1(s-1)}{6\eta_2} \frac{1}{2s^2}} \qquad (4.2)$$

where $s = \sqrt{\lambda_1/\lambda_3}$, λ_1 and λ_3 being principal quadratic elongations. It is clear from Equation 4.2 that the dominant wavelength will change with strain and the effect is illustrated in various manners in Figures 4.37 and 4.38, and 4.39. Sherwin and Chapple (1968) predict that as the deformation proceeds folds with progressively larger thickness to wavelength ratios will become those most amplified. This is illustrated in Figure 4.38.

2. The growth of Folds. It is important to note that all of the theoretical treatment discussed above is only true for small amplitude folds but the ques-

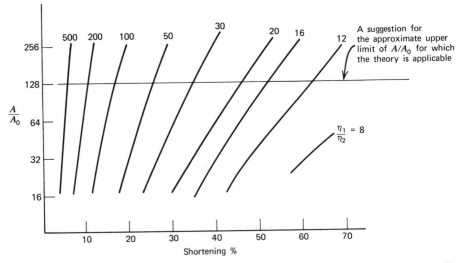

FIGURE 4.37 Amplification of the dominant wavelength as a function of the bulk shortening based on the Sherwin-Chapple analysis. For example, we see a fold whose initial deviations from straight are 1/30 of the amplitude observed; that is, $A/A_0 = 30$. Then if the viscosity ratio is 200 we postulate shortening of about 8 percent whereas if the viscosity ratio is 20 we have to postulate shortening of 35 percent.

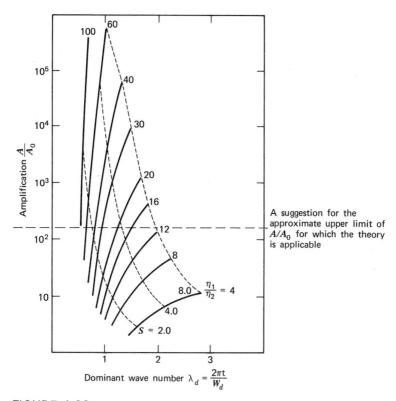

Dominant wave number $\lambda_d = \dfrac{2\pi t}{W_d}$

FIGURE 4.38   Amplification of the dominant wavelength as a function of the dimensionless ratio λ_d, the dominant wave number. The thickness of the layer is t and W_d is the dominant wavelength. The amplification is presented for various values of the viscosity contrast (heavy lines) and of the ratio, S, of the principal quadratic elongations λ_1 and λ_3, associated with bulk shortening of the specimen. For example, consider the case for a viscosity ratio of 4. As the ratio λ_1/λ_3 increases from 2 to 8, λ_d changes from 1.6 to 2.8. At the same time, the amplification increases from 2 to about 10. Thus, for a viscosity ratio of 4, a fivefold change in amplification is associated with a 1.75-times change in the ratio of thickness to dominant wavelength. [From Sherwin and Chapple (1968), *Amer. Jour. Sci.*, v.266.]

tion arises: At what stage in the growth of a fold will these treatments begin to fall down? Bayly (1964) and Chapple (1964 and 1968c) have extended Biot's viscous theory to situations where the amplitudes of folds are large. For the single layer situation Chapple shows that a 15° dip for the limbs is about the limit in amplitude for where the dominant wavelength type of analysis discussed in (1) above is likely to be operative. Chapple (1968c), Dieterich (1969,

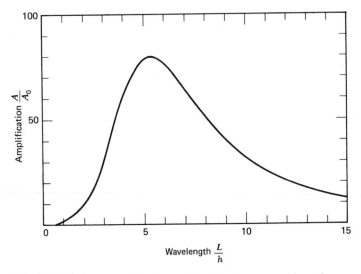

FIGURE 4.39   Amplification of the dominant wavelength as a
function of wavelength for a viscosity ratio of 18 and a quadratic
elongation of a layer of 3.2. For these conditions, the dominant
wavelength is indicated by the peak in the curve. [From Sherwin
and Chapple (1968), *Amer. Jour. Sci.*, v.266.]

1970), Stephansson and Berner (1971), and Hudleston and Stephansson
(1973) have all examined the ways in which the folds grow using numerical
methods and assuming the materials to be ideally viscous. Some aspects of
Dieterich's results are depicted in Figure 4.40. They are essentially the same as
those of Chapple (1968c) regarding the enclosing medium, but show, in addition, interesting suggestions about strains inside the folded layer. An important point illustrated by these results is that as the viscosity contrast decreases,
layer shortening increases and folding becomes less important in the deformation. Chapple (1969) has also considered some nonlinear relationships between stress and strain and has shown that folds with long, straight limbs and
sharp hinges can develop. Other nonlinear materials have been considered by
Parrish (1973); his treatment, like those of Dieterich, leads to folds that are
not very far from parallel in style. Hudleston and Stephansson (1973) have
produced folds for the viscous single-layer situation, but where there is not
very great viscosity contrast between the layer and the surrounding material.
In addition to their work based on energy minimization, Hudleston and
Stephansson also offer a more explicit mechanical description in which the
layer buckles incrementally and an increment of homogeneous shortening is
added separately to the deformation. The fold shapes produced resemble
natural folds more than the shapes of folds developed by most other workers

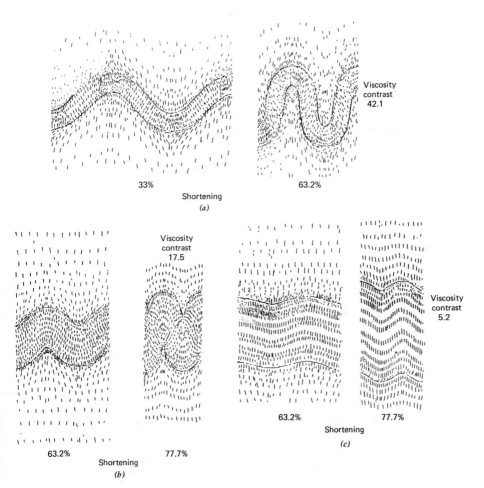

FIGURE 4.40 Computer simulation of single layer folding for three different sets of conditions. In each diagram the short line is drawn normal to the principal axis of shortening at each point. (a) Viscosity ratio is 42.1. Initial thickness equals dominant wavelength/12. (b) Viscosity ratio is 17.5. Initial thickness equals dominant wavelength/9. (c) Viscosity ratio is 5.2. Initial thickness equals dominant wavelength/6. [From Dieterich (1970). Reproduced by permission of the National Research Council of Canada from the *Candian Journal of Earth Sciences*, Vol. 7, No. 2, pp. 467–476.]

to date. The shape of the fold can vary from class IB through class IC to nearly class II depending on the viscosity ratio, the amount of shortening and the wavelength thickness ratio.

The mathematical theory for the viscous situation has now reached a stage where the nucleation process and the variation of dominant wavelength with

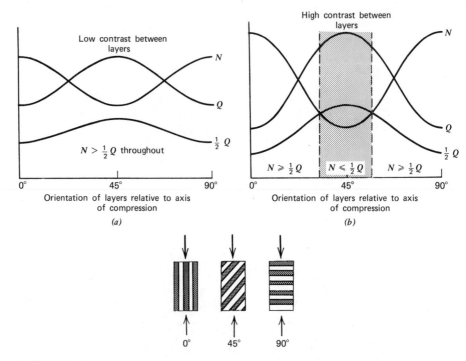

FIGURE 4.41   Sketch of the variation of N and Q with orientation of multilayered specimen relative to the axis of compression. Individual layers are assumed to be isotropic and ideally viscous. (*a*) Alternating layers of low viscosity contrast. (*b*) Alternating layers of great viscosity contrast.

layer shortening in Newtonian materials is fairly well understood. Numerical techniques have enabled the growth of folds to be investigated for various viscosity ratios and layer thicknesses. The ideal would be to treat the growth of folds from an analytical rather than a numerical point of view but the pressing need is to develop analytical approaches for materials with nonlinear relationships between stress and strain rate. Bridwell (1974), however, has considered some of the basic assumptions used in all of these computer simulation studies. He points out that considerable errors can result in such studies if the sides of the basic deforming element are not allowed to deform in a manner compatible with the buckling deformation. Other nonlinear effects associated with rotation of the basic deforming element can introduce substantial errors. In additon, de Caprariis (1974) has considered the effect of non-Newtonian behavior and indicates that the dominant wavelength selection process of Biot and Ramberg may need modification. The work of Bridwell and of de Caprariis is an important indication of the problems that arise once various kinds of nonlinear behavior are considered. The problem needs much more work.

Most of the discussion presented above applied to the single-layer situation (Fig. 4.36*a*). Biot (1965, pp. 184–204), Johnson (1970, pp. 322–327), and Cobbold et al. (1971) have considered the important question raised in Figure 4.36*b*: What mechanical properties decide if buckling or kinking will be important in the deformation of a multilayered sequence? Cobbold et al., following Biot (1965), have considered the situation at the instant instabilities are about to form in the multilayered specimen. They use two quantities, N and Q, which are measures of the compressive and shear moduli, respectively, of the bulk material at the instant of instability formation for compressive stress acting parallel to some reference direction. That is, N is a measure of the ratio of compressive stress to elongation of the bulk specimen (both measured parallel to the reference direction) whereas Q is a measure of the ratio of shear stress to shear strain of the bulk specimen (again, both measured parallel to the reference direction). Notice that if both of the materials making up the layered specimen are highly viscous then both N and Q are large, but if one material is highly viscous and the other has low viscosity, one or other of N and Q is likely to be small.

The relative values of N and Q change for different orientations of the layering with respect to the reference direction:

1. For layering parallel to the reference direction, Biot (1965, pp. 186) shows that

$$N \geq Q \qquad (4.3)$$

if the materials making up the layers are themselves isotropic and viscous. In this orientation N is a maximum and Q a minimum.

2. For layering at 45° to the reference direction,

$$N \leq Q \qquad (4.4)$$

if both materials are isotropic and viscous (Biot, 1965, pp. 187), and N is a minimum whereas Q is a maximum.

3. For layering normal to the reference direction, relationship (4.3) holds once again. The variation of N and Q with orientation is sketched in Figure 4.41.

Biot (1965) has shown that the behavior of a laminated material depends not only upon the moduli, N and Q, but also upon the state of stress in the specimen just before instabilities develop. He defines two other moduli, M and L, related to N and Q by

$$M = N + P/4$$
$$L = Q + P/2 \qquad (4.5)$$

where P is a measure of the initial stress in the material, that is, the stress present in the increment before instability develops.

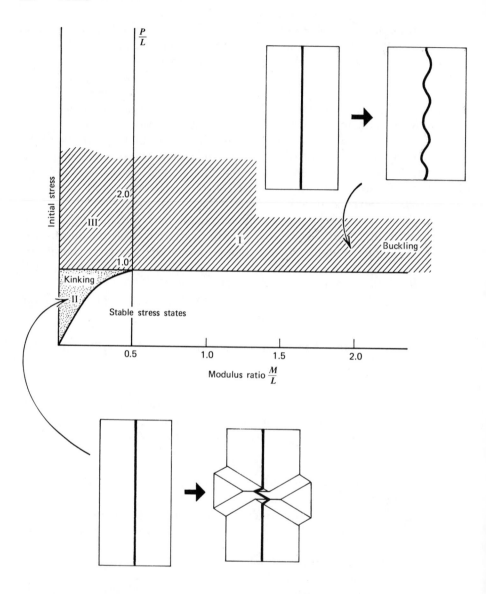

FIGURE 4.42   Plot of stress in the increment of strain prior to development of an instability against the modulus ratio M/L for a laminated specimen. In the region marked "stable stress states," no instabilities develop. In region I, the specimen deforms by buckling. In region II the specimen deforms by developing kinks oblique to the axis of loading. No specimen can reach region III because it enters region II first. [After Cobbold et al. (1971).]

Using the two quantities M and L, Biot (1965), Johnson (1970), and Cobbold et al. (1971) delineate two types of mechanical behavior where mechanical responses are contrasted. They show that, for principal stress parallel to the reference direction:

I: If $M < \tfrac{1}{2}L$ and $P < L$ one type of response is expected, while if

II: $M < \tfrac{1}{2}L$ and $L < P < 4M(L-M)/L$ a different type of response is expected.

These relationships are shown graphically in Figure 4.42, but are difficult to express in a more meaningful manner since they involve the stress P, which is in the specimen prior to instability formation.

However, regardless of P, the critical condition $M = \tfrac{1}{2}L$ is equivalent to $N = \tfrac{1}{2}Q$. In alternating layers of little contrast, this condition cannot be satisfied (Fig. 4.41a). However, in alternating layers of greater contrast, $N \leqslant \tfrac{1}{2}Q$ for a range of orientations close to 45° (Fig. 4.41b).

Experiments on multilayered materials [cf., for instance, Paterson and Weiss (1966); Ghosh (1968)] show that for alternating layers of little contrast, buckling tends to develop rather than kinking. On the other hand, for alternating layers of high contrast, kinking and buckling are observed; hence it is tempting to equate the two theoretical types of response with buckling and kinking. However, an important difficulty is that in experiments where kinks develop, the kinking process is best developed for orientations where the layering is parallel to the axis of compression and, in fact, tends not to develop at all in the 45° position [see Paterson and Weiss (1966) and the discussion of their experiments below]. Thus, although the temptation is strong to equate type II behavior with kinking, the theory predicts the best response in the 45° orientation where experiment shows it to be least developed. Moreover, the theory says that in the 0° orientation, $N \geqslant \tfrac{1}{2}Q$ and hence type II behavior cannot occur. This again is directly opposed to the experimental results.

Cobbold et al. (1971, Fig. 6) recognize this difficulty and suggest, plausibly, that it may be resolvable by postulating small irregularities upon which kinks nucleate; even tiny regions in the 45° orientation embedded in a sample that otherwise has the 0° orientation would be sufficient to promote type II behavior. However, it is necessary then, that every time a kink forms, it relies on a small inhomogeneity for nucleation. Clearly, this remains to be established and the exact extent to which the Biot equations for type II behavior apply to kinks in rocks is still open to study.

Another way over this difficulty arises from Biot's discussion of instability development in materials that are not ideally viscous but show behavior similar to perfect plasticity. Biot (1965, pp. 403) shows that, in such materials which show relatively little flow until some critical value of the stress is reached, it is possible for $M < \tfrac{1}{2}L$ for increments of strain close to this critical stage of flow. It could then be possible to initiate type II behavior in the

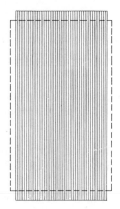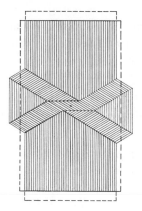

FIGURE 4.43 Kinking of a foliated specimen. The dotted lines indicate the prescribed strain. [From Paterson and Weiss (1966), *Geol. Soc. Amer. Bull.*, v.77.]

material close to the 0° orientation without relying on the presence of local inhomogeneities to nucleate the process. Suggestions such as this reinforce the need for theoretical treatments of materials with nonlinear mechanical properties; and by *nonlinear* is meant not only *power-law flow* (Equation 1.27) but more complicated relationships between stress and strain rate.

The behavior of laminated materials during kinking has been extensively studied by Paterson and Weiss (1966) and Weiss (1969). Their results are summarized below and in Figures 4.43 and 4.44.

1. Constrained specimens compressed parallel to the foliation deform by producing conjugate kinks inclined to the axis of shortening at between 55 and 65° (Fig. 4.43).
2. The angle between the foliation and the kink plane is about the same inside and outside the kink (Fig. 4.43).
3. A pair of conjugate kinks intersect to produce a fold whose axial plane is normal to the axis of shortening (Fig. 4.43).
4. The number of kinks increase as the shortening increases, there being just a few at 5 percent shortening. As the numbers of kinks increases, the number of intersections increases and so also do the number of folds with axial planes normal to the axis of shortening. By 50 percent shortening kinks are no longer obvious and the rock consists solely of these folds that have developed in intersection regions (Fig. 4.44).
5. By 50–70 percent shortening the folds are tight and have the appearance of many natural folds. [See Paterson and Weiss (1966, Plate 3E.)]
6. For specimens loaded obliquely to the foliation at angles up to 30° the same sequence of events occurs except that one family of kinks tends to be better developed, even to the exclusion of the other family. This

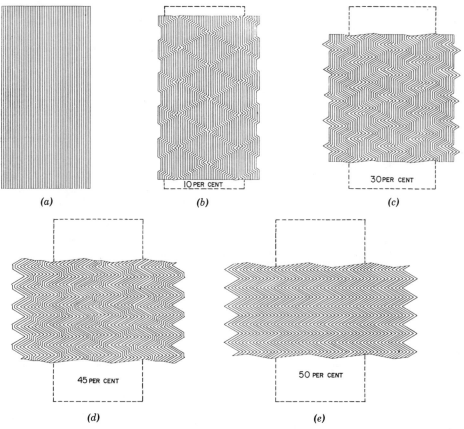

FIGURE 4.44   Sequence of kinking of a foliated specimen with increase in strain. [From Paterson and Weiss (1966), *Geol. Soc. Amer. Bull.*, v.77.]

means that folds with axial planes normal to the axis of shortening, if developed, have one limb longer than the other [See Paterson and Weiss (1966, Plate 6)].

7. Specimens loaded at 30–45° to the foliation deform by sliding on the foliation planes. No kinks develop. [See Paterson and Weiss (1966, Plate 7.)]

8. Specimens loaded at greater than 45° to the foliation normally deform by shearing across the foliation planes. These shear planes sometimes resemble crenulation cleavage. [See Means and Williams, 1972.)]

It is to be expected that a deformation behavior as geometrically simple as kinking would be capable of straightforward mathematical treatment. To date, however, most treatments have not progressed very far [see, for instance, Jaeger (1969, pp. 255–257)].

5

FOLIATIONS

5.1 INTRODUCTION

In metamorphic rocks there are commonly pervasive surfaces defined by discontinuities, preferred orientation of inequant minerals, laminar mineral aggregates, or some combination of these microstructures (see Fig. 5.1). In many cases the surfaces are inclined to bedding and since they have no counter-part in undeformed sedimentary rocks they must be a product of deformation. Elsewhere however, bedding cannot be identified with certainty and it is not obvious whether the surface, or the earliest surface if there is more than one, is of sedimentary or of metamorphic origin; in some instances it may even be a sedimentary surface that has been modified by metamorphism. For this reason there is a need for a nongenetic, general term to cover all surfaces found in deformed metamorphic rocks; we use the word *foliation* in this sense and describe some of these structures in this chapter. Some geologists use the word foliation in a more restricted sense to refer only to surfaces produced by deformation and metamorphism. We do not subscribe to the latter usage because in many areas of deformed rocks it is difficult to decide just which surfaces are a product of deformation or accompanying metamorphism. The term *s surface* is used, by some writers, with the same meaning as foliation, as used here.

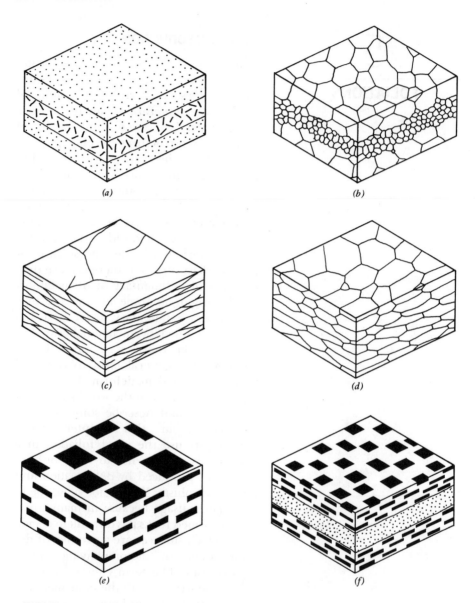

FIGURE 5.1   Diagrammatic representation of various types of foliation. The foliations are defined by: (*a*) compositional layering; (*b*) grain-size variation; (*c*) closely spaced, approximately parallel discontinuities such as microfaults or fractures; (*d*) preferred orientation of grain boundaries; and (*e*) preferred orientation of platy minerals or lenticular mineral aggregates. These various microstructures can be combined and (*f*) shows a combination (*a* + *e*) that is very common in both sedimentary and metamorphic rocks.

5.2   AXIAL PLANE FOLIATIONS: DESCRIPTION

5.2.1   Introduction

In many areas of folded rocks there is a foliation inclined to the folded surface and systematically oriented with respect to the fold. This surface is, generally, approximately parallel to the axial plane of the fold in the hinge area and is therefore referred to as the *axial plane foliation*. However, it is important to realize that axial plane foliations commonly are not strictly parallel to the axial planes of folds but may depart from parallelism in systematic ways. For example, such foliations commonly fan (Fig. 5.2) and generally change orientation where they cross layers of different composition (Fig. 5.3). In certain layers of some folds (e.g., Fig. 5.3*b*) it is impossible to find any foliation that is actually parallel to the axial plane. Furthermore, even in folds in which the traces of foliation and axial plane as seen in profile are parallel, it is not uncommon for the foliation to be slightly inclined to the hinge line in the third dimension. However, axial plane foliation, or the more specific terms *axial plane cleavage, axial plane schistosity,* and so on, are useful and well-established terms and are, therefore, used *despite the fact that such foliations are commonly not exactly parallel to the axial surface* as can be seen in Figures 5.2 and 5.3.

The microscopic morphology of axial plane foliations varies considerably, even where only one foliation is being considered. For example, there are variations within a single foliation associated with differences of metamorphic grade [White (1949); Olesen et al. (1973)], with position in larger fold structures [Talbot (1965); Williams (1972)], and with rock type [Balk (1936); Talbot (1965); Hoeppener (1956)].

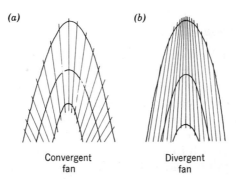

(a)   *(b)*

Convergent   Divergent
fan   fan

FIGURE 5.2   Diagrammatic representation of axial plane cleavage fans.

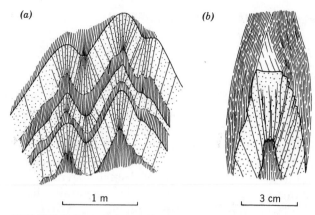

(a)   *(b)*

⊢ 1 m ⊣   ⊢ 3 cm ⊣

FIGURE 5.3   Folds in alternating greywackes (stippled) and shales (Bermagui, Australia) illustrating the fanning and "refraction" of axial plane foliations. Note that in (*b*) the foliation from one limb intersects the foliation from the other limb, despite the fact that both can be considered part of the same axial plane foliation.

A number of morphological foliation types are generally recognized and named but there is no complete agreement in the literature as to the meaning of some of these names. This confusion arises partly because (due to morphological gradation between various end members) the various foliations cannot be fitted into distinct pigeon holes. Thus, the types of foliation described below are simply useful end members and a spectrum of intermediate forms exists.

5.2.2   Fracture Cleavage

Fracture cleavage is used here in the sense of Knill (1960) and is defined as a cleavage consisting of closely spaced microfaults or fractures that divide the rock into a series of tabular bodies or *microlithons* [de Sitter (1964, p. 268)]; within these microlithons, earlier surfaces, if present, are essentially planar. The term has been used by some writers to describe cleavages referred to here as crenulation cleavage (see below) and, morphologically, fracture cleavage commonly grades into such foliations. However, it is useful to distinguish the two end members. It also grades through foliations defined by microfaults, with a few micas concentrated in the fault plane, into a definite layering of the type depicted in Figure 5.4. This particular type of layering, because of its gradational relationship with fracture cleavage is also referred to as fracture

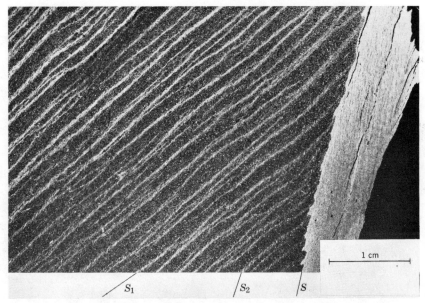

FIGURE 5.4   Differentiated layering (S_1) in a thin section of a specimen from the Upper Vicdessos Valley, Central Pyrenees, France. A bedding plane (S) can be seen at the right side of the photograph, separating a pelite (light area) from a quartz-rich bed (main part of photograph). The layering occurs in the quartz-rich bed; narrow layers, consisting almost exclusively of mica (light zones in the photograph), alternate with broader, quartz-rich zones. Mica in both zones is mostly aligned approximately parallel to the length of the zones, although the orientation pattern is complicated by a later crenulation cleavage (S_2) that is just visible in the photograph. (Nicols at 45°, negative print by W.C. Laurijssen.)

cleavage by some European geologists [e.g., see Lapré (1965)]; we, however, prefer to call it differentiated layering.

Fracture cleavage is generally formed in low to medium metamorphic grade rocks and is common within sandstone beds in folded sequences of alternating sandstone and pelites where the cleavage in the pelite is a crenulation cleavage [e.g., Balk (1936)].

5.2.3   Crenulation Cleavage

Crenulation cleavage varies considerably in morphology but the various forms share one diagnostic feature: in all cases an earlier foliation, generally defined by preferred orientation of layer silicates, is folded (crenulated) on a micro-

FIGURE 5.5 Crenulation cleavage parallel to axial plane of a small fold in quartz mica phyllite. (a) Cleavage seen in hinge of fold. Here the crenulations are symmetrical and the cleavage is defined principally by the limbs of the tighter crenulations. Locally, however, it is defined by microfaults that are developed within the limbs of the crenulations and oriented parallel to their axial planes. Plane-polarized light. (Photograph by W.C. Laurijssen.) (b) Cleavage seen in limb of fold. Here the crenulations are asymmetrical and the cleavage is defined principally by microfaults, which are particularly well developed at the left-hand side of the photograph. Plane-polarized light. (Photograph by W.C. Laurijssen.)

scale. The folds may be symmetrical (Fig. 5.5a) or asymmetrical (Fig. 5.5b), but the latter are the most common. The cleavage may be defined simply by the parallel limbs of the microfolds (as in much of Fig. 5.5a) or it may be defined by microfaults developed parallel to the fold limbs (as in much of Fig. 5.5b and parts of Fig. 5.5a).

Very commonly, rocks bearing crenulation cleavage are differentiated so that what was initially homogeneous in composition and composed of a uniform distribution of quartz and mica, for example, is now layered with mica-rich and quartz-rich layers alternating. The mica-rich layers coincide with both limbs of symmetrical crenulations and with alternate limbs of asymmetrical crenulations (Fig. 5.6); the micas within these layers are aligned approximately parallel to the length of the layer. Careful examination of very thin sections (10 μ thick for fine-grained rocks), however, will commonly reveal a

(b)

0.5 mm

FIGURE 5.5 (Continued)

slight difference of orientation as in Fig. 5.6. The micas, in fact, are still parallel to the earlier foliation but have been rotated toward parallelism with the new foliation. In areas where the progressive development of crenulation cleavage can be observed the early foliation may be enhanced by development of the later cleavage. Thus, even in the microlithons, the early foliation may show a much-stronger preferred orientation of layer silicates where it is crenulated than it does where the crenulation cleavage is not developed.

Some areas are characterized by two crenulation surfaces [see Knill (1959); Rickard (1961)], which intersect at angles generally between 60° and 90° and which are believed to be contemporary. These surfaces are described as *conjugate crenulation cleavages* and may be associated with conjugate folds [e.g., Roberts (1966)] although folds tend to be rare in such rocks. The cleavage surfaces are generally less regularly spaced than their counterpart in areas where only one family of surfaces exists [e.g., see Rickard (1961, Fig. 3)] and tend to be more localized in their distribution. Conjugate cleavage is common in areas where folds predating the crenulation cleavage are tight or isoclinal and possess a well-developed axial plane cleavage or schistosity. In such an environment they are commonly, symmetrically related to the early foliation so that the earlier foliation bisects either the acute or the obtuse angle between the two surfaces of the crenulation cleavage.

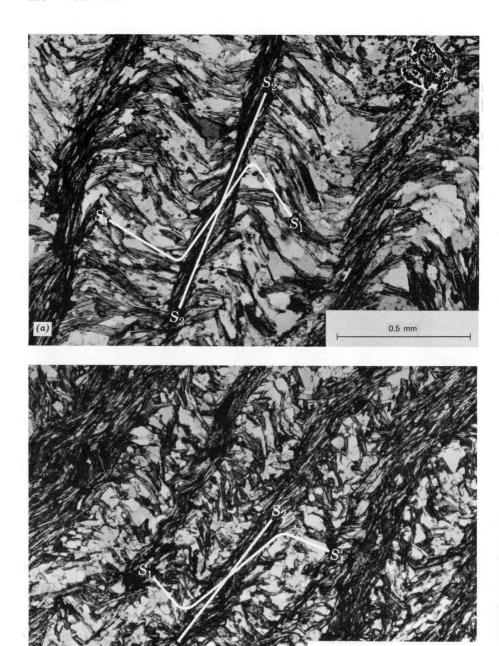

FIGURE 5.6

FIGURE 5.6   Differentiated crenulation cleavages (S_2). Note that the earlier foliations (S_1) can be traced through the mica-rich layers; micas in the mica-rich layers are not strictly parallel to the layer as might appear at first sight. (*a*) Schist from Cooma, N.S.W., Australia. Plane-polarized light. (*b*) Greenschist facies phyllite from the Seve complex, Marsfjällen, Sweden. Note that the mica-rich domains are weakly kinked due to a later deformation. Plane-polarized light. (photograph by W.C. Laurijssen.)

Crenulation cleavage is found in layer silicate-bearing rocks of all metamorphic grades and in low-grade rocks it passes morphologically into fracture or slaty cleavage. It is most spectacularly developed in medium and high-grade rocks rich in mica.

This type of cleavage is also referred to as *strain slip cleavage* [Bonney, (1886)] but we prefer the name crenulation cleavage [Rickard, (1961)].

FIGURE 5.7   (*a*) Cleavage in slate from the Rheinisches Schiefergebirge, Germany. Dark lines running from bottom right to top left are layer-silicate-rich films and they define the cleavage. Between these films are quartz-rich domains in which the layer silicates vary considerably in orientation and are generally inclined to the foliation. Plane-polarized light. (Photograph by W.C. Laurijssen.) (*b*) Cleavage in slate from the Ribagorzana Valley area, Spanish Pyrenees. Layer-silicate-rich domains (black) anastomose around large quartz grains and aggregates consisting of two or three quartz and randomly oriented layer-silicate grains. The large quartz grains show only weak undulatory extinction. Plane-polarized light. (Photograph by W.C. Laurijssen.)

5.2.4 Slaty Cleavage

The word *slate* originated as a quarryman's term for fine-grained rocks that were sufficiently fissile to be split into very thin, planar slabs suitable for roofing. The term slaty cleavage was adopted by early workers to describe the fabric responsible for this fissility but subsequent work has revealed that the fabric varies considerably. One feature, however, is common to all slates and that is a planar preferred orientation of inequant grains. [In the case of layer silicates this means a preferred orientation of (001).] This statement is an adequate description of the fabric of some slates but in general it is incomplete.

Contrary to what their appearance in hand specimen might seem to suggest, many slates, when viewed under the microscope, are found to be domainal, that is they can be divided into many small regions (domains) that are distinguished from their neighbors by differences of composition and fabric (see Fig. 5.7). Generally the domains are of two types: one is lenticular and is surrounded by the other, which is film like [see Voll (1960)]. There is a strong preferred orientation of the long dimensions of the lenticular domains parallel to the cleavage and in sections cut perpendicular to cleavage the filmlike domains appear as an anastomosing network (e.g., Fig. 5.7a and b). The lenticular domains are rich in the major constituents of the rock other

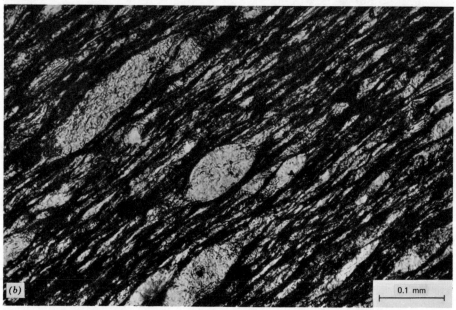

(b)   0.1 mm

FIGURE 5.7 (Continued)

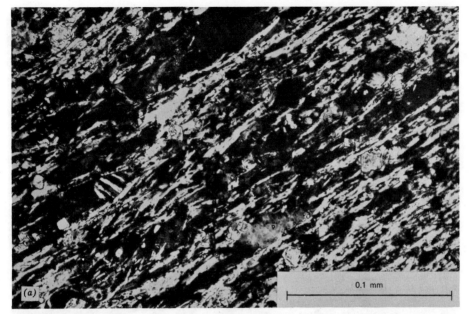

FIGURE 5.8   (a) Cleavage in Vermont slate from Lake Bomoseen, Vermont, U.S.A..
Layer silicates form films that anastomose around single grains of quartz, carbonate,
and plagioclase and occasionally around small aggregates of quartz and/or feldspar
and/or layer silicate. Crossed nicols. (Photograph by W.C., Laurijssen.) (b) Cleavage in
slate from Bermagui, Australia. This is a rock very rich in layer silicates with a very
strong preferred orientation. Plane-polarized light. (Photograph by W.C. Laurijssen.)

than layer silicates (usually quartz) and any layer silicates that they may contain generally show little or no preferred orientation. The domains vary in
size and in the number of grains that they contain; they may comprise aggregates of many grains, but there is a complete gradation through to domains
that comprise just several, or even one large lenticular grain (e.g. Fig. 5.7b and
5.8a). The filmlike domains are commonly accentuated by oxides and are rich
in layer silicates which have a strong preferred orientation so that their long
dimensions are parallel to the long dimension of the film. Minerals such as
ilmenite and zircon also tend to be concentrated in these domains (see Fig.
5.9). There is commonly a marked difference in the appearance of layer
silicate grains in the different domains: in the films they are very elongate
parallel to (001) while in the lenticular domains they tend to be more
equidimensional or even elongate perpendicular to (001) [e.g., see Hoeppener (1956); Etheridge (1971 and 1973)].

There is a complete gradation between this type of domainal fabric and
crenulation cleavage. There is also a gradation from the domainal type of

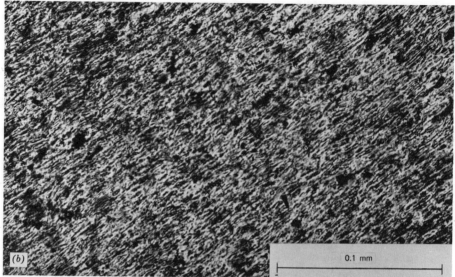

FIGURE 5.8 (Continued)

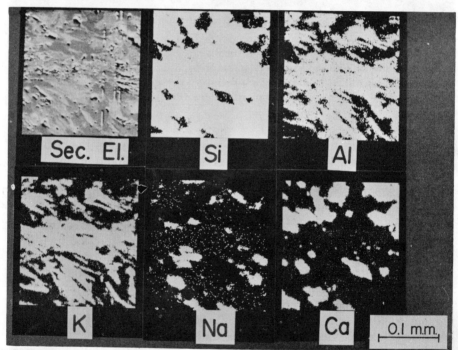

FIGURE 5.9 Electron microprobe scan showing the chemical distribution across a slaty cleavage lamella. The mica-rich domain is horizontal and centrally placed in the images and is approximately 0.02 mm wide. The secondary electron image shows surface topography only. (Reproduced by permission of R.J. Holcombe.)

cleavage through rocks in which the lenticular domains become less prominent and the films broader, into rocks in which there is a strong preferred
orientation of all mica and no domainal distribution (Fig. 5.8*b*). This fabric
grades into one in which there is still no domainal arrangement but the layer
silicates define two orientation maxima. Although there are no discrete surfaces passing through the rock these maxima can be thought of as two statistically defined surfaces and the angle between the two varies from less than 10°
up to at least 80°. In rocks in which the two orientation maxima are readily
recognized the fabric is best described as a bimodal mica fabric rather than as
a slaty cleavage.

Slates are typically rich in layer silicates but the term slaty cleavage is also
used to describe foliations in low-grade metamorphic rocks of entirely different composition. Sorby (1856a) used the term to describe the fabric of Devonian marbles in which the foliation is defined by a strong preferred dimensional orientation of elongate carbonate grains, grain aggregates, and/or fossils, and this usage has been adhered to by some more recent writers in
describing carbonate and quartz-rich rocks. It is also used to describe low-
grade, fine-grained micaceous quartzites in which the foliation is defined by a

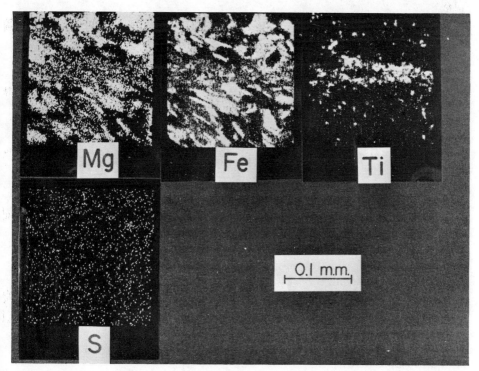

FIGURE 5.9 (Continued)

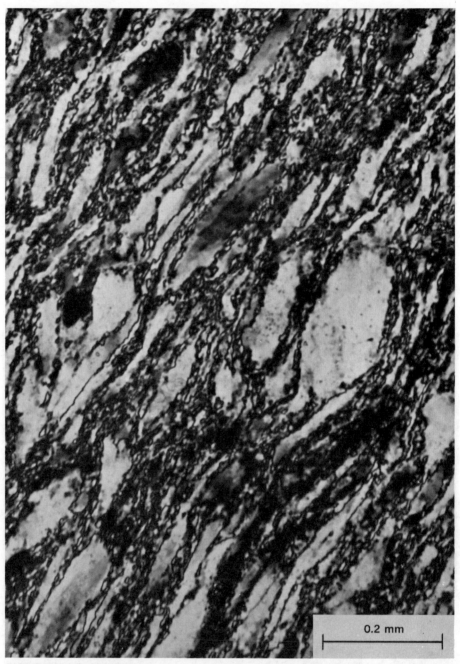

FIGURE 5.10   Foliation defined by elongate quartz grains in a quartz mylonite from Mt. Isa, Australia. Crossed nicols.

strong preferred orientation of sparse, isolated layer silicates [e.g., Billings (1972, p. 389)]. We favor use of the term specifically for rocks rich in layer silicates that possess the marked fissility of roofing slate. The other types of foliation (e.g., Fig. 5.10) are then referred to simply as axial plane foliations and the term amplified by a description of the fabric. This type of approach has been adopted by Cloos (1971) who describes a foliation defined by preferred dimensional orientation of ellipsoidal ooids in oolitic limestone and refers to the surface simply as "cleavage."

Every gradation between the above end members, including those found in rocks poor in layer silicates, is possible but domainal fabrics appear to be very common in rocks commercially described as slates. Thus, the commonly held belief that a slate can be split again and again until the laminae are only one grain thick is true of some slates but not of all.

A feature commonly found in slates is a steeply plunging lineation, that is generally inclined at a high angle to the fold axes. This lineation is generally believed to be parallel to a direction of finite elongation [e.g., Hobbs and Hopwood (1969); Borradaile (1973); Ramsay and Wood (1973)].

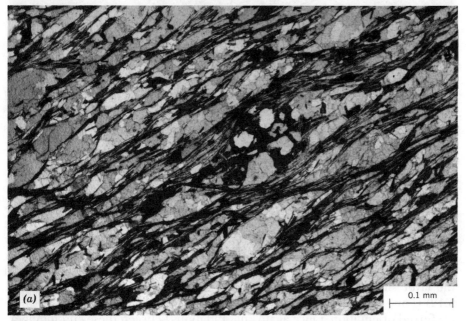

(a)   0.1 mm

FIGURE 5.11   (a) Domainal schistosity in schist from Ducktown, Tennessee, U.S.A. Micas form films that anastomose around aggregates composed principally of quartz. A weakly developed later foliation crosses the picture horizontally. Plane-polarized light. (Photograph by W.C. Laurijssen.) (b) Strong preferred orientation of mica in schist from the Chester Dome, Vermont, U.S.A. Crossed nicols. (Photograph by W.C. Laurijssen.)

FIGURE 5.11 (Continued)

Slaty cleavage is a term that is used to describe the foliation in fine grained rocks belonging generally to the greenschist facies. At higher metamorphic grades similar fabrics are generally coarser and the foliation is referred to as schistosity.

5.2.5 Schistosity

Like the other foliations described schistosity varies in morphology. There are three common end members; one (Fig. 5.11a) resembles the domainal slaty cleavage of Fig. 5.7 but is coarser; the second (Fig. 5.11b) common in layer silicate rich rocks, is a coarser version of the slaty cleavage fabric in which all the layer silicates have preferred orientation (cf. Fig. 5.8b); the third (Fig. 5.12) is one in which layer silicates are sparsely distributed but have a single preferred orientation. Once again, all gradations between these end members are possible.

Schistosity is a term used to describe the foliation in rocks that have a grain size coarse enough to be called schists, the arbitrary boundary commonly being placed at the coarseness at which individual layer silicates just become discernible to the unaided eye. The fabric of rocks called phyllites, which are

FIGURE 5.12   Schistosity defined by preferred orientation of mica (mostly biotite) in quartz-biotite-feldspar gneiss, from Ducktown, Tennessee, U.S.A. Plane-polarized light. (Photograph by W.C. Laurijssen.)

intermediate in grain size between slates and schists, can be adequately described as slaty cleavage or schistosity. Schistosity is most common in high grade metamorphic rocks but also occurs in low-grade rocks, particularly in greenschist facies rocks of retrograde origin.

5.2.6 Differentiated Layering

In some rocks there is a foliation that is defined by a layering visible in hand specimen. Where it can be demonstrated that this layering has developed by some differentation process the foliation is referred to as *differentiated layering*. This can be done, for example, where the layering can be seen to cut across the bedding (e.g., Fig. 5.4). It has been indicated that slaty cleavage, crenulation cleavage, and schistosity can all be differentiated but the distinction between these cleavages and layering is one of scale and of persistency of domains. Thus, slaty cleavage is generally too fine and too lenticular to be considered as layering. Differentiated crenulation cleavage may or may not be too

FIGURE 5.13   Gneissic layering in Lewisian rocks near Bettyhill, Scotland.

fine to be called a layering. Domainal-type schistosity is generally described as schistosity as long as the mica-rich domains anastomose markedly, otherwise it becomes gneissic layering. Once again the gradational nature of these fabrics must be stressed.

Differentiated layering is found in metamorphic rocks of all grades. In high-grade rocks it is normally described as *gneissic layering* although not all gneissic layering need be formed by differentiation. The layers can be of various compositions but gneissic layering is most commonly defined by alternating mafic and felsic layers (Fig. 5.13).

Gneissic layering may be bedding, modified bedding, or a foliation introduced during deformation, that is, it may reflect initial sedimentary compositional differences or may be due entirely to differentiation. Coincidence of gneissic layering with bedding should not be assumed but should be demonstrated. Demonstration generally requires well-preserved sedimentary structures and is commonly impossible.

In low-grade rocks, differentiated layering has been demonstrated parallel to bedding [in lower greenschist facies rocks by Williams (1972)] and as an axial plane foliation [e.g., Turner (1941); Hobbs and Talbot (1968); Williams (1972)]. Figure 5.14 shows a typical example of axial plane type of differentiated layering in lower greenschist facies rocks. It can be seen that there is a

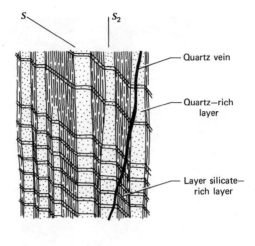

FIGURE 5.14   Differentiated layering
(S_2) in psammopelite from the limb of a
second-generation fold from Bermagui,
NSW Australia. Note the correlation
between the composition of the layering
and its position with respect to the folds in
bedding (S). Assuming that S_2 was initiated
approximately perpendicular to S, the
layer-silicate-rich layers coincide with
zones of relatively high shear strain.

relationship between strain and composition; the angle between the folded
layering and the differentiated layering is large in the quartz-rich domains
and small in the mica-rich domains. This relationship is very common in
differentiated layering but does not always hold true [see Fig. 5.15, see also
e.g., Hobbs and Talbot (1968)].

5.3   ORIGIN OF AXIAL PLANE FOLIATIONS

5.3.1   Introduction

The origin of axial plane foliations has been a topic of interest for over a
century now but is still very much a problem. There is a lack of suitable theory
and commonly a lack of adequate observation on rocks themselves. In this

FIGURE 5.15   Differentiated layering (S_1) in folded geneiss, Västervik, SE Sweden. An earlier, folded layering (S) is intersected by coarse differentiated layers (S_1). S_1 is represented in the line drawing (b) by the alternating stippled and unornamented zones. (Photograph and information provided by H.J. Zwart.)

section we seek mainly to outline the problems, some of which are indicated below.

1. The significance of axial plane foliations in terms of strain and fold geometry. Is foliation throughout its development parallel to a principal plane of strain or to a plane of maximum shear strain? Or can the orientation change with respect to both of these during development of the fold? Is there a minimum amount of strain required for the first appearance of axial plane foliations?
2. The development of preferred orientation of layer silicates. Are layer silicates rotated or do they grow in a preferred orientation?
3. Development of preferred dimensional orientation of grains other than layer silicates. Does this result from rotation, flattening, or diffusional processes including growth?

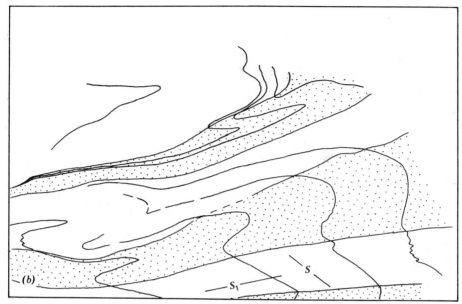

FIGURE 5.15 (Continued)

4. Development of compositional layering parallel to the foliation. What is the driving force(s) for differentiation? Is this layering controlled by discontinuities in strain? Are there local or gross volume changes associated with this process?

Much research has been concerned with cleavage in low-grade metamorphic rocks and such rocks are therefore emphasized in this chapter. Interpretations of these foliations do not necessarily apply directly to high-grade rocks where the mechanisms of deformation on a granular scale may be different.

In the following discussion of these problems the strain ellipsoid referred to is always the representation of the strain in regions that are small compared to the fold.

5.3.2 Orientation of Foliation with Respect to Strain

Analysis of strain in foliated rocks indicates that some foliations are perpendicular to the direction of maximum finite shortening, λ_3 [e.g., Badoux (1970); Cloos (1971); Wood (1973)]. The belief that the orientation of foliation is, *in general*, normal to λ_3 has increased in popularity in recent years, even to the stage of reaching theory status in many people's minds. Other evidence

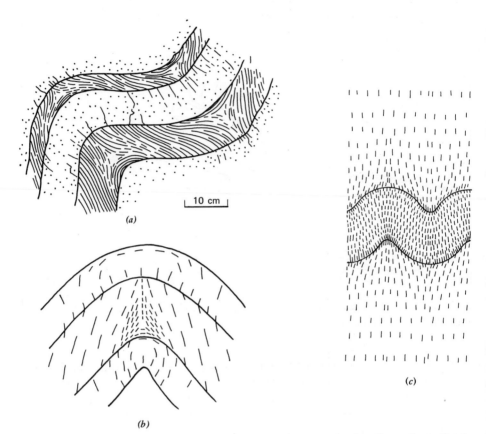

FIGURE 5.16   (a) Unusual "slaty cleavage" orientations in pelites from Kongsfjord, northern Norway. [From Roberts and Strömgård (1972), with permission of Elsevier Scientific Publishing Co.] (b) Orientation of λ_1 in experimentally deformed layers of rubber (one soft layer and two hard layers) showing same orientation and distribution as natural cleavage in (a). [From Roberts and Strömgård (1972), with permission of Elsevier Scientific Publishing Co.] (c) Orientation of λ_1 in a computer-simulated fold in a viscous layer flanked by layers of lower viscosity. Viscosity contrast is 17.5 : 1 and the layer has been shortened 63.2 percent. [From Dieterich (1969), with permission of the *American Journal of Science.*]

cited in favor of this hypothesis comes from theoretical considerations of strain. Folds have been modeled numerically [e.g., Dieterich, (1969)] and by using viscous or elastic materials [e.g., Roberts and Strömgård (1972)] as rock analogues. It is found that orientations of the trace of the $\lambda_1 \lambda_2$ plane of the strain ellipsoid, on the fold profile, can correspond closely to the orientations of axial plane foliations observed in natural folds (cf. Fig. 5.16a with 5.16b and 5.16c with 5.3a).

1 mm

FIGURE 5.17   Faultlike offsets of bedding, parallel to cleavage, in slate from the Ocoee Gorge, Tennessee, U.S.A. Plane-polarized light. (Photograph by W.C. Laurijssen.)

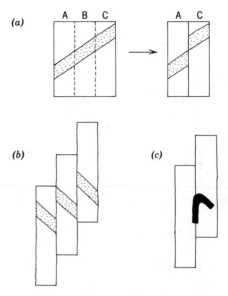

FIGURE 5.18 Diagrammatic representation of solution effects in the development of cleavage. Part (*a*) demonstrates the way in which an apparent shear displacement can be produced by solution. The slice of rock labeled B has been removed in solution, bringing A and C into juxtaposition and thereby giving rise to the apparent shear displacement. In (*b*) the displacement cannot be achieved without shear parallel to the cleavage. The magnitude of the shear must be at least as large as the apparent displacement. It can be argued that this structure was produced by selective removal of alternate limbs of a series of asymmetrical folds but then the folding implies the same shear strain. Part (*c*) shows a common situation in which large detrital micas have been folded during the development of the cleavage. The situation illustrated indicates that there has been real shear strain associated with the development of the cleavage plane.

There is little doubt that some foliations, such as those defined be penetrative preferred orientaion of micas or those defined by flattened objects, can develop perpendicular to λ_3, and such foliations are discussed in Sections 5.3.3 and 5.3.4, respectively. However, there are other foliated rocks in which it is quite common to observe that the foliation is a plane of shear strain. Where good markers can be found they are commonly offset along a succession of foliation planes (see Fig. 5.17). Where the latter is a crenulation cleavage the system of crenulations (see Figs. 5.5 and 5.6) itself indicates that the foliation is parallel to a plane of high shear strain except in special cases (e.g., in the hinge of some large folds where the crenulations are symmetrical). Sometimes the shear displacement may be only apparent, for example, if a slice of rock is removed in solution as depicted in Fig. 5.18a an initially planar marker becomes offset across the solution plane despite the fact that there has been no shear displacement parallel to the surface. However in many actual cases the displacement of the marker has the wrong sense of movement to be explained simply by solution (see Fig. 5.18b and c).

We are thus faced with the problem of explaining how a foliation can be a plane parallel to which shearing displacements have occurred and yet be parallel, or approximately parallel, to a principal plane of the strain ellipsoid (a principal plane of strain is a plane of no shear strain (see Section 1.3.1). In the following discussion we consider two types of models. The first assumes that cleavage is normal to λ_3 and attempts to explain away the shear displacements. The second assumes that cleavage develops parallel to a plane of shear strain and attempts to explain away the apparent parallelism with the $\lambda_1 \lambda_2$ plane.

Models Assuming That Cleavage Is Normal To λ_3.   A possible answer to the problem is that on the scale of individual domains the foliation is parallel to planes of finite shear strain but that on the scale of a sample containing many foliation planes, it is parallel to the principal plane of the mean strain ellipsoid for the specimen as a whole (see Section 1.3.2). This is possible if the sense of shear parallel to the cleavage varies systematically so that on the scale of the area considered the strains associated with individual domains cancel one another out (see Fig. 5.19). The strain pattern shown in Fig. 5.19a is perhaps common in the hinges of folds with crenulation cleavage. The patterns shown in Fig. 5.19b and c are qualitatively similar to patterns seen on the limbs of folds with crenulation cleavage, in that there is shear strain of opposite senses in neighboring domains.

Dieterich (1969) proposes that foliation develops perpendicular to λ_3 and has suggested another solution to the problem of reconciling this with offset parallel to the foliation. He points out that under the conditions of a noncoaxial strain history, prevailing during the development of a fold, there will at times, be shear stress on planes parallel to the principal plane of the finite

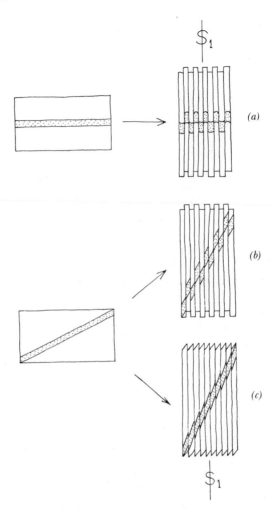

FIGURE 5.19   Diagram illustrating how cleavage (S_1) can be a shear surface but still be parallel to a principal plane of the mean strain ellipsoid. In the right-hand diagrams the shear strains cancel one another so that on the scale of many cleavage planes a surface initially perpendicular to the cleavage (represented here by the upper and lower edges of the rectangles) is also perpendicular after deformation, and the cleavage is, therefore, parallel to a principal plane of the mean strain ellipsoid.

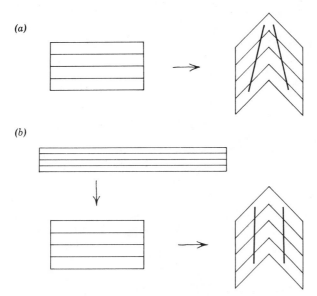

FIGURE 5.20   Diagrammatic representation of the
development of two folds of identical appearance from
bodies of different initial shape. For simplicity an
intermediate state in (*b*) is taken as the initial state in (*a*).
Despite their identical appearance the finite strain for
the two folds is different as can be seen from the
orientation of the lines of maximum extension which are
represented for each limb by the heavy line. Note that
the lines representing maximum extension in (*b*) are not
parallel although they may appear so at first sight.

strain ellipsoid, since the stress and finite strain ellipsoids are differently
oriented. Thus, he argues that foliation develops parallel to a principal plane
of the strain ellipsoid and that shearing then occurs parallel to this foliation
because resistance to shear on the foliation is low. This argument can explain
the existence of occasional faults parallel to a contemporary axial plane foliation; in such circumstances, since the fault is nonpenetrative, the foliation is
still parallel to the principal plane of the ellipsoid, as determined for a sample
that is not faulted. However, if, as is commonly observed, the offsets parallel
to the foliation are sufficiently common to constitute a penetrative feature of
the fabric, and if these offsets are not cancelled out by equivalent shear displacements of opposite sense (in the manner shown in Fig. 5.19), then by
definition the foliation cannot have developed and remained exactly parallel

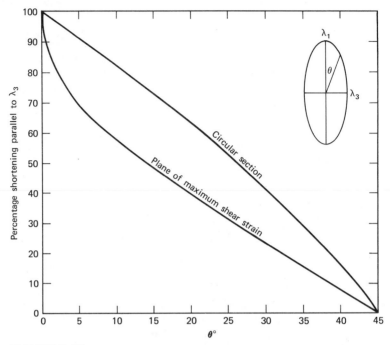

FIGURE 5.21 Orientation of circular sections and planes of maximum shear strain in plane strain, for various values of shortening parallel to λ_3. Volume change is assumed to be zero.

to a principal plane of strain. There is one possible exception to this statement and it is the special case considered in the previous paragraph (see Fig. 5.19).

Models Assuming That Cleavage Develops Parallel To A Plane of Shear Strain. Perhaps the most likely answer to this problem is that not all foliations do develop parallel to the $\lambda_1 \lambda_2$ plane. That they are commonly approximately parallel to that plane in tight folds seems fairly well established, but that they are perpendicular to λ_3 throughout their development does not seem to be a generally established fact. Strain analyses based on the use of fossils are commonly two dimensional and can only be made three dimensional, in general, by making such assumptions as foliation is parallel to a principal plane [see Hobbs and Talbot (1966)]. Obviously such an assumption is not acceptable for present purposes and the analyses simply indicate that the respective *traces* of the foliation and the $\lambda_1 \lambda_2$ plane, on the bedding plane, are closely parallel. Strain analyses based on such objects as reduction spots [Wood (1973)], ooids [Cloos (1971)] and lapilli [Oertel (1970); with discussions by Helm and Siddans, (1971), and Mukhopadhyay (1972)] seem to indicate that the foliation in these particular slates and marbles is strictly perpendicular to λ_3. This conclu-

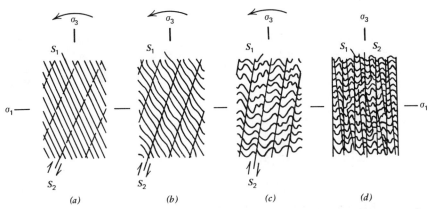

FIGURE 5.22   Diagrammatic representation of stages in the development of crenulation cleavage (S_2). [From Hoeppener (1956), with permission of *Geologische Rundschau.*]

sion, however, need not apply to all foliated rocks or even to all slates and marbles.

Comparison of foliation orientation with orientation of the strain ellipsoid in the various model analyses is, at best, only qualitative since the models considered, so far, have involved only elastic, ideally viscous and power law materials (see Section 4.8). Without much better knowledge than we presently have, of material properties and fold mechanisms, we cannot make quantitative comparisons. Furthermore, in order to compare a natural fold with a model fold it is necessary to know the complete finite strain and incremental strain history for both, since both must have undergone the same strain history in order to make the comparison. It is not sufficient that two folds have the same appearance after deformation, since their total finite strain distribution and strain history may still be different. For example, in Fig. 5.20, the products of deformation are the same but the initial shapes are different; therefore the shape and orientation of the finite strain ellipsoids are also different. Thus comparison of natural folds with modeled folds provides no compelling evidence for the development of all foliations perpendicular to λ_3.

The fact that many foliations finish approximately perpendicular to λ_3 does not necessarily indicate that all started that way. Becker (1893) considered axial plane foliations to be phenomena related to shear strain and, erroneously believing that planes of maximum finite shear strain were parallel to surfaces of no finite strain, he suggested that foliation developed parallel to the circular sections of the strain ellipsoid. Although the two planes mentioned are not parallel (see Fig. 5.21) there remains the possibility that foliation is parallel to a plane of high shear strain. For an infinitesimal strain the planes of maximum shear strain are inclined to the $\lambda_1 \lambda_2$ plane at 45° but this

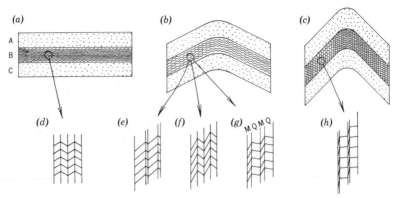

FIGURE 5.23   Diagrammatic representation of a possible mechanism for the development of differentiated crenulation cleavage and certain types of differentiated layering. For explanation see text.

angle decreases with increasing strain, and at the magnitudes of strain that one might expect in many folds [see Hobbs (1971)] the angle will be very small. This point is demonstrated in Fig. 5.21; it can be seen that at strains in excess of 60 percent shortening the angle is less than 10° and is decreasing rapidly. In view of the errors to be expected, in the measurement of the orientation of both the foliation and the strain axes, this angle may not be detectable. It is important to realize in connection with this argument that to obtain *local* strains in excess of 60 percent shortening the overall shortening of a sequence of rocks need be much less [see Hobbs (1971)].

A similar type of mechanism was invoked by Hoeppener (1956) as an explanation of crenulation cleavage in the slates of the Middle Moselle region of Germany [see also Turner and Weiss (1963, p. 464); Talbot (1965); Roberts, (1966)]. He suggested that this cleavage developed parallel to a plane of high resolved shear stress and that concomitant shear on the preexisting foliation gradually rotated the new cleavage toward parallelism with the $\lambda_1 \lambda_2$ plane of the strain ellipsoid (Fig. 5.22). There are two special features of the cleavage that Hoeppener has described. First, although the foliation does occur parallel to the axial plane of second-generation folds, such folds are rare and, second, the sense of shear on the foliation is almost invariably constant in sense [see Talbot (1965)]. It may be that what Hoeppener is describing is basically different to the axial plane crenulation cleavage of tightly folded areas where the sense of rotation varies across fold hinges.

Another model for crenulation cleavage is outlined in Fig. 5.23. It is assumed that at the onset of deformation some layers (A and C) shorten, initially without folding, while other layers rich in layer silicates fold on a small scale (Fig. 5.23a). Then as large folds develop in the other layers (A and C) the small folds are constrained to become asymmetrical as shown in Fig. 5.23*b* [cf. Ramberg (1963a)]. They can do this in two ways:

1. Fold hinges can migrate (Fig. 5.23*e*).
2. Folds can become asymmetrical in the sense that the angle between the limbs and the axial plane is different in alternate limbs (Fig. 5.23*f* and *g*).

The difference between Fig. 5.23*f* and *g* is that, in *f*, the volume of rock between two folded surfaces, in each limb of the microfolds, has been preserved by changing the dimensions of the folded surface, whereas in *g* the dimensions of the folded surface have been preserved and volume allowed to change within individual limbs, but not necessarily within the system as a whole. The steps represented by Fig. 5.23*a* and *b* have been observed in experimentally produced folds (see Fig. 1.17) and the microfolds appear to become asymmetrical, mainly by the processes represented by Fig. 5.23*e* and *f*; however, in natural folds there is evidence for volume changes compatible with Fig. 5.23*g* so that it is perhaps more common as a mechanism for the development of natural crenulation cleavage. If the volume change is achieved, in a quartz mica schist, by migration of quartz from the limbs marked M, in Fig. 5.23*g* into the limbs marked Q, or out of the system, then the M limbs will become rich in mica and the Q limbs will be relatively rich in quartz. The result (Fig. 5.22*h*) can be a typically differentiated crenulation cleavage or, on a larger scale, a common type of differentiated layering. Figure 5.24 compares the orientation of a cleavage developed according to this model and the orientation of the $\lambda_1 \lambda_2$ plane for model folds as determined from fold geometry. Three fold models have been considered; two assume that volume in the incompetent layer, in which the cleavage is developing, is constant; the other allows the volume to change. For simplicity the large folds are assumed to have perfect kink geometry. The angle between the cleavage and the $\lambda_1 \lambda_2$ plane is plotted for various fold limbs, dihedral angles, and for varying amounts of assumed homogeneous flattening. It can be seen that the angle may be large or small for open folds, depending on the amount of homogeneous flattening. In tight folds the angle is generally small; for example, in folds with a limb dihedral angle of 30°, the angle between cleavage and the $\lambda_1 \lambda_2$ plane for one of the constant volume models (Fig. 5.24*b*) is less than 2° regardless of the amount of homogeneous flattening.

In general the perfect kink geometry, assumed in Figure 5.24, is an oversimplification since most folds have more rounded hinges. If the mechanism of folding is one involving buckling (see Section 4.8) of competent layers, the simple situation represented in Fig. 5.23 will be modified slightly. There will be a tendency for the axial planes of the crenulations to be rotated [see Bayly (1965)] in such a way as to define the divergent cleavage fan (see Fig. 5.2) so common in rocks. Such a mechanism will result in a correlation between the angle of divergence and the thickness of folded beds as demonstrated by

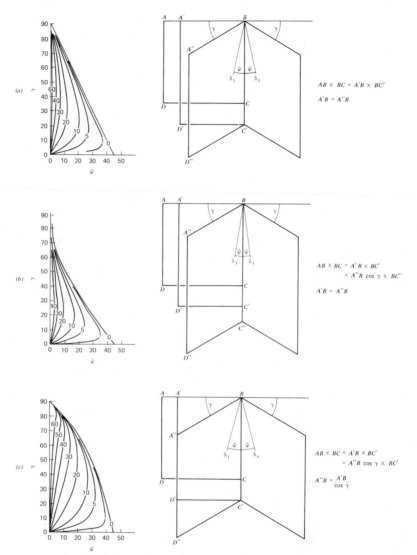

FIGURE 5.24 Curves showing the relationship between the axial plane, and therefore between cleavage developed parallel to the axial plane, and the principal, finite extension axes for the three simple folds models [(a), (b), and (c)] represented by the line drawings accompanying each graph. Plane strain is assumed for all three models and the curves are calculated for variable amounts of prefolding, constant volume, homogeneous flattening parallel to AB; these values, expressed as percentages $(ab - a'b)/ab \times 100$ are indicated on each curve. In models a and b the area of the folded surface, after the initial homogeneous flattening, is preserved, whereas in model c the area of the folded surface varies throughout the deformation. Model a involves a volume decrease and models b and c are constant volume.

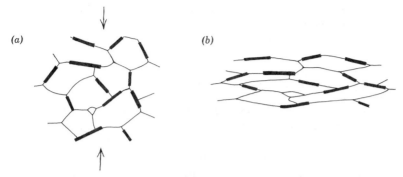

(a)   *(b)*

FIGURE 5.25   Diagram showing way in which a preferred orientation
can develop by the flattening of an aggregate. Part (*b*) is sketched from an
aggregate of experimentally deformed salt (NaCl) and mica. This material
was shortened 60 percent and, prior to deformation, had an appearance
similar to the reconstruction of (*b*) represented in (*a*).

Bayly (1965) and will also be compatible with a volume decrease on the concave side of competent layers and/or a volume increase on the convex side of
the competent layers, a situation that can be demonstrated in some rocks
[Williams (1972); Durney (1972)].

The model described above, and the model of Hoeppener preceding it, are
not necessarily mutually exclusive. It seems likely that the morphology of
crenulation cleavage simply indicates shear strain parallel to the cleavage
rather than indicating parallelism of the cleavage and a particular plane of the
strain ellipsoid.

If this is so, it is reasonable to think in terms of at least two models, for the
formation of crenulation cleavage; one involving modification of small crenulations in response to large-scale folding (Fig. 5.23) and the other involving
the development of conjugate or single cleavages initially parallel to planes of
high resolved shear stress [Hoeppener (1956)]. The latter explanation introduces a further problem as to why only one Hoeppener-type cleavage should
develop in some areas, as opposed to the conjugate pair. Hoeppener (1956)
points out that in the Moselle area the original foliation is close to a plane of
maximum shear stress and, therefore, it is likely that slip can occur on that
surface at stresses too low to give rise to another foliation of only slightly
different orientation. The original foliation, therefore, acts as the slip plane
and there is no new cleavage formed in that orientation. The other plane of
maximum shear stress is more steeply inclined to the original foliation and,
therefore, it does give rise to a new surface; the new crenulation cleavage.
Becker (1893) has offered another explanation which may be applicable, particularly to rocks in which the old foliation and the planes of maximum shear
stress are more symmetrically oriented [see Jaeger (1969, p. 87)].

Furthermore foliations defined by discontinuities, such as microfaults or layering, can only have developed and remained parallel to a principal plane provided that the strain history was coaxial. Such a situation is possible during the development of folds but is believed to be unlikely in the general case.

Foliation and Strain Magnitude

Although strain is believed to be the important factor in the development of many, if not all, axial plane foliations it is impossible at present to say much about the minimum strain required for the development of a foliation. Wood (1973) reports that strain in slates, from Wales and Vermont, is in excess of 60 percent shortening and Cloos (1947) states that in the South Mountain area, foliation first appears in the limestones at "20 percent deformation." There are thus indications that foliations are associated with a wide range of strains but there is need for much more systematic work.

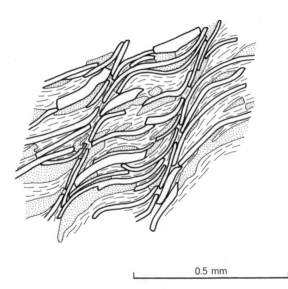

|⎯⎯⎯⎯⎯⎯⎯⎯⎯⎯ 0.5 mm ⎯⎯⎯⎯⎯⎯⎯⎯⎯⎯|

FIGURE 5.26   Development of preferred orientation parallel to narrow shear zones. Micas, initially oriented parallel to S, have been rotated locally, into parallelism with narrow shear zones, developed in an experimentally deformed aggregate of salt (NaCl) and mica. [From Means and Williams (1972). Copyright © 1972, the University of Chicago. All rights reserved. Reproduced with permission of the University of Chicago Press.]

5.3.3 Preferred Orientation of Layer Silicates

Much of the controversy concerning the development of foliations has centered around the manner in which layer silicates become oriented [e.g., Sedgwick (1835); Sorby (1853, 1856b); Becke (1913); Leith (1905); Colette (1958); Maxwell (1962); Oertel (1970)]. Some of the mechanisms involved in the development of a preferred crystallographic orientation have already been discussed (Section 2.3). However, a few points can be added here with special reference to foliations.

It was suggested by Sorby (1853) very early in the study of slates that preferred orientation can develop by rotation of existing grains. The idea became unfashionable for a while but is now generally accepted as one possible orienting mechanism. That it does occur in some rocks is readily demonstrated in areas where detrital layer silicates are oriented parallel to bedding in the absence of foliation and parallel to the foliation where it is developed [Williams, (1972)]. Experiments on aggregates containing layer silicates or other platy grains have shown that two kinds of fabric may develop by rotation. The first is a homogeneous fabric characterized by general preferred orientation of the platy grains (Fig. 5.25) with their long dimensions parallel to the $\lambda_1 \lambda_2$ plane. This mechanism was first demonstrated by Sorby (1856b). The second (Fig. 5.26) is a heterogeneous or domainal fabric in which reorientation is localized along narrow shear zones. This type of fabric has been produced experimentally in various materials [e.g. Weymouth and Williamson (1953); Raleigh and Paterson (1965); Borg and Handin (1966); Means and Paterson (1966); Means and Williams (1972)].

Mathematical models for rotation have been discussed by a number of writers and there are two principal models that may be applicable to rotation of micas. One, the "March analysis," is concerned with rotation of passive markers in a homogeneous body [March (1932); see also Owens (1973)] and the other is concerned with rotation of rigid bodies in a viscous fluid [Jeffery (1923); Gay (1968a)]. Tullis (1971) has shown that the preferred orientations predicted by both models are essentially the same, the Jeffery analysis differing from that of March by a factor that describes the shape of the rotating particles. Tullis has also shown that the degree of preferred orientation of micas in Welsh slates containing deformed reduction spots is consistent with the strain indicated by the spots, assuming an initial random fabric for the micas. The principal limitation of this model, however, is that it does not explain the domainal fabric observed in many slates. Nevertheless, it provides one possible model for the development of preferred orientation in micaceous rocks. It also provides a possible explanation of the bimodal mica fabric (Section 5.2.4). Since, if it is assumed that the initial mica fabric comprises a preferred orientation parallel to bedding, then the March analysis indicates that, if the principal axis of shortening initially lies approximately parallel to the bedding plane, the micas are rotated in such a way as to produce a bimodal distribution symmetrical about the $\lambda_1 \lambda_2$ plane of the finite strain ellipsoid (Fig. 5.27).

Maxwell (1962) has invoked rotation, under rather special conditions, to explain foliation in some of the Appalachian slates. He believes that the foliation and folds developed while the sediment was still very wet and unconsolidated. This hypothesis may well be correct for some rocks [e.g., see Williams et al. (1969); Tyler (1972); Corbett (1973)], but similar fabrics are found in rocks that have been previously deformed and metamorphosed, and which

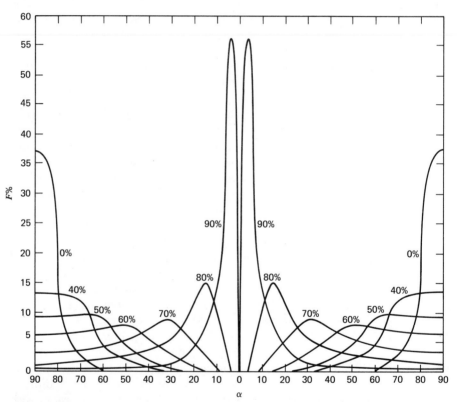

FIGURE 5.27   Diagram illustrating the effect of homogeneous strain on the preferred orientation of a micaceous rock, assuming that the micas behave as passive markers and assuming that the strain is symmetrical with respect to initial fabric. The preferred orientations are represented by curves showing frequency (F) as a function of orientation (α), relative to λ_3, for various amounts of strain. The strain is shown on each curve and is the amount of shortening perpendicular to λ_3. The initial fabric is given by the 0 percent curve. An initial preferred orientation parallel to the principal direction of shortening is split into two maxima, which increase in intensity and approach parallelism with the principal direction of extension with increasing strain.

were therefore presumably lithified prior to the development of the new foliation [Williams (1972)].

Although rotation of preexisting layer silicates is important in the formation of many foliations, there can be no doubt that new grains commonly form *during* the development of a foliation; that is, some or all of the grains may be syntectonic. It is, therefore, necessary to consider what influences their orientation. Metamorphic grains that develop before or during the deformation, if not parallel to the foliation, may be rotated by strain in the same way as detrital grains. It has been suggested, however, that new grains develop to produce a foliation under the influence of stress and that grains in other orientations are preferentially dissolved [Kamb (1959b)]. These processes may be capable of producing a preferred orientation of mica basal planes and, if so, it is predicted that the foliation will be perpendicular to σ_1 [e.g., see Kumazawa (1963); Flinn (1965)]. Analysis of viscous models, however, suggests that there need be no such relationship between stress and foliation [Dieterich (1970)]. However, knowledge of stress in geological materials is inadequate to allow the hypothesis to be completely overruled at this time. Even if it could be proved that, for certain geological situations, the foliation did develop approximately perpendicular to σ_1, it would still be difficult to prove a causal relationship because of the difficulty of distinguishing grains that have crystallized in a given orientation and metamorphic grains that have been rotated into that orientation. Furthermore, even if it can be demonstrated that a grain has not been rotated, there are factors, other than stress, that can influence its growth orientation. One possibility is that the orientation of the new grains is controlled by, and is parallel to, the orientation of preexisting grains or aggregates [e.g., see Oertel (1970)]. That is, the growth may be *mimetic*. Etheridge (1971) has demonstrated the importance of mimetic growth in some experimentally produced foliations and has pointed out that the variation in grain shape in different domains in slates (see Section 5.2.4), is consistent with the operation of such a mechanism in the development of the slaty cleavage. In further support of the mimetic growth mechanism it can be pointed out that layer silicates replacing, or growing adjacent to, detrital micas are commonly parallel to the relict grains, even though the latter vary in orientation.

During deformation of an aggregate the grains of mica aligned with (001) more nearly parallel to the direction of maximum shortening tend to kink whereas micas more steeply inclined tend to rotate, possibly as almost rigid bodies, toward parallelism with the $\lambda_1 \lambda_2$ plane of the strain ellipsoid. If recrystallization then takes place it is possible that the kinked grains, having maximum strain energy density, will be preferentially consumed [see Green (1968); Tullis (1971)] thus decreasing the number of micas inclined to the $\lambda_1\lambda_2$ plane at large angles, and thus providing another mechanism that can contribute to the development of foliations.

5.3.4 **Preferred Dimensional Orientation**

Preferred dimensional orientation of grains, other than layer silicates, is a feature of many foliated rocks and there are various ways in which it can be achieved. The different mechanisms can be classified as follows:

1. Rigid body rotation of preexisting inequant grains.
2. Modification of grain shape by such processes as crystal slip or diffusion.
3. Growth of inequant grains in a given preferred dimensional orientation.

Rigid body rotation can be demonstrated in some deformed clastic sediments where detrital grains, showing practically no evidence of strain and preserving their clastic appearance, are oriented parallel to the axial plane foliation [Yagishita (1971); Dallmeyer (1972); Williams (1972)]. The process governing the development of a preferred orientation by this mechanism has been discussed by Gay (1968a). Rotation is not restricted to rigid bodies but can be combined with other processes such as strain or growth of the rotating body.

There are a number of ways in which grain shape can be changed. A grain may be flattened in the plane of the foliation as a result of plastic deformation or by a purely diffusion-controlled process such as that operative in Herring-Nabarro creep (Section 1.4.3). Alternatively, the shape of the grain may be changed by solution on sides parallel to the foliation with or without concomitant addition to the grain on boundaries inclined to the foliation. Obviously by looking at the shape of a grain it is difficult if not impossible to determine which of these mechanisms has operated, although the latter has been recognized in detrital rocks [Elliot (1973)] where original grain boundaries can be recognized by the presence of lines of dusty inclusions within the existing grains. Wilson (1973) has described yet another mechanism. In large quartz grains deformation bands develop parallel to the foliation. When the quartz begins to recrystallize new grains nucleate along deformation band boundaries and subdivide the original grain into several very elongate grains (see Fig. 5.10). Finally, in some rocks the quartz is believed to grow as elongate grains due to the orientation and distribution of micas [see Hobbs (1966a)].

5.3.5 **Differentiation**

As pointed out in Section 5.2 many foliations are defined partly by a compositional layering. This layering is believed to develop by some process of differentiation that accompanies development of the foliation. One of the problems, therefore, associated with the origin of foliations is how the layering develops.

DeVore (1969) has demonstrated that in a stressed polycrystalline material, arrangement of the component minerals into monomineralic layers, parallel to the $\sigma_2 \sigma_3$ plane of the stress ellipsoid is favoured thermodynamically. As with the effect of stress on preferred orientation, it is difficult to evaluate the importance of this effect in rocks. It could be that, in some or all cases, the layering develops by nonequilibrium processes such as diffusion of some material from the system, so that such classical thermodynamic arguments would not apply.

Where the foliation is a crenulation cleavage or a differentiated layering of the type visible in a hand specimen, it is often possible to trace markers across the differentiated layers. Where this is so, the layer silicates are generally found to be concentrated in the zones in which the angle between the foliation and the marker is most acute, as in Figure 5.14. One possible explanation of this structure is that strain has been higher in these zones and that, as a result, there is a greater degree of lattice disorder or concentration of microfractures relative to intervening zones. This would facilitate diffusion of materials to or from these zones either by solid diffusion or in interstitial fluids. Another possibility, that is more consistent with the model suggested above for crenulation cleavage (see Fig. 5.23), is that layer silicates act as strong fibers and cause the rock to strain in such a way as to establish the type of pressure gradients shown in Figure 5.28, where a small element of rock is sandwiched between two mica layers or two large micas. If it is assumed that the continuity of these layers must be maintained, then as the element is folded the area of $ABCD$ tends to decrease so that there is a pressure gradient from this area into the adjacent rectangle. Such a gradient will encourage migration of material away from the area $ABCD$ either in solution or by solid diffusion. The ability of layer silicates to behave in this manner has to be demonstrated; in support of the hypothesis it can be said that in experimentally deformed salt/mica aggregates a mechanical differentiation takes place involving flow of salt away from zones analogous to the area $ABCD$ [Means and Williams (1972)]. This is believed to indicate the existence of a pressure gradient and demonstrates

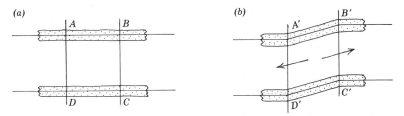

FIGURE 5.28   Diagrammatic representation of asymmetric kinks or crenulations in a material in which the micas behave as strong fibers. The length of the mica remains constant ($\therefore AB = A'B'$ and $CD = C'D'$) and continuity is preserved along the boundaries AD & BC and therefore the area of the rectangle $ABCD$ decreases with an increase in strain ($AB \times CD > A'B' \times C'D'$). Thus development of the fold tends to establish a pressure gradient as indicated by the arrows in (b). For further discussion see text.

that in some aggregates the micas can behave in the required manner, but whether this mechanism will work or not, when salt is replaced by quartz, is open to question until more is known about the behavior of such materials under natural conditions.

The layer silicate rich films of slates also appear to be a product of differentiation. Plessmann (1964) has demonstrated that they develop by removal of carbonate from carbonate rich slate with resulting concentration of layer silicates. In quartz-rich slates it can be shown that quartz is removed selectively [Williams (1972); Durney (1972); Holcombe (1973)].

5.3.6 Summary

Axial plane foliations are common in all grades of metamorphic rocks above lower greenschist facies conditions; differentiated types are increasingly obvious as grade increases but are also common in the lower grade rocks. Axial plane foliations can form at lower grades of metamorphism or even in sediments but are less common in such rocks.

Strain is believed to be the important factor in the development of many if not all axial plane foliations. Some may develop in other ways, such as in direct response to stress but even then the process is almost certainly accompanied by straining.

There may be a number of mechanisms involved in the development of various foliations but all are tending toward the same result. That is, all are tending to produce a preferred dimensional orientation of inequant grains and/or aggregates of grains, that define a planar structure perpendicular, or approximately perpendicular, to the axis of maximum shortening, λ_3, at any given point in the rock. Some axial plane foliations, such as those produced by flattening of particles, are probably parallel to the $\lambda_1\lambda_2$ plane of the strain ellipsoid throughout their development. Others may be parallel to a plane of maximum shear strain or, like crenulation cleavage as interpreted in Figure 5.23, may be related to strain in a more complex way. However, as the magnitude of strain increases all axial plane foliations tend toward parallelism with the $\lambda_1\lambda_2$ plane of the strain ellipsoid.

5.4 TRANSPOSED FOLIATIONS

5.4.1 Introduction

Transposition layering is still another type of foliation that occurs parallel to the axial plane of folds, but it differs from most axial plane foliations, in that it is largely defined by a layering that predates the folding.

In many rocks there is a layering that closely resembles bedding, may contain genuine sedimentary structures, and appears to represent a simple stratigraphic sequence; some if not all of the interfaces between layers may in fact be original, sedimentary surfaces. Nevertheless, such a layering may have no real stratigraphic significance; inferences drawn in the usual way, concerning the stratigraphic sequence, the gross disposition of the stratigraphic units and direction of younging may be entirely misleading. This is the situation in areas where the layering is a product of transposition and such areas are not uncommon in deformed rocks [e.g., Sander (1911); Greenly (1919; 1930); King and Rast (1955); Weiss and McIntyre (1957); Weiss (1959b); Baird (1962); Christensen (1963); Hobbs (1965); Bishop (1972)].

The concept of transposition is simple but carries with it too many implications to be defined by a short statement. An essential part of the transposition process is rotation of a preexisting foliation by folding into an orientation approximately parallel to the axial plane of the folds. Thus, it is instructive to first consider rocks that are isoclinally folded or approximately isoclinally folded.

In an area of tight folding, the predominant orientation of bedding is approximately parallel to the axial surface of the folds and is inclined to the enveloping surfaces for the various beds [e.g., see Balk (1936)]. This point is

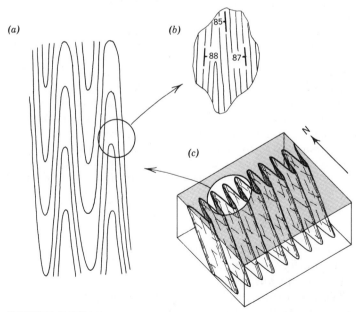

FIGURE 5.29   Diagrammatic representation of a system of tight folds showing how bedding, as seen in outcrop (b), may strike N/S, in areas where the gross distribution of a given lithology is E/W(c).

(a)

(b)

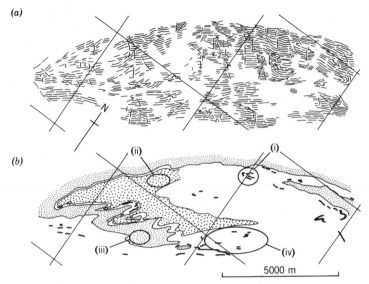

5000 m

FIGURE 5.30   Little Broken Hill, Australia: an area of complexly
deformed, locally retrogressed, amphibolite facies rocks. (a) Form
surface map showing trend of foliations. High-grade schistosity and
layering are generally parallel and trend approximately NE/SW.
Cross-cutting trends represent retrograde schistosity, which is generally
restricted to narrow shear zones. (b) Map of same area showing gross
distribution of different lithologies. The heavily stippled area is occupied
principally by a granitic gneiss ("Potosi gneiss") and the lightly stippled
area is occupied largely by lenses of similar gneiss, abundant amphibolite,
and a sillimanite-rich schist or gneiss. The unornamented area is
occupied principally by a sillimanite-rich schist or gneiss, with occasional
lenses of amphibolite and banded ironstone. The latter is represented by
heavy black lines. (c) Equal area projection of 2355 poles to layering.
Contours are 1, 2.5, 5 and 10 percent per 1 percent area. (d) Details of
outcrop from areas (i)–(iv) indicated in (b). Banded ironstone is indicated
by solid black areas, granitic gneiss by heavy stippling, amphibolite by
light stippling, and sillimanite-rich rocks by fine lines that also indicate
the trend of the foliation. [From Williams (1967), with permission of the
Geological Society of Australia.]

demonstrated in Figure 5.29 where, in outcrop, bedding generally strikes
North/South and dips steeply East or West (Fig. 5.29b). However, the gross
distribution of the various beds is represented by zones that, in this example,
strike East/West and dip parallel to the plunge of the folds (Fig. 5.29c). In
areas where the folds are readily seen this point is trivial but if, by one means
or another, the fold hinges are obscured, it becomes a very significant point
for two reasons. First, since there is then practically no variation in the orien-

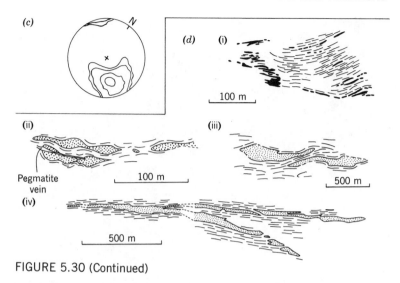

FIGURE 5.30 (Continued)

tation of bedding the complexity of the area may not be recognized. That is, the sequence of fold limbs may be mistaken for a normal stratigraphic sequence. Second, even if the existence of folding is recognized, the bedding information collected from small outcrops (e.g., Fig. 5.29b) tells nothing of the gross distribution of stratigraphic horizons, direction of younging on a regional scale, or stratigraphic thickness. To obtain this information we have to be able to piece together the regional structure. *It is this development of a new orientation of bedding in such a way that bedding, as seen in small outcrops, no longer provides information concerning stratigraphy or gross distribution of stratigraphic horizons that is the essence of transposition of bedding.*

The development of a transposed foliation has been described here with specific reference to bedding, but any foliation may be transposed and, although the situation in the case of other foliations cannot be described in terms of sedimentary stratigraphy, it is strictly analogous.

An important part of the transposition process is the obscuring of the fold closures. It is achieved by such processes as extreme flattening, development of discontinuities parallel to axial surfaces, development of axial plane foliation, and segmentation of marker beds or layers. The same end, of course, is always aided by paucity of outcrop.

5.4.2 Example of A Transposed Foliation

Figure 5.30b is a map showing the gross distribution of various lithologies in an area of deformed rocks [Williams (1967)]. The rocks have been metamorphosed to amphibolite facies. Figure 5.30a is a form surface map and

FIGURE 5.31   (a) Transposition of carbonate-rich layers in slate near Ducktown, Tennessee. Isolation of the fold hinges may have been achieved in part by metamorphic differentiation (R.J. Holcombe, personal communication). Note that to the left of the transposed layer bedding in the slate is much less folded and shortening has apparently been achieved by a more homogeneous flattening. (b) Transposition of quartz-rich layer in slate from Devon, England. (Photograph by H.J. Zwart.)

it shows the trends of foliations recognized in the field. The maps of small-scale structures (Fig. 5.30b, i-iv) show what is seen in representative outcrops and the stereogram (Fig. 5.30c) shows the distribution of poles to layering for the whole area. The picture is apparently simple, there are few obvious folds and the first impression is one of lenticular layers of various lithologies, all with a similar orientation and all parallel to schistosity. Some early workers believed that the structure was simple and that the area occupied the limb of a regional syncline. The few folds recognized were interpreted as minor "drag folds" related to the regional structure and the layering was interpreted directly as bedding with all the stratigraphic connotations that the term implies. Repetition of lithologies across the area was interpreted as a product of cyclic sedimentation and the lithologic distribution patterns represented in Figure 5.30b were interpreted as a product of sedimentary facies variation. However, detailed mapping reveals many small, tight or isoclinal folds with their axial planes parallel to the general orientation of the layering and shows that the

FIGURE 5.31 (Continued)

whole area is intensely folded. Hence, the boundaries drawn in Figure 5.30*b* are the traces of enveloping surfaces and the structure can be seen to be much more complex than the previously suggested limb structure.

The folds in this particular area are difficult to recognize for several reasons. Distinct lithologies such as the amphibolites, banded ironstones, and granitic gneiss might be expected to make folds readily recognizable at map scale and sometimes they do, but in general they are so discontinuous that the structure is far from clear. For example, the banded ironstone in Figure 5.30*d*, i defines a fold (one of the most obvious in the area), but only two of the lenses can actually be traced around closures. Existence of a lenticular, differentiated layering, parallel to the axial surface of folds, is another factor in this problem. This layering is common in the gneisses, schists, and amphibolites, and it tends to obscure folds, particularly in the amphibolites where, for the most part, there does not appear to have been a well-developed earlier layering. Thus, for example, in the fold illustrated in Figure 5.30*d*, iv, the only layering in the amphibolite and the prominent layering in the adjacent silimanite gneiss is the differentiated layering, and it is parallel to the axial plane of the fold. For this reason the fold was previously interpreted as a sedimentary structure resulting from the lenticular nature of the beds. However, a folded surface can be found in the gneiss defining the fold hinge so that there is no doubt that the structure is a fold.

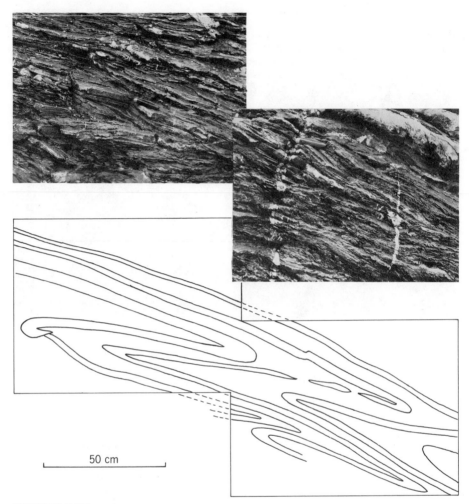

FIGURE 5.32 Transposition of bedding in flysch, below the Segnes Pass near Elm, Swiss Alps. Continuity of many beds is fairly well preserved as is shown in the line drawing; nevertheless, many of the folds are not obvious as can be seen from the photographs.

5.4.3 Examples of Individual Folds From Transposed Rocks

Transposition is found in rocks of all metamorphic grades as is apparent from the examples illustrated in Figures 4.21, 5.30, 5.31, 5.32, and 5.33. These drawings and photographs illustrate the types of folds common in areas of transposition. It is important to note that in one of the examples (Fig. 5.33c),

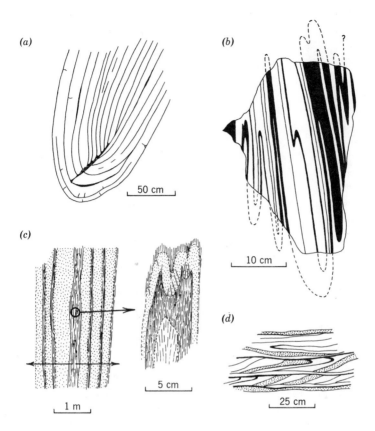

FIGURE 5.33   (*a*) Tight fold in amphibolite facies gneiss, Mary
Kathleen, Queensland, Australia. (*b*) Tight and isoclinal folding
in a block of Murphy Marble from Tennessee, U.S.A., sketched
from a photograph. Note that one layer is repeated many times
across the full width of the specimen. Dotted completion is based
on observations made on the sloping sides of the block. (*c*)
Isoclinal fold in lower greenschist facies, meta greywacke,
Bermagui, Australia. Arrows indicate the direction of younging
as determined from graded bedding and microcross lamination
in the greywacke beds (stippled). (*d*) Transposed bedding in lower
greenschist facies meta sediments, Wyangala, New South Wales,
Australia. Sketched from a photograph. (*e*) Isoclinal fold in
marble and quartz feldspar gneiss (the latter comprises the
boudins), Taylor Valley, Victoria Land, Antarctica. Note that the
right-hand boudin comprises the fold closure and the other two
boudins belong to opposite limbs of the fold as is indicated by the
fine layering in the marble.

(e)

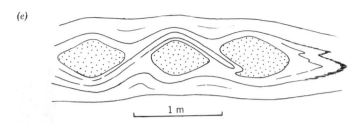

1 m

FIGURE 5.33 (Continued)

despite the tight nature of the fold, graded bedding and microcross bedding are preserved in the greywacke beds in the fold limbs; strain during the formation of this fold was apparently concentrated in the pelitic layers. Attention is also drawn to the cross bedding-like deformation structure (Fig. 5.33*a* and *d*) that is a common feature of many transposed sequences.

5.4.4 Processes Involved in the Development of Transposed Foliations

Transposition starts with tight folding and from then on the process is one of elimination of fold closures. So long as closures exist, there are portions of the folded surface that have not been rotated parallel to the axial surface of the folds and transposition is therefore incomplete. Some possible processes involved in the completion of transposition are now outlined.

Sander (1911), [see also Ingerson (1938) and Knopf and Ingerson (1938)] pointed out that as folds become tight, discontinuities commonly develop parallel to the layering in the limbs of the folds. Such surfaces, if persistent, sooner or later cut a fold closure and thereby obscure its presence. This process commonly results in a structure resembling cross bedding (see Section 5.4.3).

Turner and Weiss (1963, p. 94) have shown how transposition can be achieved by attenuation of alternate fold limbs in areas of asymmetrical folding (Fig. 5.34). This process produces a sequence of rocks with more or less constant facing but with repetition of the stratigraphic sequence. A similar process involves boudinage of competent layers and their rotation and translation in an incompetent matrix. This can also produce a strongly transposed sequence, as is illustrated diagrammatically in Figure 5.35; Figures 5.30*d* and 5.33*e* depict examples of the resulting structures.

Heim (1900) described transposition with specific reference to crenulation cleavage (*Ausweichungsclivage*). His is a rather special case and differs slightly from the structures already considered. In the development of a crenulation cleavage an existing foliation is folded and rotated towards parallelism with the axial plane of the folds. Thus far the process is one of transposition, but it is commonly accompanied by differentiation such that the limb areas become

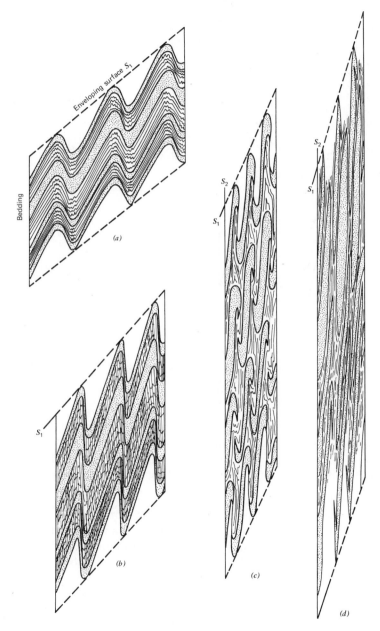

FIGURE 5.34   A possible sequence of events in the development of a transposed foliation (S_2) by folding of S_1. [From F. J. Turner and L. E. Weiss (1963), *Structural Analysis of Metamorphic Tectonites.* Copyright © 1963, McGraw-Hill Book Co. Used with permission of McGraw-Hill Book Co.]

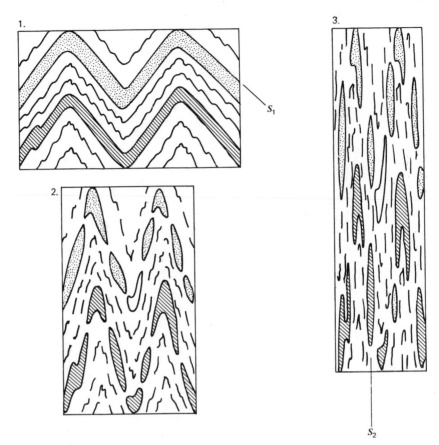

FIGURE 5.35 A possible sequence of events in the development of transposed layering (S_2) by folding and boudinage of S_1. [From Williams (1967), with permission of the Geological Society of Australia.]

layer silicate rich. Thus, there is a new layering that is defined by alternating microlithons, in which fold hinges are preserved, and mica-rich domains, in which the preexisting foliation is aligned approximately parallel to the length of the domain (Fig. 5.36a). The result, as demonstrated by Heim, may be a very strong layering parallel to the axial plane of the folds (see Section 5.2.3), but it differs from the structures described above in that:

1. The boundaries of the new layers are inclined to the folded foliation; the layering is, in fact, new and not simply an old one rotated [cf. Knopf (1931)].

2. The new layers are defined by compositional variation even in rocks that prior to development of the crenulations were compositionally homogeneous.

(a)   *(b)*

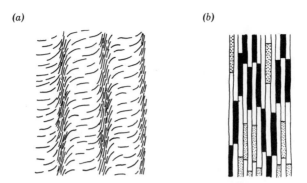

FIGURE 5.36   (*a*) Diagrammatic representation of
crenulation cleavage; lines represent individual layer
silicate grains that are aligned parallel to the folded
foliation. (*b*) Diagrammatic representation of gleittbrett
folding showing the development of a new layering
parallel to the axial planes of the folds.

3. In general, crenulation hinges, in terms of volume, represent a large
   percentage of the rock (commonly 50 percent) and are easily recognized.
   This type of foliation might be considered as morphologically intermediate between transposition layering as described above and the lenticular layering that can be produced in the development of *gleittbrett
   folds* [Schmidt (1932)]. Such folds develop by shear on discontinuities
   parallel to their axial plane. Thus, the folded layers become divided into
   many lenses that may become isolated and thereby define a new layering
   (Fig. 5.36*b*). Thus, if a graded bed is segmented in this way the resulting
   layers are graded along their length. This structure does not appear to
   be common in rocks and differs from the much more common transposition layering in that the new layering is not parallel to the old one.

5.4.5 Recognition of Transposition

In general, recognition of transposition depends on observation of folds of
the type depicted in Figures 4.21, 5.30, 5.31, 5.32, and 5.33. In some areas
younging data may be used to identify transposed layering and in areas where
younging evidence is good, it may be easier to recognize that the sequence is
transposed than it is to recognize actual folds (e.g., see Fig. 5.33*c*).

Parallelism of cleavage or schistosity and layering is not diagnostic of transposition. However, since it is a feature of transposed layering its occurrence
should make an observer cautious and on the lookout for more definitive
evidence. Similarly, occurrences of certain cross bedding-like structures should
make an observer wary, especially where these structures involve sediments

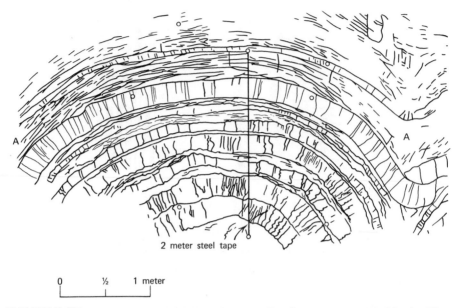

2 meter steel tape

0   ½   1 meter

FIGURE 5.37   Fold in interbedded sandstone and pelite near Leon, NW Spain. The sandstone is jointed approximately perpendicular to bedding. In the pelite beds (see especially bed A-A) a foliation defined by preferred orientation of micas is parallel to bedding in the fold hinge and is inclined to bedding in the same manner in each of the fold limbs. [From Savage (1965), with permission of I.T.C. Publications, Delft.]

not normally found cross bedded. For example, micro cross bedding is common in flysch deposits, but large-scale cross bedding involving several greywacke beds is atypical and would suggest some other sedimentary explanation or, possibly, transpositon.

5.5 OTHER METAMORPHIC FOLIATIONS

Not all secondary foliations necessarily develop as axial plane foliations. There is the possibility that some metamorphic foliations mimic an earlier sedimentary foliation, and there seems little doubt that, in some rocks where a group of structures fold an earlier foliation defined by preferred orientation of layer silicates, the folded surface is of direct or indirect sedimentary origin. It has also been suggested [e.g., Billings (1972, p. 399)] that during folding a new foliation may be produced that is parallel to the existing surface in which the folds are developing. This interpretation was invoked [Lewis et al. (1965)], for example, for folds at Broken Hill, Australia, which fold schistosity

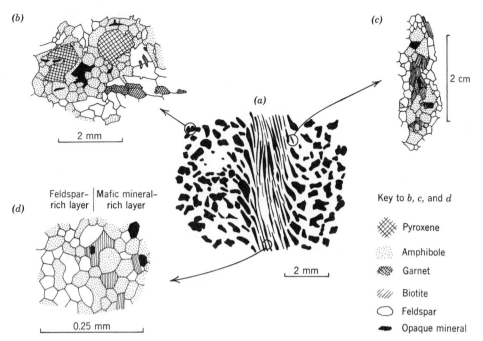

FIGURE 5.38   Diagrammatic sketch of a small shear zone in a metagabbro from the Adirondack Mountains, New York. In (*a*) the dark areas are rich in mafic minerals and the light areas rich in feldspar Parts (*b*). (*c*) and (*d*) show the change in mineralogy and microstructure between the country rock and the shear zone. The country rock is unfoliated (*b*). At the edges of the zone (*c*) there is a foliation defined principally by the shape of mineral aggregates and, to a lesser extent, by the preferred orientation of biotite that is found only in and close to the shear zone. In the shear zone (*d*) there is a well-defined foliation in hand specimen which, in thin section, is seen to be defined principally by variation in composition (i.e., it is a layering) and in biotite-rich areas by the preferred crystallographic orientation of biotite. (Drawings prepared from a specimen provided by J.F. Dewey.)

and gneissic layering. Subsequent work [Ransom (1968)], however, has shown that the folded surface is an axial plane foliation of an earlier group of folds. We know of no well-documented evidence to support the hypothesis that a pervasive foliation can be produced, by folding, parallel to the layers being folded. As stated above (Section 5.2.3), enhancement of the existing foliation in the development of crenulation cleavage is known and *localized* development of a foliation parallel to the surface being folded is also known. For example, a foliation defined by anastomosing layer silicate-rich layers has been described from the A horizon [see Bouma (1959)] of a sequence of folded greywackes and shales [Williams (1972)]. This foliation is found in the limbs of some folds and is more or less equally developed in adjacent limbs and absent in intervening hinges. It is not likely that this foliation is of sedimentary origin since shaly intercalations are not found in A horizons and their symmetrical positioning on opposite limbs of folds suggests that they are a product of the folding. The layer silicate-rich layers are believed to be domains that have undergone a large shear strain relative to adjacent quartz-rich domains and their composition is believed to be due to selective removal of quartz in solution. This type of foliation may be common but is, of course, difficult to recognize because of its parallelism with the folded surface which may have a similar fabric.

A foliation has been described [Savage (1965, 1967); van Veen (1965)] from northwest Spain that is curviplanar and inclined to both the axial surface and the folded surface (Fig. 5.37). The foliation, described as a slaty cleavage, is defined by a penetrative preferred orientation of micas and is believed to have developed during folding. It is referred to as concentric cleavage.

A somewhat similar foliation has been described by D. Roberts (1972) from Norway as arcuate hinge cleavage and Roberts and Strömgård (1972) have indicated the similarity between the cleavage pattern and the distribution of principal elongation axes in some experimentally deformed materials (see Fig. 5.16). The inference is that the foliation has developed parallel to the direction of maximum elongation.

There is another type of foliation common in some rocks (especially in high-grade gneisses and deformed igneous rocks), that is not necessarily parallel to the axial plane of folds. It is found in structures commonly referred to as *shear zones*. These structures are narrow zones across which markers are displaced as by faulting but in which faults, in the brittle sense, if present at all, are incidental. They are simply zones of large ductile shear strain with or without a component of shortening perpendicular to the zone. The foliation may be approximately parallel to the zone in part but generally transgresses it, as shown in Figure 5.38. Restriction of the foliation to the zone and immediately adjacent rocks indicates that it is a product of the deformation that produced the zone. This is particularly clear in areas where the country rock surrounding the zone is an isotropic igneous rock [see Ramsay and Graham (1970)] so that there is no earlier foliation that could be simply rotated. The foliation described by Ramsay and Graham is not planar but has either been folded or, as seems more likely, developed as a curviplanar surface.

Ramsay and Graham (1970) present evidence indicating that the new foliation in the zones that they have described is parallel to a principal plane of the strain ellipsoid and is perpendicular to the direction of maximum shortening.

6

LINEATIONS

6.1 INTRODUCTION

The word *lineation* is used to describe any linear structure that occurs repetitively in a sample of rock; for example, it may refer to an array of elongate pebbles, oriented with their long dimensions mutually parallel, or it may refer to the lines of intersection of two foliations (see Fig. 6.1). It should not be confused with the word *lineament,* which is used to describe linear topographic features of regional extent, that are believed to reflect crustal structure [Dennis (1967); Hills (1972)]. Lineation may be defined in a more precise way as any linear fabric element (see Section 2.1); it may be a primary igneous [Balk (1937)] or sedimentary structure [Potter and Pettijohn (1963, p. 114)] or a secondary structure related to deformation. Here we are concerned only with the latter.

Lineations are ubiquitous in deformed rocks and it is common for more than one lineation to be visible on a given foliation plane. If related to a given group of folds they are generally, but not invariably, either parallel to the fold axis or inclined to the fold axis at an angle approaching 90°. Many lineations are associated with a foliation and actually lie in that surface, but this is not necessarily so, and lineated rocks that are not foliated are fairly common, particularly in areas of gneissic rocks. It is commonplace for a single deformation to give rise to lineations of more than one orientation. Thus, for example, in slates, a lineation forms more or less parallel to the fold axes while another forms in the cleavage plane at a high angle to the first (see Section 9.10).

Various types of lineation are described below (Section 6.2) and the processes involved in their formation are then discussed (Sections 6.3 and 6.4).

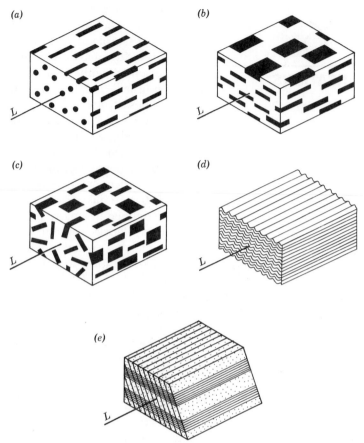

(a)

(b)

(c)

(d)

(e)

FIGURE 6.1 Diagrammatic representation of lineation (*L*). The elements represented by the idealized black shapes in (*a*), (*b*), and (*c*) may be individual grains, metamorphic mineral aggregates, or any other bodies such as fossils, pebbles, and so on. (*a*) Simple linear fabric defined by preferred orientation of linear bodies. (*b*) Combined lineation and foliation defined by preferred orientation of elongate tabular bodies. (*c*) Linear fabric defined by common axis of intersection of variably oriented, tabular bodies. (*d*) Linear fabric defined by penetrative folding. (*e*) Lineation defined by intersection of two foliations.

6.2 DESCRIPTION OF LINEATIONS

6.2.1 Slickenside Striae

Slickenside striae are a common linear structure in many rocks but they are not generally a penetrative feature and therefore not a fabric element. Excep-

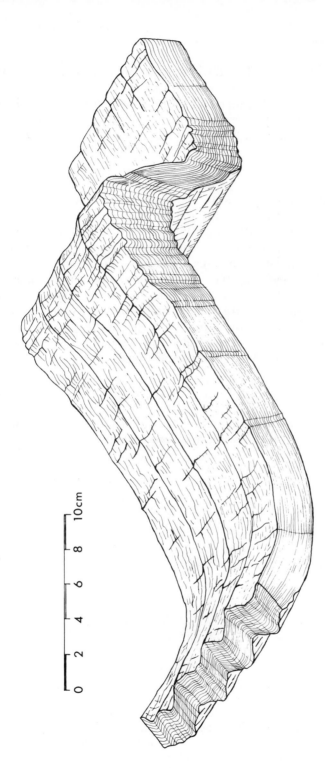

FIGURE 6.2 Intersecting lineation defined by microfolds in schist from Mt. Robe near Broken Hill, Australia. [From Vernon (1969), with permission of the Geological Society of Australia.]

tions to this statement are found where striae occur in wide fault zones or on closely spaced foliation planes where, during the development of folds, there has been slip on the folded surface, as for example in some kinked slates; the striae then are generally oriented perpendicular to the fold axis. Since these structures are considered in Chapter 7 they are not discussed further here.

6.2.2 Folds as Lineations

Fold hinges are a linear structure and may or may not be a fabric element depending on the size of the folds and the size of the area under consideration. On a regional scale, the folds in a strongly folded area commonly do constitute a lineation, and small crenulations (see Fig. 6.2), including those associated with crenulation cleavage, can form a prominent lineation in outcrop or hand specimen. This type of lineation can be very regular and spectacular examples are found in many schists. In Figure 6.2 two such lineations intersect one another; both are defined simply by small-scale folding and the specimen as a whole is bent on a larger scale, about the same axes. Where the small folds or crenulations are associated with a crenulation cleavage this type of lineation belongs equally well in the next type.

6.2.3 Lineations Due To Intersection of Foliations

Lineations defined by the intersection of two foliations are a feature of most rocks that are folded and that have an axial plane foliation. For example, in

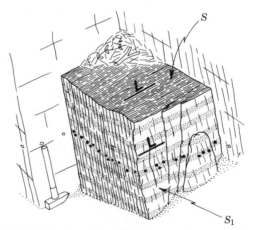

FIGURE 6.3   Sketch of slate outcrop in Ocoee Gorge, Tennessee, U.S.A.. Note lineation on bedding plane (S) due to trace of cleavage S_1 and lineation on S_1 due to trace of S. Note also "downdip" lineation on S_1. Black squares represent large pyrite crystals.

slates, the trace of the bedding appears on the cleavage plane as a lineation, and vice versa (Fig. 6.3). However, lineations due to the intersection of surfaces do not always involve bedding. For example, the intersection of crenulation cleavage and an earlier secondary foliation produces a lineation that is very common in layer silicate-rich rocks. The trace of the cleavage on a joint surface produces a linear structure that may or may not be penetrative. However, this feature is generally of no value in structural analysis; it throws no light on the structural history and it tells less about the regional situation than does direct measurement of the joint surface and the foliation.

6.2.4 Mineral Lineations

Mineral lineations can be defined by the preferred dimensional orientation of inequant grains (Figs. 6.4 and 6.5) or by elongate mineral aggregates (Figs. 6.5 and 6.6).

There are many examples of lineations defined by crystals of inequant habit that are oriented parallel to the axes of minor folds. Sturt (1961) has described a lineation from Sørøy, Norway, defined partly by the long dimension of nepheline crystals that are aligned parallel to the minor fold axes. At Broken

3 cm

FIGURE 6.4   Lineation defined by hornblende crystals in Garbenschiefer from Klimpfjäll, Västerbotten, Sweden.

FIGURE 6.5   Lineated Dogger marble in a quarry at Andermatt, Swiss Alps. The lineation is defined by the preferred dimensional orientation of inequant grains and also by elongate aggregates in impure layers.

Hill, Australia, sillimanite and amphiboles are oriented parallel to Group I fold axes in the Willyama Complex [Hobbs (1966b); Ransom (1968)]. Similarly, in the Hudson Highlands of the Appalachians, sillimanite and hornblende define lineations parallel to the axes of both early and late folds [Dallmeyer (1972)].

Mineral lineations are not always parallel to the axes of related folds. For example, in the Mount Robe area north of Broken Hill, Australia, a lineation defined by alignment of such minerals as sillimanite, andalusite, and biotite is generally inclined to the related fold axes at angles varying from 0–90° [Anderson (1971)]. Similarly in a sequence of folded and boudinaged migmatites in Saskatchewan [Schwerdtner (1970)] the hornblende lineation, which lies in the folded surface, in a given fold may be one of the following:

1. Parallel everywhere to the fold axis.
2. Perpendicular everywhere to the fold axis.
3. Parallel to the axis in the hinge region, approximately perpendicular to the axis at the inflexion line and of intermediate orientation in intermediate positions.

Lineations defined by elongate *mineral aggregates* are common both parallel and at a high angle to the axes of related folds. The nephelines described by Sturt [(1961), see above] occur in lenticular or pencil-shaped aggregates that are aligned parallel to the fold axes. Similarly, at Broken Hill, in addition to the lineations defined by alignment of individual crystals (see above), many minerals occur in aggregates that define a lineation. For example, aggregates of various minerals, including biotite, sillimanite, hornblende, quartz, and feldspar are aligned parallel to Group I fold axes [Hobbs (1966b); Ransom (1968)], whereas aggregates of biotite, muscovite, quartz, and garnet are inclined, at various angles, to Group III axes [Vernon and Ransom (1971)].

In slates there is generally a lineation on the cleavage surface, which, because of its common orientation, is referred to as the *down dip* lineation. This lineation is inclined to the fold axes at angles commonly approaching 90° and is partially defined by elongate minerals and mineral aggregates. Commonly, lenticular quartz-rich aggregates are separated by mica films and these two structures, which also define the cleavage (see Section 5.2.4), are largely responsible for the lineation. The quartz-rich lenses are elongate in the plane of

FIGURE 6.6   Strongly lineated Monte Rosa gneiss in a quarry at Villadóssola, Italian Alps. Slabs are split parallel to the lineation. Inset photograph shows the appearance of a surface cut perpendicular to the lineation. The lineation is defined principally by elongate aggregates of feldspar, quartz, and biotite.

the cleavage thus contributing to the lineation which is further enhanced by the trace of layer silicate-rich films outcropping on the cleavage plane, by the parallel orientation of elongate grains such as quartz and rutile [see Behre (1933, p. 181)], or by rock and fossil fragments [Hobbs and Hopwood (1969)].

"*Pressure shadow*" or "*pressure fringe*" structures [see Spry (1969, p. 240)] are another type of mineral aggregate that can contribute to the definition of a lineation. Such structures generally comprise aggregates of new grains growing on opposed sides of a host porphyroblast or detrital grain, thereby producing an elongate structure (Fig. 6.7). This structure is generally aligned parallel to a foliation and may define a lineation [see Magnée (1935); Fairbairn (1950); Hoeppener (1956); Zwart and Oele, (1966); Elliot (1972)].

6.2.5 Pebbles, Boulders, and Ooids

Detrital grains or fragments of any size may be deformed and/or rotated, to define a lineation. Such lineations are particularly impressive when they occur in pebble or boulder beds. Ooids are generally approximately spherical before deformation and they therefore must be deformed, rather than simply rotated, before they can define a strong lineation. Lineations defined by pebbles may be parallel [e.g., Agron (1964); Gay (1969); Ramsay and Sturt (1970)] or inclined [e.g., Brace (1955); Dewey and McManus (1964); Hossack (1968)] to the axes of related folds, but in a given area are commonly one or the other. However, in some areas, pebbles oriented with their short dimension perpendicular to bedding, have their long dimension perpendicular to the fold axis in limb areas and parallel to the fold axis in hinge areas [Billings (1972, p. 415)].

In the South Mountain area of Maryland [Cloos (1947, 1971)] oolitic limestones have been deformed so that individual ooids, that were once approximately spherical, are now ellipsoidal. These ellipsoids are commonly triaxial and are oriented such that their long axes define a lineation and their long and intermediate axes combined, define an axial plane foliation (see Section 5.2.4 and Fig. 1.25). The lineation is generally inclined to the related fold axes at an angle close to 90° but locally it is parallel to the fold axis [Cloos (1947, 1971)]. Badoux (1970) has also described oolites in which the ooids define an axial plane foliation and a lineation that is perpendicular to the related fold axes.

6.2.6 Rods, Mullions, and Boudins

The terms rod, mullion, and boudin are treated together here because the structures that they describe grade morphologically into one another rather than falling into three distinct groups.

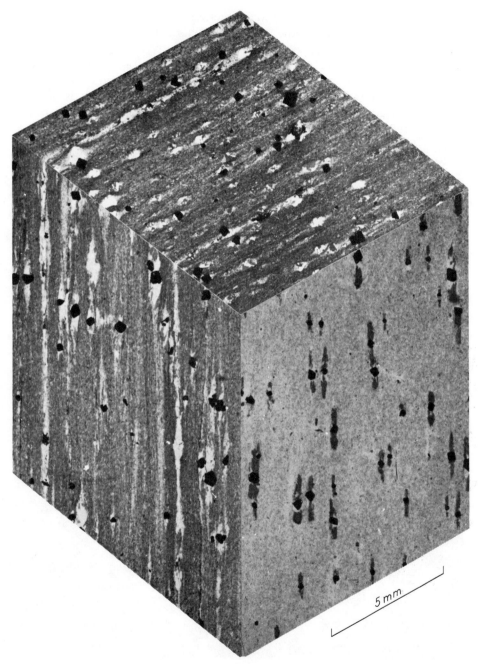

FIGURE 6.7   Lineation defined by pressure shadows associated with magnetite crystals in phyllites from the Rocroi massif, Ardennes. Ordinary light. (Photo by W.C. Laurijssen.)

Rods are a common feature of many metamorphic areas of all grades and mostly comprise elongate bodies of quartz. Wilson (1953) in his early work used the term to describe bodies of this type, which he believed to have formed by metamorphic segregation of the quartz. More recently, Wilson (1961), in recognition of the problem of identifying the origin of the quartz, suggested that the term should be purely descriptive. Following his usage a *rod* is any very elongate aggregate that is essentially monomineralic and is not demonstrably formed by disruption of constituent layers of the original rock (see Fig. 6.8). In profile, rods may have any outline — from elliptical to irregular to that of a dismembered fold. They are generally believed to be elongate parallel to the local fold axes.

Mullions [see Wilson (1953, 1961)] are structures that generally form in the original rock material as opposed to segregated or introduced material (see Fig. 6.9). They may form in a competent layer where they occur as elongate bodies bounded partly by bedding planes or other pre-existing surfaces and partly by new surfaces, such as joints or cleavages. Alternatively, they may form in a surface rather than a layer. The mullion is then simply a corrugation of the surface of a competent bed or preexisting competent layer. Whatever

FIGURE 6.8   Quartz rods in Moine schists on Ben Hutig, Scotland.

FIGURE 6.9   Mullion structure in bedding surface between sandstone and pelite, North Eifel, Germany.

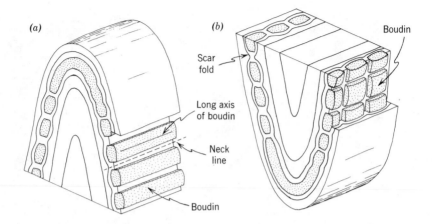

FIGURE 6.10   Diagrammatic representation of normal boudinage (a) and chocolate block boundinage (b), showing the normal relationship, of both, to associated folding. Useful terms associated with boudinage are illustrated.

their form they share certain diagnostic features: they are remarkably cylindrical as well as being complex in profile so that they have a ribbed or grooved appearance, and individual surface features are very persistent along the length of the mullion. Thus, they were described by Holmes (1928, p. 16) as resembling the mullions of Gothic church windows. Characteristically mullions are coated by what Wilson (1961) describes as a "veneer of mica." Like rods, they are generally believed to be oriented parallel to related fold axes.

Boudinage is a structure that forms by the segmentation of preexisting bodies, that are generally more competent than the rock surrounding them. Typically a body such as a given bed or dyke is broken up into a series of elongate bodies aligned parallel to one another (Fig. 6.10a and 6.11). Less commonly, other objects such as pebbles [e.g., Forman (1971)], minerals [e.g., Misch (1970)], or fossils [e.g., Heim (1878); Badoux (1963)] are deformed into small boudinagelike structures. The boudins are separated by material that originally lay on either side of the boudinaged layer or by mineral aggregates that have grown in situ as individual boudins moved apart (e.g., quartz veins concentrated between boudins in Fig. 6.11). The appearance of boudins as seen in profile varies considerably [e.g., see Ramberg (1955)] and there is a complete range of outlines from rectangular or rhomboid to elliptical (Fig. 6.12). In some rocks incomplete boudinage can be seen where preexisting bodies have necked but not broken through. This structure (see Fig. 5.30dii) is known as *pinch and swell* [see Ramberg (1955)] and may be combined with true boudinage (see Fig. 5.30dii).

Boudins are commonly aligned parallel to the axes of related folds but

FIGURE 6.11   Boudinage structure in Paleozoic limestone, Forsyth Quarry, Maryland, U.S.A.. Note the quartz veins between individual boudins.

exceptions are known [see Wilson (1961)] and in some areas a given body of rock may be segmented in two directions to produce fairly equidimensional boudins (Fig. 6.10b) rather than the typical elongate forms [e.g., Wegmann (1932); Sylvester and Christie (1968)]. Such boudins are referred to as *choco-late block boudinage* following Wegmann (1932).

   Mullions and boudins tend to be large in size and are commonly restricted to certain layers or, in the case of mullions, are restricted even to certain surfaces in a deformed sequence. Thus at outcrop scale they are quite commonly a nonpenetrative feature. Rods are generally smaller in size and tend to be more evenly distributed.

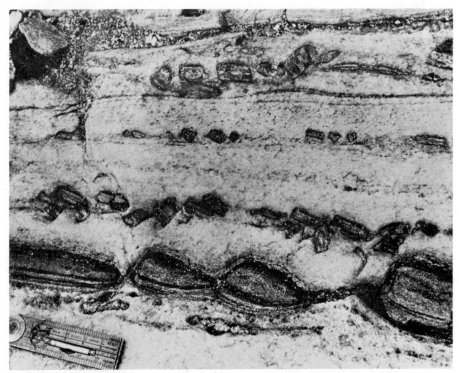

FIGURE 6.12   Gneiss boudins in marble matrix, Nussbaum Riegel, Taylor Valley, Antarctica. Note variation in shape of boudins. [From Williams et al. (1971), with permission of the Geological Society of Australia.]

6.3   ORIGIN OF LINEATIONS

6.3.1   Introduction

In the period 1935 to 1960 lineations were used, more or less indiscriminately, as indicators of "direction of tectonic transport." The precise meaning of this term was always obscure; to some it was synonymous with the direction of maximum finite extension in a deformed rock mass, to others, the "direction of tectonic transport" was the direction that one rock mass had been displaced relative to another. Some writers believed that lineations were always parallel to the direction of tectonic transport; others believed lineations were always normal to this direction, and others again believed that they could be at any angle.

The literature of this period is very confused and the discussions are commonly in terms of Sander's kinematic, coordinate axes a, b, and c, defined for a displacement field that possesses just one plane of symmetry [Sander (1930, p.

57)]. In this coordinate system a was defined as the movement direction (or "direction of tectonic transport"), b was normal to a and was an axis of rotation, shearing, or folding, and c was normal to the ab plane. The ac plane is, therefore, the unique symmetry plane. The use of such a coordinate system for a deformed rock, assumes a very simple displacement field that amounts to simple shear. Since such a model of deformation is greatly oversimplified, the arguments that raged a few years ago [e.g., see Phillips (1937); Cloos (1946; Anderson (1948); Kvale (1953); Turner (1957); Lindström (1958b)] as to whether lineations were parallel to a or to b, now seem rather futile.

An important advance in structural geology came when discussions of deformation in terms of oversimplified displacement fields ceased and attempts to relate observed structural features, such as lineations, to finite strains or, more important still, to the history of incremental strains began.

6.3.2 Mineral Lineations

Lineations defined by elongate mineral grains present a similar problem to the one encountered in explaining the development of foliations (see Section 5.3). If the lineation is defined by a preferred orientation of grain dimensions, rather than of crystallographic directions, then a number of mechanisms can be invoked. They include deformation of grains, preferential growth, and rotation and are the same as the mechanisms already mentioned in explanation of preferred dimensional orientation of inequant grains parallel to foliations (see Section 5.3.4). Where the preferred dimensional orientation is associated with a preferred crystallographic orientation as with hornblende or sillimanite crystals that are elongate parallel to the c axis, there is again the possibility of rotation and also the additional possibility of oriented growth [e.g., see Kamb (1959b); Kumazawa (1963); Hartman and den Tex (1964)] in response to a deviatoric stress that results in an elastic strain only. Again there is the problem of distinguishing between this orientation due to stress and orientation brought about by rotation. However, since the orientations observed in lineations defined by elongate grains are also observed in lineations defined by pebbles and since at least some of the latter are due principally to rotation [e.g., see Ramsay and Sturt (1970)], it is reasonable to believe that at least some mineral lineations are due to rotation during straining.

Similarly it is not easy to tell whether mineral aggregate lineations are a strain or a growth phenomenon. Pressure shadow and pressure fringe structures are probably growth phenomena and are generally believed to develop in parts of the rock where the mean stress is low due to a shielding effect of the host grain, which is a relatively rigid body in a deforming matrix [e.g., Spry (1969, p. 240); Elliot (1972)]. Other lineations of this type are perhaps better interpreted as a product of strain or a combination of strain and growth. For example, biotite aggregates defining the lineation inclined to Group III fold axes at Broken Hill (see Section 6.2.4) are found only in retrograde shear zones and their progressive development can be traced at the edges of some zones. In the high-grade rocks outside the shear zones the aggregates are represented by garnets which generally have a coating of biotite. In the transitional interval, a few feet wide, from the high-grade rocks into a given retrograde zone, the garnets become progressively smaller and the coat of biotite progressively thicker, so that the overall size is preserved, until the garnet is completely replaced by biotite. These mineralogical changes are accompanied by a change in shape; in the high-grade rocks the garnet biotite aggregates are approximately equidimensional. As the garnet decreases in size the aggregates become elongate until the resultant biotite aggregate is very elongate and flat with aspect ratios of the order of 1:10:100.

6.3.3 Ooids and Pebbles

Ooids and pebbles may be deformed into elongate bodies that define a lineation. If these bodies were initially spherical and had the same mechanical properties as the matrix surrounding them, then after deformation their shape is the shape of the strain ellipsoid for the rock as a whole, within a homogeneously strained body. In general, however, the situation, particularly for pebbles, is more complex because the pebbles and matrix differ in mechanical properties. Under these conditions the shape of the pebble, if initially spherical, only represents the strain of that pebble and not the strain of the rock as a whole. Furthermore, when mechanical properties do differ it is very difficult to relate the strain of the pebble to that of the rock because the pebble has not only changed its shape but, in general, has rotated with respect to the rock matrix. Theory-relating mechanical properties of pebbles and matrix and type of strain history (coaxial or noncoaxial), in a rock body as a whole to the strain and rotation of the pebble is complex and incomplete [see Goldsmith and Mason (1967); Ramsay (1967, Section 5.9); Gay (1968b and c); and Jaeger (1969, p. 257)]. If the strain history is coaxial, theory predicts that elongate bodies will tend to become oriented parallel to λ_1 [Gay (1968c); Jaeger (1969)]. Under conditions of noncoaxial strain history, such as found in a viscous fluid deforming by laminar flow, Jeffery (1923) has predicted that rigid elongate bodies will align themselves in the laminar flow planes and perpendicular to the direction of flow. This prediction has been verified experimentally by Taylor (1923) who found, however, that it may take hundreds of revolutions of a prolate ellipsoid immersed in a viscous liquid for the body to reach the predicted orientation.

6.3.4 Rods, Mullions, and Boudins

Rods, mullions, and boudins overlap in morphology and there may be some overlap in the mechanisms involved in the formation of these structures.

Wilson (1953 and 1961) postulated that rods in the Moine series of Scotland formed by segregation of quartz, by the segmentation of quartz veins and by the deformation of quartz pebbles. Evidence for each of these mechanisms can be found in many other areas. Thus, there is probably no single mechanism involved in the development of these structures which morphologically are just a special, rather large variety of the group of linear structures defined by mineral aggregates as discussed above.

Mullion structures are also problematical and may be the product of more than one process. Some resemble boudinage and may develop in a similar manner. Ramsay (1967, pp. 382–386) suggests another mechanism that seems to be applicable to many mullion structures and proposes that they are the product of strong small-scale folding of the interface between rocks of different competency.

Boudinage is a structure resulting from extension and is readily reproduced in the laboratory [e.g., see Ramberg (1955); Griggs and Handin (1960); Paterson and Weiss (1968)]. In a sequence of rocks of variable competency the rate at which the various lithologies can deform in a ductile manner varies. If the rate of strain in the sequence as a whole exceeds the rate at which a given lithology can behave in a ductile manner then that particular rock type will behave in a brittle manner when its strength is exceeded. When such lithologies fail in extension boudinage results. The shape of boudins as seen in profile is a function of the ductility difference, under the prevailing conditions of deformation, between the competent layer and adjacent layers. If this difference is small, the boudins fail by *necking* (see Fig. 6.11), that is, they thin locally in a ductile manner before fracturing. The smaller the difference the greater the chance of necking and vice versa. In some rocks fracturing never occurs and individual bodies are simply necked to produce pinch and swell structure (e.g., Fig. 7.2*d*ii). The presence of elongate boudins (Fig. 6.10*a*) indicates that the direction parallel to the boudinaged layer and perpendicular to the length of the boudins was a direction of extension during at least part of the deformation. Presence of chocolate block-type boudinage (Fig. 6.10*b*) indicates that all directions within the boudinaged layer were directions of extension for at least part of their history. Sylvester and Christie (1968) used chocolate block boudinage as a means of estimating strain and demonstrated approximately 80 percent shortening perpendicular to the boudinaged layer, a figure that may not be unusual for such rocks.

6.4 PROBLEM OF LINEATIONS INDICATING EXTENSION PARALLEL TO FOLD AXES

One interesting aspect of the lineation problem is how best to interpret lineations defined by deformed or rotated bodies, such as pebbles, that are aligned parallel to fold axes. This is a common orientation, especially in the hinges of folds and if the lineation is due to deformation of the bodies under conditions of coaxial strain history then, assuming that they were initially spherical, they indicate that the fold axis is parallel to λ_1. In terms of the strain, within a body of rock, this does not present any problem; there is no reason why the fold hinge cannot be parallel to the λ_1 axis of the mean strain ellipsoid or, for that matter, parallel to λ_3 if λ_2 is also an axis of shortening (see Fig. 6.13). However, folds of the type under consideration, are commonly horizontal or gently plunging and if the hinge is parallel to λ_1 there would seem to be a space problem. It is to be expected that there may be extension parallel to the fold hinge in such areas as the culmination of doubly plunging folds [see arcuation of Cloos (1947, p. 888)] or in the outer arcs of arcuate fold belts. It is also reasonable that the lineation and fold axis should be parallel where they plunge steeply or where the parallelism is only a local feature, but for fold axes to be parallel to λ_1 over large areas would seem, at first sight, to suggest that it is often easier for fold belts to accommodate shortening by extension in a horizontal direction parallel to the fold axis, rather than in the vertical direction. This seems improbable. Furthermore, strain analyses based on

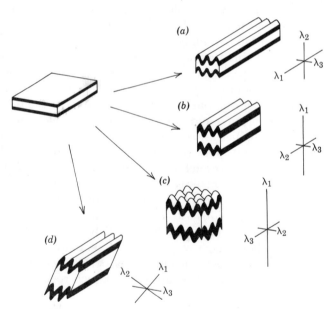

FIGURE 6.13 Various possible relationships between fold hinges and principal axes of mean strain. In (a), (b), and (c) there are hinge lines parallel to λ_1, λ_2, and λ_3, respectively. Part (d) illustrates the general case where the hinge line is inclined to all principal axes of the mean strain ellipsoid.

ooids [Cloos (1947, 1971); Badoux (1970)] and reduction spots [Wood (1973)], indicate some extension parallel to the fold axis but a maximum extension that, in general, is approximately perpendicular to the fold axis. Also, in experiments on strongly anisotropic specimens it is found that when the material deforms by folding there is very little [generally <15 percent] extension parallel to the fold axis even for shortening as high as 75 percent [observations made by the authors; see also Paterson and Weiss (1966)]. The principal extension is perpendicular to the fold axis despite the fact that the external force resisting extension (provided by the confining pressure) is the same in all directions parallel to the axial plane of the fold. Thus it appears that anisotropy of a type common in foliated rocks may determine that deformation will approximate a plane strain.

A possible solution to this problem is that the lineation is not really parallel to the principal extension axis of the mean strain ellipsoid and there are several alternative explanations that can be offered for the lineation and its orientation. One possibility is that elongate bodies owe their orientation to rotation (see Section 6.3) as predicted by Jeffery (1923). This explanation, however, seems unsatisfactory in the light of Taylor's work (1923) since the shear strain in the hinge of a fold is not likely to be nearly large enough to rotate the body sufficiently for it to reach the postulated stable orientation parallel to the fold axis. However, it may be that Taylor's work, which was carried out using viscous liquid, is not directly applicable to anisotropic material such as layered or foliated rocks. Ramsay (1967, p. 220) has· suggested another possibility. He points out that if the bodies defining the lineation are initially elongate and have a preferred orientation of their long dimension parallel to the folded surface, as might be expected of pebbles lying in the bedding plane, then a coaxial strain history with λ_2 oriented parallel to the fold axis and λ_3 perpendicular to the fold axis can produce a preferred orientation of the bodies parallel to λ_2. The point is that the initial shape of the pebble influences its final shape so that, for certain strains, preferred orientation parallel to λ_2 results. This explanation seems very satisfactory for objects that can be deformed into elongate bodies but cannot be used to explain preferred orientation of elongate mineral grains where the grains do not owe their shape to deformation. It may be, of course, that this simply means that mineral lineations parallel to fold axes do not develop by rotation but are a growth phenomenon or alternatively that they only have that orientation where the fold axis is parallel to λ_1. Yet another possible explanation is that the elongate bodies are oriented parallel to the principal extension direction for the volume of rock immediately surrounding them but that this strain axis is not parallel to λ_1 of the mean strain ellipsoid. For example, in a layered sequence that is deforming by folding, it is possible for λ_1 in the competent layers to be parallel to the fold axis and for both this local λ_1 and the fold axis to be parallel to the λ_2 axis of the mean strain ellipsoid (see Fig. 6.14). Thus if

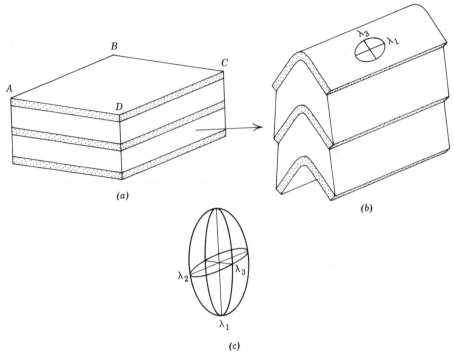

FIGURE 6.14   Diagram illustrating how the principal axis of extension, in some folded layers, may be parallel to the fold hinge even though the latter is perpendicular to the principal extension axis of mean strain. The dimension of *AD* measured parallel to bedding has decreased during folding but the area of *ABCD* has been maintained by extension parallel to *AB*. The thickness of the competent beds (stippled) is unchanged and λ_1 at any point is these beds is, therefore, papallel to the hinge line. The thickness of the incompetent beds, however, has changed and λ_1 of the mean strain ellipsoid (*c*), for a system of such folds, is perpendicular to the hinge line.

the lineation is found only in the competent layers it can be an extensional lineation without being parallel to the mean strain extension axis.

Another possibility however, is that the lineation is parallel, or approximately parallel, to the principal extension direction of the finite strain ellipsoid and that the parallelism of the lineation and fold axes simply indicates that fold axes that formed at an angle to the extension direction have been rotated into approximate* parallelism with that direction. This mechanism has been proposed for mylonites occurring along major thrust planes [e.g.,

*In general they can only be truly parallel where the strain is infinite but the discrepancy can be too small to measure or even, in view of the imprecise definition of the markers, too small to recognize.

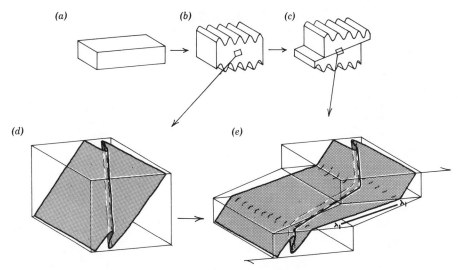

FIGURE 6.15   Diagram showing how early fold hinge lines may be rotated, within a zone of intense deformation, toward parallelism with the direction of thrusting. Parts (a)–(c) show how, on a regional scale, the crust is being shortened first by folding and then by thrusting. Parts (d) and (e) show what is happening on a smaller scale (note that in these diagrams it is assumed that, due to local variation in fold plunge, the hinge line of the fold is not parallel to the thrust, this of course will be true in general). The thrusting is assumed to be achieved solely by ductile deformation concentrated in a narrow zone and the strain is assumed to be simple shear. Since the strain is plane, the principal axis of extension, for the front face of the block (heavy line marked λ_1), is the three-dimensional, principal axis of extension for the deformed zone. As shear is increased both this axis and the fold hinge line will rotate toward parallelism with the direction of thrusting or transport, as indicated by the single barbed arrows in (e). Note that in this environment, where the thrusting may be measurable in tens of kilometers and the shear zone may be only tens of meters thick, the possible shear strains are enormous.

Johnson (1965); Bryant and Reed (1969)]. In such zones the folds are commonly parallel to a lineation that is believed to be parallel to the direction of thrusting [e.g., Balk (1952); Kvale (1953); Johnson (1965); Hooper (1968); Bryant and Reed (1969)] and that, where the movement is large, must also be approximately parallel to the principal elongation axis for the mylonite itself. This point is demonstrated diagrammatically in Figure 6.15. In the diagram, for simplicity, the strain in the mylonite is assumed to be homogeneous simple shear; in reality it is more likely to be heterogeneous and is probably not even simple shear; however, in such an environment it is to be expected that the strain history will be noncoaxial and it is believed that its principal characteristics can be represented by this simple model. It is not necessary that the folds predate the thrust; contemporary folds will also rotate in the same manner; however, as drawn here the folds and thrust can be fit into a simple scheme of continuous deformation (Fig. 6.15a–c) in which the crust, in response to continuing forces, deforms initially by folding and then by faulting.

7

JOINTS
AND
FAULTS

7.1 INTRODUCTION

Joints and faults are structures resulting from brittle behavior in which blocks of rock are displaced relative to one another across narrow and approximately planar discontinuities. The discontinuities are called *joints* if the component of displacement parallel to the structure is zero (or too small to be apparent to the unaided eye) or *faults* if the parallel component of displacement is larger.

Most joints and faults form by fracturing, that is, by development of cracks across which the original cohesion is lost. Following their inception as fractures, however, many joints and faults are partially or wholly healed by introduction of secondary minerals or by recrystallization of the original minerals. If a fracture propagates slowly enough, healing may keep pace with propagation so that only a small region near the tip is a true fracture at any given time.

7.2 JOINTS

7.2.1 Joint Sets and Systems

Joints usually occur as families of fractures with more or less regular spacing in a given rock type. We define a *joint set* as a group of joints of common origin. The joints of a set are often approximately parallel to one another, but they need not be. For example, a group of joints that are everywhere parallel to a fold hinge line and perpendicular to bedding may comprise a genetically distinct set even though the joints are not all parallel to one another. Joints of several sets commonly occur together, giving exposures a blocky or fragmented appearance (Fig. 7.1).

FIGURE 7.1   Jointing in granite gneiss at The Glen, southeastern
Adirondack Mountains, New York, U.S.A.. Three prominent joint sets are
present: one set dips steeply left and strikes toward observer; other two sets
strike parallel to plane of photograph and give rise, respectively, to the
light-colored ledges and the dark overhangs. Height of main overhang,
behind birch trees, is about 3 meters. (Photo by D. Graham.)

The whole assemblage of joints present in an exposure or map area is called a *joint system*. Typically the characteristics of a joint system, that is, the sizes, spacings, and orientations of the joints, are seen to vary in some degree across contacts between rocks of different lithology. This fact is put to use in mapping contacts, particularly in air-photo interpretation or in surface mapping of heavily weathered or inaccessible exposures. In a given area, there may also be revealing differences in the joint systems at limb and hinge positions on large folds, or at different distances from large faults. Other aspects of the use of joints in mapping folds and faults are mentioned in Section 7.2.3.

7.2.2 Joint Surfaces

The conspicuous joint surfaces in most outcrops have dimensions ranging from tens of centimeters to hundreds of meters and repeat distances of several centimeters to tens of meters. In addition, most rocks contain numerous inconspicuous joints of smaller size and closer spacing, many of them visible

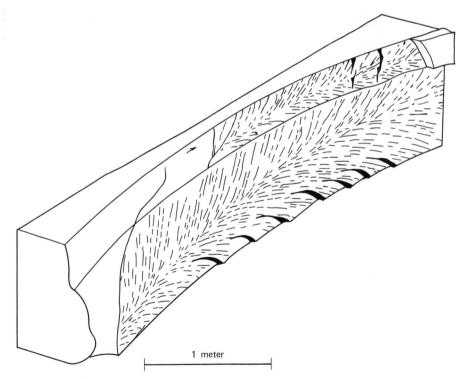

1 meter

FIGURE 7.2   Plume structure on joints in lower Carboniferous greywacke, Thuringia, Germany. [From Bankwitz (1965), with permission of Akademie Verlag, Berlin, D.D.R.]

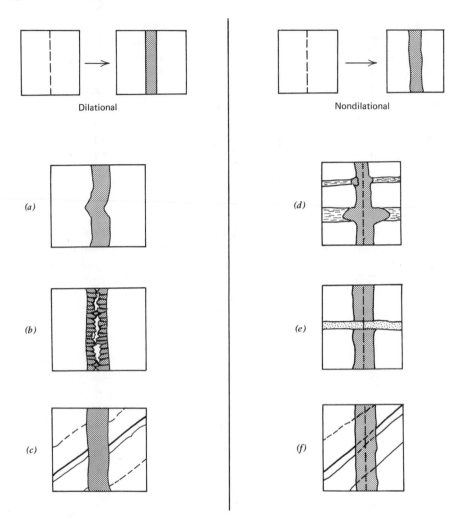

Dilational

Nondilational

(a)

(b)

(c)

(d)

(e)

(f)

FIGURE 7.3   Criteria for recognition of dilational and nondilational veins. (a) Matching walls. (b) Comb structure with vugs. (c) Early planar structures offset. (d) Vein widens in chemically favorable layers. (e) Chemically resistant layers continuous across vein. (f) Early planar structures not offset. Parts (a)–(b) are evidence for open-space filling along a dilating fracture. Part (c) indicates dilation if the shear component of displacement across the fracture is known to be zero. Part (d) indicates replacement throughout vein without dilation, or replacement in outer parts of vein with some dilation near the center. Part (e) indicates replacement without dilation if the preexisting layer (stippled) has uniform texture and mineralogy across the vein. Part (f) indicates replacement without dilation if traces of the early planar structure are seen within the vein.

only in thin section. Fractures small enough to require microscopic observation are called *microjoints* [Wise (1964)] in naturally deformed rocks or *microfractures* [Scholz (1968)] in experimentally deformed rocks. Any joints that are larger than associated joints of the same orientation are called *master joints*. The master joints seen in aerial photographs can sometimes be traced for distances of several kilometers.

Joints in massive rocks, like some sandstones, may show distinctive patterns of surface relief [Parker (1942); Hodgson (1961); Roberts (1961); Bankwitz (1965, 1966); Price (1966); Syme-Gash (1971)]. The most common type is called *plume structure* (Fig. 7.2). Markings similar to plume structure are seen on fracture surfaces in glass and other brittle materials [McClintock and Argon (1966, p. 502)] and can be interpreted in terms of the direction of propagation of the fracture front. This direction would be from left to right in Figure 7.2. Less well established is the correlation between plume structure and shearing as opposed to normal displacements on joints (see Section 7.2.4). Many workers have seen plume structure on joints identified as shear joints, but others believe plume structure is diagnostic of joints lacking a shear component and Gramberg (1961) has shown experimentally that this is possible under some conditions [see discussions by Price (1966, p. 125) and Syme-Gash (1971, p. 354)].

Some joints are barren hairline cracks or empty fissures, but many contain coatings or narrow veins of secondary minerals, very commonly quartz or calcite. These deposits along joint planes can be classified as *dilational* if the vein material occupies space between the two original fracture surfaces, or as *nondilational* if the vein material occupies space outside the two original fracture surfaces, that is, space made available by *replacement* of the original rock. Several criteria for distinguishing dilational from nondilational veins are illustrated in Figure 7.3, and others are discussed by Park and McDiarmid (1970, p. 118). Notice that it is entirely possible to have replacement together with dilation of a fracture, so that some kinds of evidence for replacement (e.g., Fig. 7.3d) are not necessarily evidence against dilation.

It is important in any study of jointing to examine thin sections *across* the joints for details of mineralogy, microstructure, and displacements because such observations are essential for working out the relative ages of several joint sets and for relating fracture history to events in the thermal history. In engineering practice, the mineralogy and texture of joint fillings is important because joints with different fillings can have different mechanical properties and different properties governing storage and flow of fluids.

7.2.3 Relations of Joints to Other Structures

In layered rocks of all kinds the most prominent joints usually intersect the layering at high angles. Where the layers are folded, joints in various special

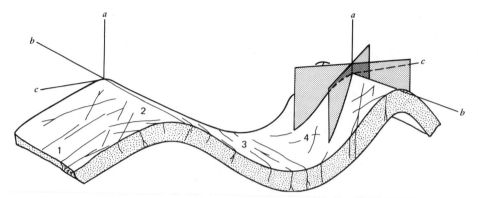

FIGURE 7.4   Joint sets commonly found in folded rocks: (1) a–c joints; (2) joints in paired sets symmetrically disposed about the a–c plane, intersecting in the pole to bedding; (3) radial or $\{h0l\}$ joints; (4) joints in paired sets intersecting in a and equally inclined to c, or $\{0kl\}$ joints. Joints types shown in separate parts of folds above for clarity only. In nature they can all occur together. References axes shown are for description of joints only. They are not recommended for other purposes.

orientations relative to the folds are seen. These joints are described with respect to orthogonal reference axes a, b, c; where b is parallel to the fold hinge line, a is normal to the hinge line and lies in the axial plane, and c is normal to the a-b plane (Fig. 7.4). Joints approximately perpendicular to fold axes are thus called *a-c joints* (Fig. 7.5). Joints parallel to fold axes and approximately perpendicular to layering are called *radial joints*. Joints in paired sets of the type represented by 4 in Figure 7.4 can be called $\{0kl\}$ joints. It is emphasized that the reference axes a, b, c, while useful for describing joint orientations, have no simple significance in terms of the displacements or strains represented by the joints or folds. The belief, widely held 20 years ago, that the a direction was necessarily a special direction of "movement" or of "tectonic transport" has been discarded (see Section 6.3.1). Excellent examples of fractures associated with folds are described by Norris (1967), Stearns (1968), and Handin et al. (1972).

Joints associated with faults may predate the faults and have no genetic relation to the faults apart from a possible control on the orientation of the fault planes. Other joints may be intimately related to faulting and useful in revealing the sense of slip on faults. The best known example here are the *feather joints* of Cloos (1932) or other *pinnate fractures*. These occur preferentially in the immediate vicinity of a fault plane and intersect the fault in an acute angle pointing in the direction of relative movement of the block containing the pinnate fractures. Examples on two different scales are shown in Figure 7.6. Pinnate fractures can evidently form both in advance of the de-

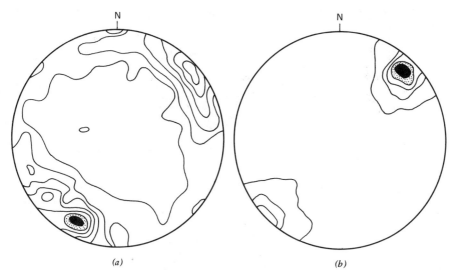

N   N

(a)   (b)

FIGURE 7.5   Poles to joint planes (a) and associated lineations (b) in an area of
highly deformed metamorphic rocks near Barstow, California; after Weiss (1954,
Figs. 4, 7). Lower hemisphere, equal area projections. Contours at 1, 2, 3, 4, 5, and 6
percent per 1 percent area in (a) and 1, 7, 13, 19, and 25 percent per 1 percent area
in (b). The joints are oriented at a high angle to the lineations (and fold hinge lines)
but few are exactly perpendicular [From Weiss (1954). Originally published by the
University of California Press; reprinted by permission of the Regents of the
University of California.]

velopment of a through-going fault plane (e.g., Fig. 9.6), or during subsequent slip on a fault. They have been observed in experiments on a wide
variety of rocks and other materials [Morgenstern and Tchalenko (1967);
Lajtai (1969); Gay (1970); Friedman and Logan (1970); Dunn et al. (1973)].

Joint systems in igneous rock bodies may be quite different from joint
systems in the surrounding rocks and they are often symmetrically related to
the contacts of the body, suggesting an origin during emplacement and cooling. One prominent joint set is commonly seen at a high angle to the nearest
contact. Balk (1937) and Price (1966) give detailed descriptions of these and
other joints in plutonic bodies. In tabular bodies, such as flows, dikes, or sills,
joints perpendicular to the contacts are again common and may display a
special configuration known as *columnar jointing*, in which the joints isolate
elongate prisms with more or less regular hexagonal cross sections. The field
worker in an area of poor exposure may be able to use well-developed columnar jointing as an indicator of the normal to the contacts of a tabular body.
This interpretation is not always reliable however because curved columnar
jointing oblique to contacts is also well known [see Spry (1962, p. 194)].

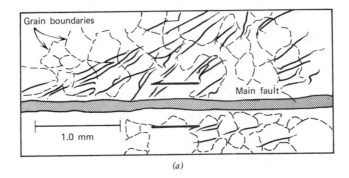

(a)

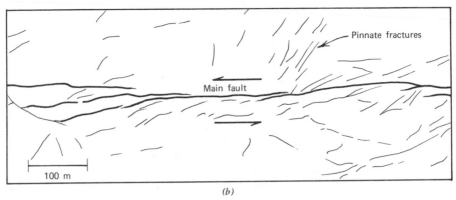

(b)

FIGURE 7.6   Two examples of pinnate fractures associated with faults. (a) Pinnate fractures in experimentally faulted quartzite, traced from photograph by Dunn et al. (1973, Fig. 8). (b) Pinnate fractures on part of fault zone associated with the Dasht-e Bayaz (Iran) earthquake of August 31, 1968. [After Tchalenko and Ambraseys (1970, Fig. 5), with permission of the Geological Society of America.] Arrows show sense of motion on faults.

7.2.4   Origin of Joints

Extension and Shear Joints.   A genetic classification of joints is based on the size of the shear component of displacement at the moment of origin. If a fracture forms with a shear component that is zero, so that the total displacement is directed normal to the fracture surfaces, the structure is an *extension joint* (Fig. 7.7a). If the shear component has some finite value, the structure is a *shear joint* (Fig. 7.7b). Following the initial displacement at the inception of a joint, there may be a long and complex history of further displacements. This is indicated by joints carrying successive secondary mineral coatings in each of which there are differently oriented slickenside striae or fibrous mineral growths (see Section 7.3.3).

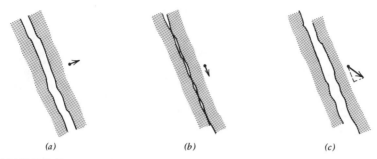

FIGURE 7.7   Classification of joints by displacement. (*a*) Extension joint.
(*b*) Shear joint. (*c*) Shear joint with finite normal displacement, alternatively
classed as an "oblique extension joint" (Dennis, 1972, p. 291).

Extension joints in isotropic rocks form normal to one of the principal
directions of stress, otherwise there would exist a finite shearing stress on the
potential joint plane at the moment before fracture and a corresponding finite
shear displacement would occur. In the usual laboratory tests and presumably
in most natural environments, extension joints form normal to the direction
of σ_3 [Griggs and Handin (1960, p. 348)]. It is not impossible, however, for
extension joints to form normal to σ_2 or even σ_1, given suitable anisotropy of
the tensile strength. Shear joints can form at any angle to the principal directions of stress other than 90°. These matters are discussed more fully in
Section 7.3.9.

Various criteria have been suggested to allow distinction between extension
joints and shear joints. These include surface markings, regional orientation
patterns, and relations to other structures. Thus, a joint set lacking slickensides and oriented normal to the hinge line of a doubly plunging anticline
might be interpreted as a set of extension joints. While a convincing case can
sometimes be built up from many lines of evidence, it is usually impossible to
determine whether a joint set *originated* as extension or shear joints although
some details of subsequent displacements may be determined. The terms
extension joint and shear joint are not suitable field terms because thin section
observations, at least, are required to determine the nature of the displacements.

Joints Due To Erosional Unloading in Isotropic Rocks.   Some joints form during
erosional unloading, on account of the greater ease with which decompressed
rock expands normal to, rather than parallel to, the free surface. Imagine that
the body shown in Figure 7.8*a* is isotropic rock at depth, in which all shearing
stresses have been relaxed by creep processes. The body will only remain in a
state of hydrostatic stress during erosional unloading if extension is equal in
all directions. Near the surface upward extension may commonly be easier
than horizontal extension because the normal stress on horizontal planes must

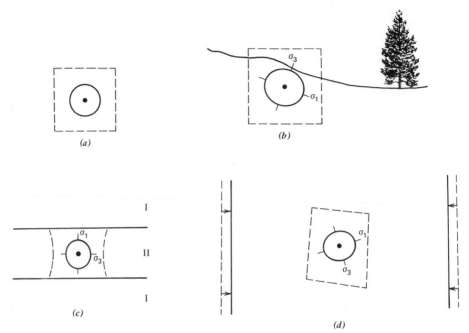

FIGURE 7.8 Schematic diagrams illustrating generation of nonhydrostatic stress required for rock fracturing. (*a*) Initial state of stress, assumed hydrostatic. Circle is stress ellipse for central point. Dashed lines outline imaginary volume element. (*b*) Nonhydrostatic stress generated by erosional unloading. Differential stresses and strains in this and other sketches exaggerated for clarity. (*c*) Nonhydrostatic stress generated by different volume changes in adjacent rock types. In example shown, layer II contracts more than layer I. (*d*) Nonhydrostatic stress in response to displacements of boundaries of a large region. Orientation of σ_1 is not simply related to the direction of boundary displacements (arrows) but depends on local structural conditions.

approach one atmosphere, whereas this is not necessarily true for the normal stress on vertical planes. The state of stress may therefore become nonhydrostatic with σ_3 approximately perpendicular to the earth's surface. Extension joints formed in this situation would be about parallel to the surface of the earth, and this is an explanation for the sets of flat-lying joints in granitic rocks referred to as *sheeting* or *sheet structure* [Jahns (1943); Johnson (1970)]. The amount of expansion to be expected from stress relief during erosion is indicated by the values of the *compressibility* of rocks, the ratio of volume change to pressure change. For near-surface conditions, typical compressibilities fall in the range 10^{-2} kbar^{-1} to 10^{-3} kbar^{-1} (Birch, 1966). Pressure changes of 2 kbar, corresponding with depth changes of about four miles, lead to volume changes ranging from a few percent to a few tenths of a percent. If such volume changes are accomplished mainly by vertical extension, and if this extension takes place fast enough, horizontal extension joints may form.

Joints Due to Differential Volume Changes in Heterogeneous Bodies. Most big rock bodies and many small ones consist of several rock types juxtaposed in layers or other configurations. When such bodies are decompressed or cooled from conditions of hydrostatic stress, local deviatoric stresses will be set up within them because of the differences in compressibilities or thermal contraction coefficients between adjacent units of different lithology. Local deviatoric stresses will also be set up on a granular scale, where adjacent mineral grains of different orientation or composition will tend to undergo slightly different strains during decompression or cooling. Local nonhydrostatic stress generated by decompression or cooling may be important in joint formation even where the original state of stress is itself nonhydrostatic or where the total state of stress is governed by regional deformation as well as by the internal makeup of a rock body.

One clear example of jointing brought about by differential volume change is provided by the columnar jointing in sills [Iddings (1886); James (1920)]. Imagine that layer II in Figure 7.8c is a sill of hot igneous rock that is contracting more than the cooler layers of country rock I on either side. Vertical contraction of the sill is accommodated by downward movement of the overlying country rock. But horizontal contraction of the sill is resisted by the country rock, assuming the sill contracts more in a given interval of time than the country rock. If the boundary between the two rock types is to remain coherent, then compressional structures must develop in the country rock or extensional structures must develop in the sill. If the sill is weaker in extension than the country rock is in compression, vertical extension joints may develop in the sill. Good columnar arrangement of these joints is enhanced if the sill contracts equally in all horizontal directions and if thermal and mechanical properties are identical in all horizontal directions in the country rock.

Joints Due to Regional Deformation. Many joints, and particularly those that cut through rocks of different lithologies, appear to be related directly to folds produced by regional deformation. The folds may be pronounced features like the example in Figure 7.4 or barely perceptible regional upwarps or downwarps. Indeed a conceivable cause of regional jointing is the very gentle flexing of lithospheric plates (Section 10.2) to be expected when a plate changes latitude and thereby its radius of curvature [McKenzie (1972); Turcotte and Oxburgh (1973)].

Joints such as *a-c* joints that are geometrically related to folds may or may not originate during the folding. If produced during the folding they may be extension joints reflecting some elongation of the rocks parallel to hinge lines of the folds. If produced later than the folding and by forces unrelated to the folding, their special orientation must be accounted for somehow by mechanical anisotropy of the folded rocks.

Other aspects of the origin of joints, which are related to the origin of shear fractures, are discussed below in Section 7.3.8.

7.3 FAULTS

7.3.1 Fault Terminology

A *fault* is a planar discontinuity between blocks of rock that have been displaced past one another, in a direction parallel to the discontinuity (Fig. 7.9*a*). A *fault zone* is a tabular region containing many parallel or anastomosing faults (Fig. 7.9*b*). A *shear zone,* in the sense preferred here, is a zone across which blocks of rock have been displaced in a faultlike manner, but without prominent development of visible faults (Fig. 7.9*c*).* Shear zones are thus regions of localized ductile deformation, in contrast to fault zones that are regions of localized brittle deformation. Another distinction is that the normal component of displacement, which is negligible for faults and fault zones, may be appreciable for shear zones. The displacement across a shear zone can be inclined at any angle, other than 90°, to the boundaries of the zone. In another common usage [Dennis (1967, p. 133)] the term shear zone refers to a tabular region of pervasively faulted rock — that is, a fault zone containing a very large number of closely spaced and anastomosing fault surfaces.

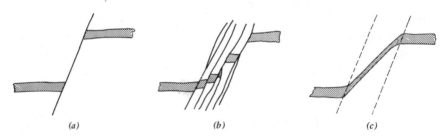

FIGURE 7.9   (*a*) Fault. (*b*) Fault zone. (*c*) Shear zone.

The rock immediately above and below any nonvertical fault is referred to, respectively, as the *hanging wall* and the *footwall* of the fault. The displacement vector connecting originally contiguous points in the hanging wall and footwall is called the *net slip.* The components of the net slip parallel to the strike and dip of the fault are the *strike slip* and the *dip slip* (Fig. 7.10).

The offset shown by a planar feature in a vertical cross section perpendicular to the fault is called the *dip separation* (Fig. 7.11). The vertical and horizon-

---

*The term shear zone as used here is synonymous with *shear belt* as used, for example, by Sutton and Watson (1962) and Ramsay and Graham (1970). The latter authors use shear belt and shear zone interchangeably.

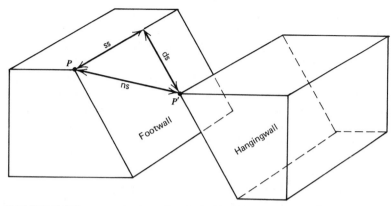

FIGURE 7.10   Terminology for fault displacements. Net slip (ns), strike
slip (ss), dip slip (ds). The fault is an oblique-slip fault with normal and
sinistral components of dip slip and strike-slip displacement, respectively.
Before faulting, points P and P' were coincident.

tal components of the dip separation are the *throw* and the *heave*. Notice, by
comparing Figure 7.11 with Figure 7.10, that the dip separation is not equivalent to the dip slip, the former depending on the orientation of the offset
surface as well as on the nature of the fault displacement.

A fault with dominant strike slip displacement is called a *strike slip fault*. A
fault with dominant dip slip displacement is a *dip slip fault*. Strike slip faults

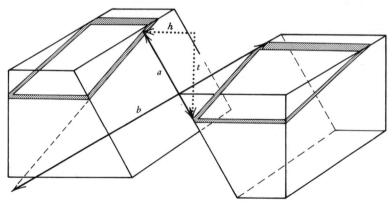

FIGURE 7.11   Same pair of blocks as in Figure 7.10 showing an offset
layer (shaded) and terminology for fault offsets. Offset of a planar feature
across a fault can be measured parallel to the dip, giving the *dip separation*
(a) or, parallel to the strike, giving the *strike separation*)b). Vertical and
horizontal components of the dip separation respectively are *throw* (t) and
heave (h).

usually have very steep or vertical dips and are then referred to as *transcurrent faults* or *wrench faults*. A large transcurrent fault that terminates in another large structure, such as an oceanic ridge or trench or triple junction, is called a *transform fault*. Transform faults are discussed in detail in Section 10.2.3.

The sense of the strike slip part of displacement on a fault is described by the terms *sinistral* and *dextral*, or alternatively, *left lateral* and *right lateral*. A fault is sinistral or left lateral if, to an observer standing on one block and facing the other, the opposite block appears to have been displaced to his left. The strike slip component of the displacement in Figure 7.10 is sinistral.

Faults dipping more or less than 45° are called, respectively, *high angle faults* and *low angle faults*. A *normal fault* is a high angle, dip slip fault on which the hanging wall has moved down relative to the footwall. A fault of similar type but with a dip less than 45° is sometimes called a *lag* [Rickard (1972)]. A *reverse fault* is a dip slip fault, either high or low angle [Gill (1971)], on which the hanging wall has moved up relative to the footwall. The terms normal fault and reverse fault, while strictly defined for faults with zero strike slip displacement, can also be used for faults with small strike slip displacements accompanying much larger dip slip displacements [Rickard (1972)]. Where the strike slip and dip slip displacements are similar in magnitude, as in Figure 7.10, the fault can be called an *oblique slip fault*.

A (relatively) downfaulted topographic trough between high angle faults is a *graben*. An upfaulted block between high angle faults is a *horst*. The faults bordering horsts and grabens are usually normal faults.

A *thrust fault* is a low-angle reverse fault, according to one common usage [see Dennis (1967)]. The term is also used by many geologists for low angle faults that are *presumed* to have involved reverse dip slip displacement but where this has not been demonstrated. A *window* (or *fenster*) is an exposure of the rock below a thrust fault that is completely surrounded by rock above the thrust. A *klippe* is an exposure of the rock above a thrust completely surrounded by rock below it (see Fig. 7.17).

7.3.2 Breccia and Mylonite

Fault planes are commonly filled with fragmental material known as *fault breccia*, or *microbreccia* if the fragments are microscopic. Some microbreccias are soft and are designated by terms such as *pug* or *gouge*. Others, particularly in metamorphic rocks, are hard and characterized by a platy or streaky "flow" structure in thin section. Such microbreccias, which occur as fault fillings and in wider zones of intense deformation (Section 9.14), are called *mylonites* [Lapworth (1885); Waters and Campbell (1935); Christie, (1960)]. Some rocks that look like mylonites in the field are seen in thin section to be highly recrystallized, so that it is not clear whether or not they were ever microbreccias. The original meaning of the term mylonite can be broadened to include any fine-grained metamorphic rock with well-developed "flow" structure [Theodore and Christie, (1969)] or the special term *blastomylonite* can be used for such rocks. Blastomylonites have not necessarily suffered any of the brittle fragmentation or "milling" envisioned for ordinary microbreccias. Their fine-grain size and distinctive microstructure may be due entirely to ductile deformation accompanied by recrystallization [Bell and Etheridge (1973)]. This subject is discussed more fully in Section 9.14. Yet another kind of fault filling, present in thin films on some large faults, is a glassy material called *pseudotachylyte*, formed by melting as a consequence of frictional heating [Francis (1972); McKenzie and Brune (1972)].

7.3.3 Slickensides and Drag Structures

A very common and diagnostic feature of fault planes are smoothed or polished surfaces of easy parting known as *slickensides*. Slickensides may be featureless, but more commonly they display a prominent parallel ribbing or striation of the type shown in Figure 7.12. The striations are believed to be parallel to the direction of relative movement during their formation.

The origin of slickenside striations is not clear. Some striae may be grooves scored on one side of a fault by hard particles drawn over it by the other side. In experiments, however we have observed apparently continuous striations that are longer than the known displacement, so that a different mechanism must be operative. Most striations are thought to be defined by as yet unknown types of lineations in the fine-grained material along fault planes. Features that resemble ordinary slickenside striations, and that in many cases contribute to the ribbed appearance of fault surfaces, are coarse fibrous grains of quartz or other minerals that appear to grow with their long axes parallel to the prevailing direction of fault displacement [Durney and Ramsay (1973)]. An example of quartz ribbing associated with and parallel to ordinary slickenside striations is shown in Figure 7.12. Quartz grains of this type are thought to grow from solutions into dilating cracks at steps on slickenside surfaces. They are valuable in the study of displacement history of faults because curved crystal fibers can preserve a record of changes in the instantaneous direction of fault displacement, as discussed for related "pressure shadow" structures by Choukronne (1971), Elliott (1972), Wickham (1973), and Durney and Ramsay (1973). By contrast, ordinary slickenside striations may often be erased and overprinted if changes occur in the direction of fault displacement so that ordinary striations may record only the latest uniform displacement.

Slickensides with striations usually contain small steps facing in one direction and oriented more or less normal to the striations (Figs. 7.12, 7.13). Prior

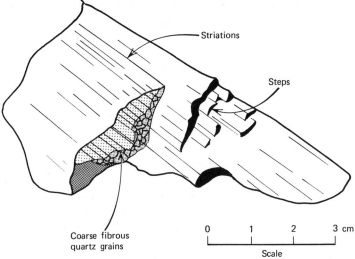

FIGURE 7.12   Fragment of a slickenside in shale from Glenmont, New York. Characteristic steps and striations are accompanied by a quartz ribbing defined by coarse fibrous crystals with strong preferred orientation of c axes parallel to the direction of fault displacement. (Photo by D. Graham.)

to 1958 it was considered axiomatic that the steps faced in the direction of movement of the opposite block. But Paterson (1958), Norris and Barron (1969), and Gay (1970) have shown experimentally that step structures can also form so that the risers of the steps would appear to oppose the movement

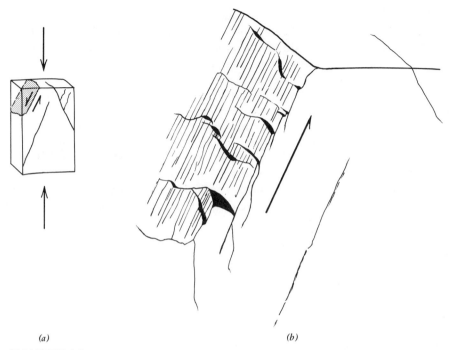

(a)   (b)

FIGURE 7.13   (a) Experimentally deformed block of Solenhofen limestone about 1 in. long, showing orientation of σ_1 (arrows) and shear fractures. (b) Detail of a fault surface [shaded in (a)], showing steps facing opposite to direction of movement of overlying block. (Unpublished experiment by Means.)

of the opposite block. Another experimental example of this type is shown in Figure 7.13, and a natural example has been described by Tjia (1964). Durney and Ramsay (1973) suggest that steps associated with striations defined by fibrous crystals should face in the direction assumed prior to 1958. Evidently, there is no general rule about the kinematic significance of the steps although they may still reveal the sense of displacement with careful microscope study of individual examples.

Another potential indicator of the sense of displacement on faults is the curvature of layering adjacent to faults, known as *fault drag*. Curvature in the same sense as the displacement is known as *normal drag*. Curvature of opposite sense to the displacement is known as *reverse drag*. Unfortunately, both kinds of drag are common, particularly in areas of flat-laying sedimentary rocks, and they may even be found together as in Figure 7.14a. There is usually a tendency in any given region for drag of one sort or the other to predominate, however. In the Colorado Plateau, for example, normal faults are consistently associated with broad reverse drag flexures on the hanging wall side of the fault [Hamblin (1965)]. In this area, and many like it, the presence of a belt of downward flexing adjacent to a fault is an indicator of relative upward displacement of the other side of the fault. In other areas, particularly those characterized by reverse faults, normal drag is more common.

Use of the term *drag* for flexures associated with faults is well established but misleading because it suggests that faulting is usually initiated first and that folding then occurs adjacent to the fault as one block is dragged past the other (Fig. 7.14*b*). This is probably not a common situation because it requires brittle behavior along the fault synchronous with ductile behavior adjacent to it. Normal drag is more likely to express an early history of limited ductile deformation followed by brittle fracture as shown in Figure 7.14*c*. Reverse drag is clearly independent of true drag effects and is not understood. A very speculative possibility is that reverse drag results from movement on curved fault surfaces (Fig. 7.14*d*).

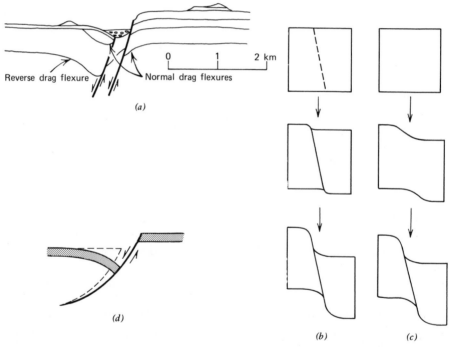

FIGURE 7.14 Drag structures associated with faults. (*a*) Normal and reverse drag flexures accompanying the Hurricane fault, Utah. [After Hamblin (1965), with permission of the Geological Society of America.] (*b*) Faulting accompanied by the development of normal drag, a less common origin for drag. (*c*) Faulting that follows development of a flexure, a more common origin for normal drag. (*d*) Reverse drag due to curvature of fault plane, very speculative. [After Hamblin (1965), with permission of the Geological Society of America.]

7.3.4 **Determination of Fault Displacement**

Complete determination of fault displacement requires knowledge of the positions on either side of the fault of two originally contiguous points. Such points may be defined by the intersection of the fault plane and a fold hinge line, or some other linear marker (e.g., the junction of a dike and a distinctive stratigraphic horizon). Where the offset linear marker is defined by the intersection of two planar structures, the planar structures must *both* predate the faulting. Offset contact lines on geological maps are generally not suitable linear markers because they are defined by the intersection of a geological plane that predates the fault and a topographic surface that postdates the

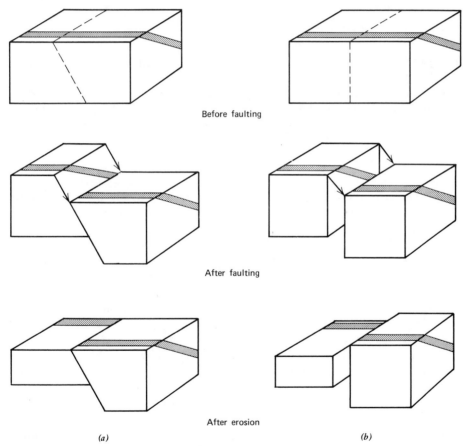

Before faulting

After faulting

After erosion

(a)   (b)

FIGURE 7.15   Examples of offset contacts on erosion surfaces that do not reflect true displacement on faults. (a) Offset parallel to strike of fault developing by purely dip-slip displacement. (b) Dextral offset developing by oblique slip with a sinistral component.

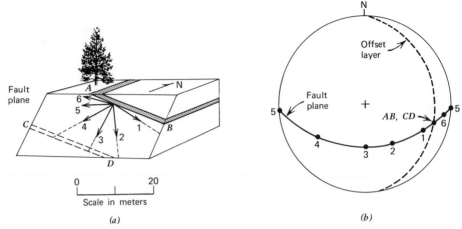

FIGURE 7.16 (a) Football block of a fault showing trace of offset layer on footwall (AB) and hanging wall (CD). Arrows show possible directions for the net slip (of the hanging wall with respect to the footwall). These correspond to the following kinds of faulting: (1) and (2) sinistral-normal, (3) pure normal, (4) dextral-normal, (5) pure dextral strike slip, and (6) dextral-reverse. (b) Lower hemisphere stereographic projection corresponding to (a). The dashed great circle represents *both* halves of the offset layer.

fault. The offset shown by such contacts may be entirely misleading (Fig. 7.15). The same may be true of offsets shown in cross sections, or observed in outcrop surfaces of any orientation.

To distinguish the offset of planar markers from offset of linear markers the former are referred to as *separations.* The separation across a fault can be measured in any direction in the fault plane. The offset layer in Figure 7.16 shows a separation of about 17 meters parallel to the dip of the fault and a separation of about 35 meters parallel to the strike of the fault. The smallest net slip possible on this fault would be about 15 meters, measured perpendicular to the trace of the marker layer on the fault plane. Where only a single planar marker is offset by a fault, all that can be established about the net slip is a lower limit on its magnitude, and a 180° range in its possible pitches on the fault plane.

Similarly, a lower limit can be placed on the displacement of a thrust fault where thrusting has brought older rocks over younger ones. Figure 7.17 shows a case like this where the younger rocks are exposed in an erosional window through the overthrust plate and where an outlier of the overthrust plate occurs as a klippe. Here the distance between klippe and fenster places a minimum limit on the displacement of the fault. Note, however, that the klippe-to-fenster method of putting a limit on thrust displacement breaks down entirely if rocks beneath the thrust are older than those above it, or if stratigraphic inversion occurred prior to thrusting (e.g., by recumbent fold-ing), or if a line drawn from fenster to klippe is not at least approximately parallel to the direction of displacement.

An unwarranted assumption commonly made in the interpretation of large, gently dipping faults is that displacement must have occurred in a direction approximately normal to the outcropping trace of the fault (e.g., E-W dis-placement in Fig. 7.17). Displacement may, in fact, have occurred in any direction within the fault plane and the correct direction should be dem-onstrated by other evidence.

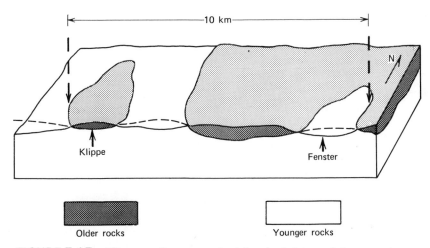

FIGURE 7.17   Klippe-to-fenster method for obtaining a minimum estimate of displacement on a thrust fault. The minimum displacement for the example sketched is 10 km, subject to qualifications mentioned in text.

7.3.5 Fault Terminations

Although it is often satisfactory to think of faults as if they had indefinite extent, they must of course have termini or edges somewhere across which the displacement drops to zero. A corollary of this is that faulting can only pro-ceed if there is some attendant deformation in the surrounding material. Consider an incipient fault, or the active segment of an established fault, that is completely enclosed within a block of rock (Fig. 7.18a). Geometric problems arise around the edges of such a segment. There is a space problem along parts of the boundary like AB, where rocks of the footwall are moving upward into a region occupied by rock that is not moving. There is a continuity problem along parts of the boundary like CD, where rock above and below the fault is moving but adjacent rock outside the edge of the fault is not moving.

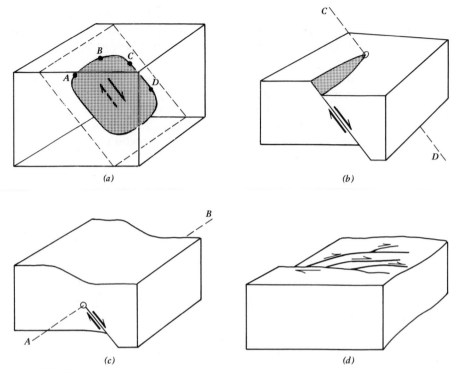

FIGURE 7.18   Accommodation strains near fault terminations. (*a*) A fault plane
(dashed) and slipped region (shaded). Displacements relative to edges of block
shown by arrows. Note similarities of boundary segments *AB* and *CD* to edge and
screw dislocations, respectively. (*b*) and (*c*) Detailed sketches of small volumes along
AB and *CD*. (*d*) *Splay faults* branching from main fault near its terminus;
accommodation strains spread over a large volume of rock.

These geometric problems near fault margins can be met or "accommodated" in various ways, depending on the ratio of fault length to fault displacement. If displacement is very small relative to the dimensions of the
active segment, space and continuity problems can be accommodated by
gradual reduction in displacement toward the fault margins and suitably distributed elastic strains (Fig. 7.18*b, c*). Where displacement is somewhat bigger
relative to fault dimensions, accommodation can still be by elastic strains, as
long as they are distributed through a large volume of rock, as, for example,
in the case of faults that branch repeatedly near their terminations (Fig.
7.18*d*). Where displacement is still larger relative to fault length, fault movement is accompanied by flow or subsidiary faulting or volume change in
transverse structures like those shown in Figure 7.19. Notice that in situations
like these it is not immediately obvious whether the transverse structures

developed to accommodate movement on the strike slip fault or whether the strike slip fault developed to accommodate displacements on the transverse structures. Figure 7.19 also illustrates the geometric similarity of oceanic transform faults (Section 10.2.3) and fault patterns on a smaller scale known in continental rocks.

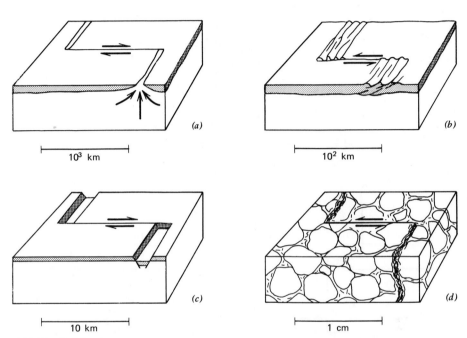

FIGURE 7.19   Faults associated with transverse structures at their termini. (a) Ridge-ridge transform fault. (b) Fault terminating in shallow fold and thrust belts (cf. *tear faults* of the Jura and elsewhere). (c) Fault terminating in block-faulted regions [cf. the Garlock fault in California as interpreted by Davis and Burchfiel (1973)]. (d) Microfault terminating in stylolitic seams, from which coarse white grains are removed in solution.

7.3.6 Shearing Resistance of Dry Rock

The *shearing resistance** of a rock is the shear stress on a potential fault plane that is just sufficient to initiate fault movement. In this section we describe how the shearing resistance of dry rock can be measured and how it is found to be related to normal stress on the fault plane.

*The shearing resistance can also be called the "shear strength," but this runs counter to the usage of some authorities [e.g., Jaeger (1962, p. 76)] who reserve the latter term for the constant c in Equation 7.1.

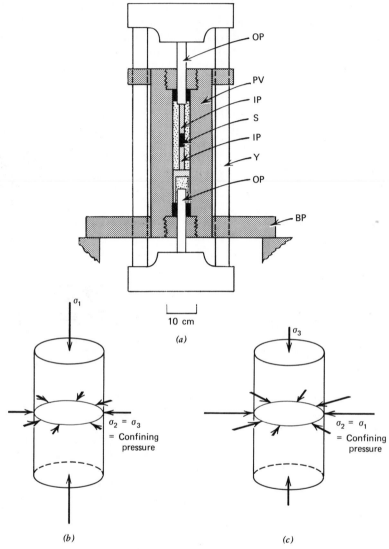

10 cm

(a)

(b)

(c)

FIGURE 7.20 *(a)* An apparatus for triaxial testing of rocks [after Paterson (1970a) with simplifications], showing outer pistons (OP), inner pistons (IP), specimen (S), yoke (Y), and pressure vessel (PV). All shaded parts are fixed relative to main base plate (BP). Yoke and other white parts are driven downward or upward by jacks (not shown) in compression and extension tests, respectively. Dotted volume is occupied by confining pressure fluid, or by confining pressure fluid and furnace for tests at high temperature. *(b)* and *(c)* Assumed orientations of principal stresses in rock cylinder during triaxial compression and extension tests, respectively. [*(a)* used with permission of the Pergamon Press.]

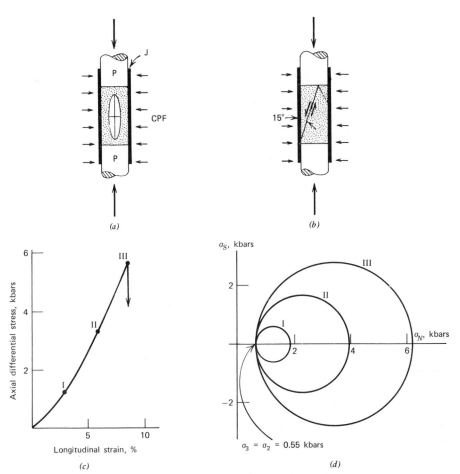

FIGURE 7.21   Experimental faulting of sandstone. (*a*) Cylindrical sample (stippled) of Potsdam sandstone, loaded axially by steel pistons (P) and laterally by confining pressure fluid (CPF). Fluid is kept out of rock pores by rubber jacket (J). Ellipse represents state of stress assumed in center of specimen at about point I on stress-strain curve (*c*). (*b*) Appearance of specimen just after faulting occurs (point III on stress-strain curve). (*c*) Stress-strain record of the experiment, showing axial differential stress ($\sigma_1 - \sigma_3$) plotted against percentage longitudinal strain. (*d*) Mohr circles representing the state of stress in the specimen at I, II, and III. The diameter of circle III is equal to the *ultimate strength* of the sandstone.

Triaxial Testing.   Figure 7.20*a* shows essential features of an apparatus used to measure the strength of rocks under geologically realistic stress conditions. Such conditions are those in which all *three* principal stresses have nonzero magnitudes, and the tests are accordingly called *triaxial tests*. A *triaxial compression test* is one in which a rock cylinder is shortened parallel to its length (Fig.

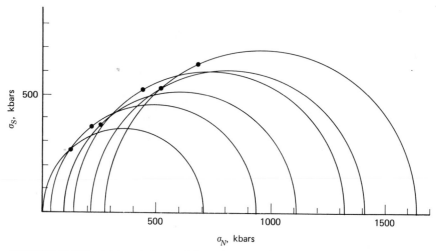

FIGURE 7.22   Mohr circles at failure for Wombeyan marble tested at six confining pressures, ranging from zero to 270 bars, room temperature. The dot on each circle corresponds to the plane on which faulting occurred. [Data from Paterson (1958).]

7.20*b*), and a *triaxial extension test* is one in which the cylinder is extended parallel to its length (Fig. 7.20*c*). A typical triaxial compression test is described below and illustrated in Figure 7.21.

A cylindrical rock sample, perhaps 2 cm long and 1 cm in diameter is enclosed in an impermeable jacket, mounted between steel inner pistons, and immersed in a fluid inside a strong pressure vessel. The fluid is then pumped up to some desired *confining pressure.* At this stage in the experiment the state of stress in the specimen is hydrostatic with $\sigma_1 = \sigma_2 = \sigma_3 =$ the confining pressure. The sample is then gradually shortened by driving the yoked outer piston assembly downward relative to the pressure vessel (Fig. 7.20*a*). This drives the inner pistons together and establishes a state of stress in the sample that is nonhydrostatic (Fig. 7.21*a*) with $\sigma_1 > \sigma_2 = \sigma_3 =$ the confining pressure.* The confining pressure stays constant because the yoked arrangement and equal diameters of the two outer pistons ensures that the volume displaced by inward movement of the upper piston is exactly compensated by outward movement of the lower piston [Griggs (1936)].

As the inner pistons are driven together, σ_2 and σ_3 remain equal to the confining pressure, but σ_1 increases. The difference $(\sigma_1 - \sigma_3)$ is called the

---

*The *symmetry* of stress is axial and for this reason the type of triaxial test described here is also called an *axial compression test.* In this usage, a triaxial test would be one in which all three principal stresses have *different* values.

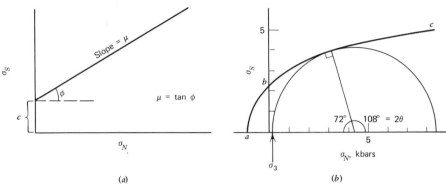

FIGURE 7.23 (*a*) Straight solid line is locus of points at which faulting is initiated, according to the Coulomb criterion. The "angle of internal friction" is ϕ. (*b*) Curve *abc* is the established Mohr envelope for a given rock type. The fault angle and ultimate strength in a new compression test can be predicted by constructing the largest Mohr circle that fits beneath the envelope and passes through the appropriate value of σ_3 (= the confining pressure). For the example illustrated, confining pressure = 200 bars, fault angle = 36°, and ultimate strength = 8.3 kbars.

axial differential stress or, commonly, just the *differential stress*. The stress history in the sample can be represented by a series of Mohr circles of increasing diameter $(\sigma_1 - \sigma_3)$, all passing through the fixed value of σ_3 as shown in Figure 7.21*d*. The relation between stress and strain is usually represented by a plot known as a *stress-strain diagram* that shows the relation between axial differential stress and percentage longitudinal shortening (Fig. 7.21*c*). As the axial differential stress increases, the shearing stress on all potential fault planes in the sample increases. Finally, the shearing resistance on one of these planes is exceeded and faulting occurs. This happens, in fully brittle materials, at the maximum value of $(\sigma_1 - \sigma_3)$ achieved during the test, which is accordingly called the *ultimate strength*. Once the ultimate strength is known for a given confining pressure, and the inclination of the fault to the specimen axis is measured, the shearing resistance can be calculated by Equation 1.7 or determined graphically with a Mohr diagram (Section 1.2.2). The ultimate strength, fault angle, and confining pressure in numerous tests on rocks have been compiled by Handin (1966).

The dependence of shearing resistance on normal stress is demonstrated if tests are run on one rock type at several different confining pressures. Figure 7.22 shows a Mohr diagram for six different tests on Wombeyan marble performed by Paterson (1958). Each circle represents the stress conditions at failure in one of the tests. The dot on each circle represents σ_N and σ_S on the plane on which faulting occurred in each test. Note that the shearing resis-

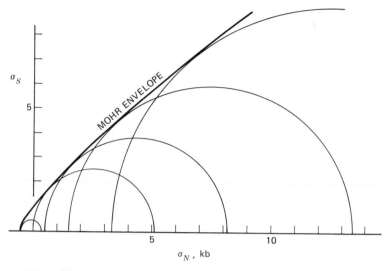

FIGURE 7.24　Mohr envelope defined by tests on Frederick diabase by Brace (1964). Note that another half of the envelope could be plotted below the axis, but this is omitted because it is symmetrically equivalent to the part shown. [From W.F. Brace (1964), "Brittle Fracture of Rocks." Fig. 22, in W.R. Judd, Ed., *State of Stress in the Earth's Crust*, New York, American Elsevier.]

tance increases with normal stress and that a smooth curve could be drawn on Figure 7.22 that passes close to each of the points. Such a curve represents, approximately, the combinations of σ_N and σ_S at which faulting is initiated in Wombeyan marble under test conditions. Combinations of σ_N and σ_S plotting below the curve are states of stress that can be supported by Wombeyan marble without faulting.

Shear Fracture Criteria　There have been several attempts to describe in general terms the characteristics of a state of stress necessary to initiate shear fracture. The simplest "criterion" of this kind was suggested by Coulomb (1773). He proposed that shear fracture would occur when the shear stress on a potential fault plane reached a critical value given by

$$\sigma_S = c + \mu\sigma_N \tag{7.1}$$

where c is a constant known as the *cohesion* or the *shear strength* [Jaeger (1962, p. 76)] and μ is another constant known as the *coefficient of internal friction*.

This criterion was based on an assumption that shear fracture in solids involves breaking some kind of cohesive bonds between particles (giving the c term), together with frictional sliding (giving the term proportional to σ_N). This physical interpretation is not entirely satisfactory for rock [see Handin (1969); Byerlee (1969)] but Equation 7.1 nevertheless provides a good fit to much experimental data. It predicts that points such as those in Figure 7.22 should lie on a straight line with slope μ and intersecting the ordinate at c as in Figure 7.23a. The points on Figure 7.22 can be seen to lie fairly close to a straight line and this is characteristic of many rocks tested at moderate confining pressures. At higher confining pressures where ductile behavior begins, or at very low values of σ_3 where other special effects enter, plots of σ_N against σ_S at failure are characteristically somewhat concave toward the σ_N axis. A tendency in this direction is seen in Figure 7.22.

A second suggestion regarding stress conditions required for fracture was proposed by Mohr (1900). According to this hypothesis σ_S and σ_N are related in general by a nonlinear function that is different for each material. The function in each case represents the shape of the envelope to a series of Mohr circles at failure for that rock type. Such an envelope, called the *Mohr envelope,* is drawn in Figure 7.24 for some tests on diabase by Brace (1964). Note that the equation for a linear Mohr envelope would have the form of Equation 7.1. Once a Mohr envelope is established by a series of tests on a given rock type, this curve can be used to predict both the ultimate strength and the fault angle in tests at other confining pressures (Fig. 7.23b). If one constructs, in Figure 7.22, an envelope to the Mohr circles (representing conditions for failure predicted by Mohr's hypothesis) and the best-fitting straight line through the dots (representing the Coulomb criterion based on the same data), one can see that these two curves are very similar.

A physically more realistic theory than the Coulomb theory derives from a hypothesis of Griffith (1924) to explain the anomalous weakness of glass in tension. It is proposed that even apparently homogeneous materials like glass contain numerous tiny flaws, or *Griffith cracks,* some of which become enlarged and propagate under the influence of applied stress. The macroscopic weakness of brittle solids is ascribed to high stress concentrations existing around the edges of the cracks. Griffith cracks in rocks may be original or induced openings along grain boundaries or within grains [Brace (1964)]. Figure 7.25 shows cracks along a grain boundary in granite as revealed by scanning electron microscopy. McClintock and Walsh (1962) have derived an expression for the shear strength of rock based on the Griffith model and including the effect of closure of cracks at high pressures. For compressive values of σ_N, the equation is

$$\sigma_S = 2\,T_0 + S_0\sigma_N \tag{7.2}$$

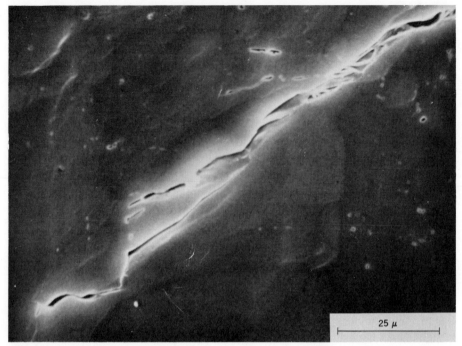

FIGURE 7.25   Scanning electron micrograph of cracks along grain boundary between two feldspar grains in Westerly granite. Scale bar is 25 microns. (From Brace et al., Science, 178, p. 162–64. Copyright 1972 by the American Association for the Advancement of Science.)

where T_0 is the uniaxial tensile strength* of the rock and S_0 is a coefficient of friction. This equation has the same form as the original Coulomb equation, indicating that shearing resistance depends on two material constants and is linearly proportional to normal stress on the fault plane. There remain serious questions, however, about the physical basis for Equation 7.2 [Brace and Byerlee (1967)] because it gives the shearing resistance of a simplified mathematical model of a rock and not the shearing resistance of a real rock, in which the microscopic behavior is determined by a complex array of anisotropic crystals and their grain boundaries [Jaeger (1971, p. 103)]. The reader is referred to Brace (1964), Handin (1969), Jaeger and Cook (1969), Lajtai (1971), and Jaeger (1971) for further discussion of fracture criteria.

*The *uniaxial tensile strength* of a rock is the tensile stress at which a rock cylinder breaks when it is stretched parallel to its length at zero confining pressure. It is represented in a Mohr diagram by the point at which the Mohr envelope crosses the σ_N axis.

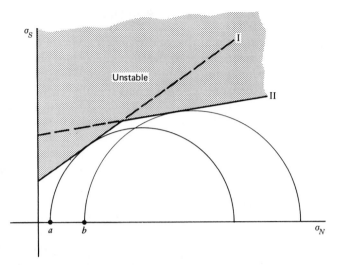

FIGURE 7.26   Mohr diagram illustrating change in mode
of failure with increasing depth in dry rocks. σ_3 is assumed
to be vertical and equal to the lithostatic pressure. Curve I is
plot of shearing resistance versus normal stress for failure
by shear fracture. Curve II is similar plot for failure by
crystal slip. The depth for the transition from faulting to
flow corresponds to σ_3 values between a and b.

Maximum Depth for Faulting in Dry Rocks.   The above dependence of shearing resistance on normal stress leads to the prediction that the resistance of
dry rock to faulting should be greater at greater depths. This leads to a
maximum depth for faulting if the shear strength of a given rock type becomes so high that deformation can occur by other mechanisms at lower
stresses. Figure 7.26 illustrates a case like this, where the other mechanism is
taken to be crystal slip. The critical stress conditions required for faulting are
represented by curve I, with a slope of 0.7, a common value for the coefficient
of internal friction. The stress conditions required for flow by crystal slip are
represented by curve II with a much flatter slope, in accordance with the
known insensitivity to normal stress of the critical stress for slip in single
crystals of at least some minerals [see Paterson (1970b)]. At a depth where the
value of σ_3 is a, the Mohr circle at failure is tangent to curve I and the rock
faults. At a greater depth, where σ_3 is b, the Mohr circle is tangent to curve II
when stress conditions reach the boundary of the unstable region and the rock
yields by crystal slip. The maximum depth for faulting in this rock would be
between the depths at which σ_N is a and b.

7.3.7 Shearing Resistance of Rock with Pore Fluid

Rocks deformed under most natural conditions are expected to contain pore fluid at pressure. Its influence on both the mechanical and chemical behavior of rock systems is extremely important. We consider here the influence of pore fluid pressure on shearing resistance and the ability of high pore pressures to promote faulting under conditions where it would not occur in dry rocks. The "pores" in rocks include features like vesicles in volcanic rocks and intergranular spaces in sedimentary rocks that are primary openings; and microcracks, joints, and faults that are secondary openings induced by deformation. Scanning electron microscope pictures of pores in sedimentary rocks and granite are given by Timur et al. (1971).

The "normal" or *hydrostatic pressure* to be expected in pore water at any depth is the pressure in a column of water extending from this depth to the surface. Thus, the hydrostatic pore pressure at any depth will be about 0.4 times the lithostatic or rock pressure, if we take the mean density of the rock column to be 2.5 g/cc. Pore pressures greater than hydrostatic can be generated by several mechanisms, including rapid compaction of sediments by burial or tectonic deformation and dehydration of mineral assemblages during metamorphism [see Bredehoeft and Hanshaw (1968); Hanshaw and Bredehoeft (1968)]. Ratios of pore pressure to lithostatic pressure in excess of 0.8 have been measured in oil fields, and ratios approaching 1.0 are considered possible.

Effective Stress and Shearing Resistance. Terzaghi (1923) proposed that the shearing resistance of saturated soils is given by a simple modification of the Coulomb criterion,

$$\sigma_S = c + \mu \left(\sigma_N - p \right) \tag{7.3}$$

where p is the pore pressure and the difference $(\sigma_N - p)$ is defined as the *effective normal stress* (σ_N'). An equation of this form has since been found to fit experimental data for rocks as well as soils [Hubbert and Rubey (1959); Handin et al. (1963); Handin (1968)]. The term $(\sigma_N - p)$ is called the effective normal stress because it is this function, rather than the total stress (σ_N), that is found experimentally to be "effective" in controlling the shearing resistance. A rock will thus exhibit essentially the same shearing resistance when $\sigma_N = 1$ kbar, $p = 0$ and when $\sigma_N = 2$ kbar, $p = 1$ kbar, because the effective normal stress is 1 kbar in both cases.

High pore pressure can be thought of as having a lubricating effect on faults in the sense that it reduces frictional resistance to movement, but it does not have a lubricating effect in the sense that μ is changed by pore pressure. The coefficient μ is, in fact, reasonably constant for a wide range of pore pressures [Handin et al. (1963)]. The effect represented here is one in which increase in the fluid pressure is associated with a change in the state of stress

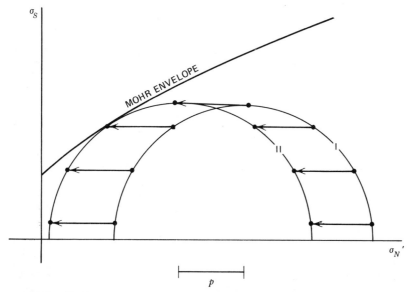

FIGURE 7.27   Mohr diagram with effective normal stress plotted against shear stress. Circle I represents state of stress in stable rock at zero pore pressure. Circle II represents state of stress in unstable rock if total stresses are the same as in I but if pore pressure is raised by an amount p, represented by bar.

throughout the rock that somehow reduces the shear stress required for faulting.

Equation 7.3, like the Coulomb equation for dry rocks (7.1) provides a reasonable fit to much experimental data, but a complete understanding of the physical and chemical mechanisms involved in faulting of wet rocks is lacking. Equation 7.3 relates the shearing resistance entirely to the *pressure* of the pore fluid when chemical properties of the fluid may also be important, for example, in controlling the rate of stress corrosion at the tips of cracks [Scholz (1972)]. Furthermore, the state of stress is itself complex in porous materials, and there is as yet no general agreement on how best to write the effective normal stress, whether as a simple difference between σ_N and p as proposed by Terzaghi, or as a more complex function of σ_N, p, and material and geometric properties of the rock, as proposed by others [see Nur and Byerlee (1971); Garg and Nur (1973)].

Pore Pressure and Faulting.   The way in which pore pressure promotes faulting can be illustrated by a Mohr diagram on which effective normal stress rather than total normal stress is plotted along the horizontal axis (Fig. 7.27). Circle I represents the state of stress in a rock mass when the pore pressure is

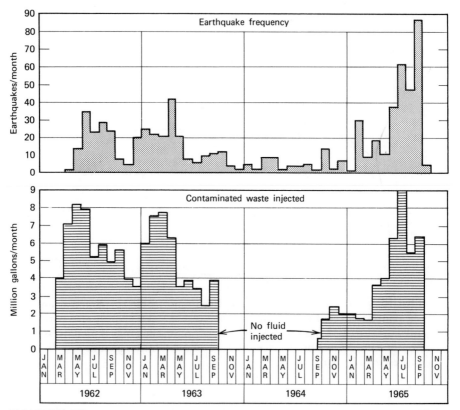

FIGURE 7.28   Correlation between local earthquake frequency and rate of fluid injection observed near Denver, Colorado. [From Evans (1966), with permission of the Rocky Mountain Association of Geologists.]

zero, so that the effective normal stress equals the total normal stress. The rock is stable under these conditions. Now suppose that the pore pressure is increased gradually to a level p while the total normal stress is held constant. This decreases the effective normal stress on all planes by an amount p but does not alter the shear stresses. The stress circle is shifted to the left in Figure 7.27 from its original position I to a new position II. If p is large enough, the circle in position II will intersect the Mohr envelope, as shown, and faulting will occur. The pressure of the pore fluid allows faulting to occur *even though the shear stresses present would not have been adequate for faulting in dry rock,* or in wet rock at lower pore pressures.

Recent research on earthquake control is much concerned with pore pressure effects because pore pressure is the one critical variable that can be manipulated to some extent by man. The basic idea is that suitable local increase in pore pressure, brought about by pumping fluids into drill holes, could lower the shearing resistance of rocks sufficiently to trigger local faulting and small earthquakes. Many such small earthquakes might gradually release the stored energy that would otherwise accumulate for a single large and catastrophic earthquake. A correlation between the rate of fluid injection and earthquake frequency (Fig. 7.28) has been observed near Denver, Colorado [Evans (1966); Healy et al. (1968)], and more detailed studies of the relation between pore pressures, rock stresses, and faulting frequency and orientation are being carried out near Rangely, Colorado [Raleigh et al. (1972)]. These studies tend to confirm that small earthquakes can be triggered or suppressed by judicious control of pore pressure, but it is not yet clear what effect such control of small earthquakes will have on the occurrence of large ones.

High pore pressures are considered to play a vital role in some examples of low-angle thrust faulting, as first discussed by Hubbert and Rubey (1959) and Rubey and Hubbert (1959). Thrust sheets 30–100 km wide pose serious mechanical problems if they moved dry, because friction along their bases would seem to require either impossibly high shearing stresses in the sheets (if they are pushed from behind) or unreasonably steep slopes (if they slide downhill under the influence of gravity). Hubbert and Rubey point out how pore pressures higher than normal could allow thrust sheets to be pushed more easily or to slide down slopes with dips as low as a degree or two. There has been much subsequent discussion and refinement of this idea [see Raleigh and Griggs (1963); Davis (1965); Heard and Rubey (1966); Hsu (1969); Forristall (1972); Roberts (1972)].

7.3.8 Maximum Depth for Extension Joints

Extension joints form perpendicular to σ_3 under several different conditions: the total value of σ_3 may be tensile, in which case the joints are true *tension joints;* or the total value of σ_3 may be compressive but tensile stresses exist locally on a granular scale; or the total value of σ_3 may be compressive but pore pressure is sufficient to make the principal effective stress $(\sigma_3 - p)$ tensile.

In principal, there is a limiting depth for true tension joints that can be estimated from the Mohr envelope for a given rock type as shown in Figure 7.29. We assume that σ_1 is vertical and due entirely to the weight of the overlying rocks. The state of stress just prior to tension jointing is represented by a circle that passes through point A (the uniaxial tensile strength) but is otherwise wholly within the stable region. If we now construct the largest

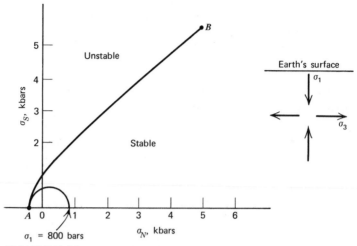

FIGURE 7.29 Mohr diagram illustrating method for estimating maximum depth for tension joints in dry diabase. Mohr envelope from Brace (1964, Fig. 22). σ_1 is assumed vertical. For further explanation see text.

Mohr circle that satisfies these requirements, its σ_1 value will correspond to the maximum depth at which tension jointing can occur. The difficulty with this procedure in practice is that it will tend to give excessive maximum depths

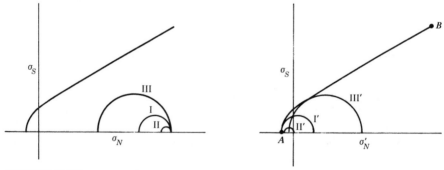

FIGURE 7.30 Mohr diagram showing how the pore pressure and the stress difference ($\sigma_1 - \sigma_3$) determine whether or not extension fractures form at depth. Total normal stresses are plotted in the diagram at left; effective normal stresses plotted at right. In case I-I′ an extension fracture forms. In case II-II′ there are effective tensile stresses but no fracture forms. In case III-III′ there are effective tensile stresses but a shear fracture forms.

because the Mohr envelope is based on short-term laboratory testing and does not express the time-dependent weakening effects expected in natural environments. Thus, Figure 7.29 suggests a maximum depth of 2.7 km for tension joints in dry diabase, whereas Griggs and Handin (1960, p. 351) suggest that macroscopic tensile stresses are inconceivable, even in stronger rocks, at depths below a few hundreds of meters.

Extension joints associated with compressive values of the total stress σ_3 can form to indefinitely greater depths than true tension joints as long as pore pressure is high enough and the stress difference $(\sigma_1 - \sigma_3)$ is in the right range [Secor (1965, 1969)]. This is illustrated in Figure 7.30 in which circles I, II, and III represent the total stresses and circles I', II', and III' represent the corresponding effective stresses. In the situation represented by I and I', the pore pressure $p = 0.8\,\sigma_1$ and an extension fracture develops because the circle I' touches the Mohr envelope at point A only. In situation II', no fracture forms even at $p = 1.0\,\sigma_1$ because the difference $(\sigma_1 - \sigma_3)$ is less than the tensile strength A. In situation III' the stress difference $(\sigma_1 - \sigma_3)$ is greater than the tensile strength but a shear fracture forms instead of an extension fracture because the stress difference $(\sigma_1 - \sigma_3)$ is so big that the circle impinges on the Mohr envelope at a point between A and B.

7.3.9 Shear Fracture Orientation Relative To Principal Stress Directions

In triaxial tests shear fractures form at angles less than 45° to the direction of σ_1. Angles of about 30° are particularly common [see compilation in Handin (1966, p. 244)]. Where $\sigma_2 = \sigma_3$ (as in the usual axial compression test) there are an infinite number of equally favored planes for faulting, all equally inclined to σ_1. Where $\sigma_2 \neq \sigma_3$ there are just two equally favored planes, one either side of σ_1, which intersect in σ_2. The direction of maximum shearing stress in both of these planes and, therefore, the direction of displacement, is parallel to the σ_1-σ_3 plane. Where faults or joints develop more or less synchronously in both of the equally favored orientations, the structures are called *conjugate faults* or *conjugate joints*. Figure 7.31 shows conjugate shear fractures, and extension fractures, developed experimentally in a block of Solenhofen limestone.

The Coulomb criterion predicts that shear fractures should form at less than 45° to σ_1, as follows. Figure 7.32 shows the shearing resistance calculated from the Coulomb criterion and the shear stress for all planes parallel to σ_2 at the moment of failure. It can be seen that the plane for which the shear stress equals the shearing resistance must always be inclined to σ_1 at less that 45°, because of the positive slope of the shearing resistance curve and the symmetrical shape of the shear stress curve.

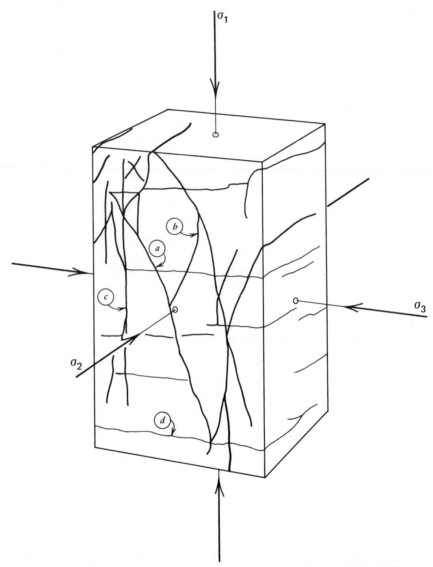

FIGURE 7.31   Fractures formed experimentally in a block of Solenhofen limestone shortened about 1 percent at room temperature, with $\sigma_1 > \sigma_2 > \sigma_3$. Fractures of sets (a) and (b) are conjugate shear fractures. Fractures of set (c) are extension fractures formed during loading. Fractures of set (d) are extension fractures formed during unloading, when the position of σ_3 becomes vertical in the apparatus used. (Means, unpublished experiment.)

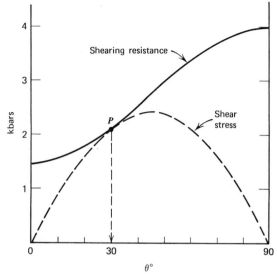
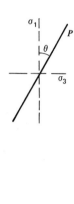

FIGURE 7.32   Shearing resistance and shear stress at the moment before failure in Solenhofen limestone, for planes parallel to σ_2 and inclined at $\theta°$ to σ_1. The Coulomb criterion, from which the shearing resistance curve is calculated, predicts faulting on plane P at about 30° to σ_1. Data from Solenhofen limestone test by Handin (1969, p. 5346) as follows: $c = 1.05$ kbars, $\mu = 0.53$, σ_1 and σ_3 at failure 5.5 kbars and 0.75 kbar, respectively. Shear stress and shearing resistance calculated from Equations 1.7 and 7.1, taking into account that θ used here is the complement of θ in Equation 1.7.

The observed orientation of shear fractures can be correlated with the shape of the Mohr envelope, as shown in Figure 7.33. Failure is expected to occur approximately on the planes P for each of the stress circles shown. The angle is less than 45° in each case because the slope of the envelope is positive. Notice that a complete transition is indicated from faults like P_1 inclined at 30° to σ_1, through shears with a dilational component like P_2 inclined at lower angles to σ_1, to extension fractures like P_3 that are parallel to σ_1. The occurrence of shear fractures at very low angles to σ_1 is corroborated by experiments [Griggs and Handin (1960); Brace (1964)] and by field evidence [Muehlberger (1961)].

Since the surface of the earth must be a plane across which there is no shearing stress, it can be assumed that one of the principal directions of stress will be approximately vertical at the surface (in areas of low relief) and for some distance beneath it. This consideration led Anderson (1951) to propose

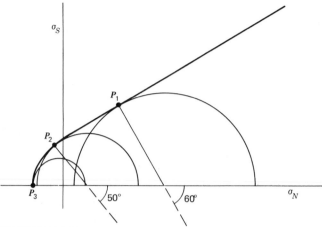

FIGURE 7.33   Fracture orientation correlated with shape of
the Mohr envelope. Fractures parallel to planes P_1, P_2, and P_3
form, respectively, at 30°, 25°, and 0° to the direction of σ_1.

the threefold classification of faults shown in Figure 7.34. Normal, reverse,
and transcurrent faults are considered to be the commonest types of faults
and to represent the response of isotropic rock bodies to each of the three
possible stress orientations in which one of the principal directions is vertical.

Most rock bodies will be anisotropic with respect to one or both of the
constants in Equation 7.1, especially when considered on a scale large enough
to include joints or contacts between rocks of different lithology. In such cases
all the simple relations given above between shear fractures and stress directions can break down. A fault can be inclined at any angle to σ_1 (other than
zero or ninety degrees); it need not parallel σ_2; and the displacement need not
occur parallel to the σ_2-σ_3 plane. An example of possible behavior in anisotropic rock is given in Figure 7.35. A plane of weakness is oriented at 70° to σ_1.
σ_3 is kept constant while σ_1 increases until faulting occurs. The Mohr diagram
shows separate Mohr envelopes for shear fracture parallel to the plane of
weakness and for shear fracture in other orientations. The point P on each
Mohr circle represents the stresses on the plane of weakness. For states of
stress represented by circles I and II the rock is stable, but for circle III P lies
on the appropriate envelope and faulting occurs parallel to the plane of
weakness even though this lies at a very high angle to σ_1. Planes Q and R, at
45° and 30° to σ_1 feel higher shear stresses than P, but these stresses are still
well below the critical stress defined by the upper envelope. Cases of this type
have been investigated theoretically by Jaeger (1960) and experimentally by
Donath (1961) and Lajtai (1969a).

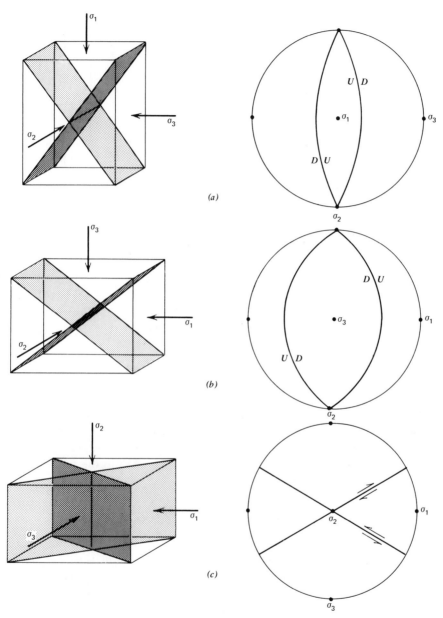

FIGURE 7.34   Fault planes (shaded) and principal stress directions in isotropic rock, with one principal stress vertical. (a) Normal faults with σ_1 vertical. (b) Reverse faults with σ_3 vertical. (c) Transcurrent faults with σ_2 vertical. Lower hemisphere stereographic projections, with the primitive circle representing the horizontal plane.

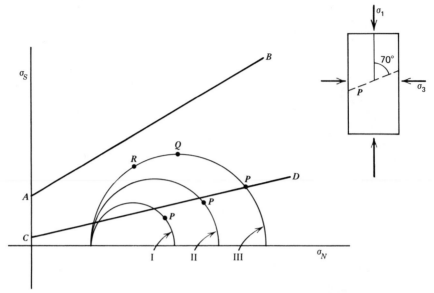

FIGURE 7.35   Faulting in anisotropic rock. P is a plane of weakness at 70° to σ_1. AB is the Mohr envelope for shear failure parallel to planes other than P. CD is the Mohr envelope for P. For further explanation, see text.

Where a preexisting plane of weakness controls the orientation of a fault, and where this orientation is not parallel to any of the principal stress directions, the direction of fault displacement depends on the relative magnitudes of the principal stresses [Bott (1959); McKenzie (1969a)] because the relative magnitudes of these stresses determine which direction in the potential fault plane feels the greatest shear stress.

7.3.10   Principal Stress Orientations From Field Data

Determination of the orientations of σ_1, σ_2, and σ_3 just prior to faulting requires, in general, that one know the orientation of a fault, the direction and sense of initial displacement, and the angle θ between the fault and the direction of σ_1. Since this much information is generally not available, reliable stress interpretations are usually not possible. There are circumstances, however, under which some reasonable guesses can be made. Some of these are reviewed below.

The easiest situation to interpret is the case of conjugate faults in an isotropic-looking rock body. The faults can be regarded as conjugate if there

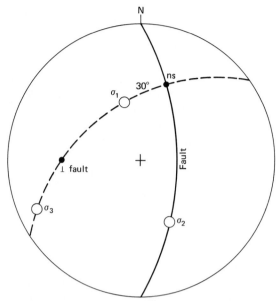

FIGURE 7.36   Determination of principal stress directions in isotropic rock from knowledge of the orientations of the fault and the net slip (ns), and the sense of displacement. The fault represented dips 60 to 90° and has a net slip plunging 30 to 20°. The hanging wall block is assumed to have moved down relative to the footwall. σ_1, σ_3, the pole to the fault, and ns are coplanar. For further explanation see text.

is evidence of movement in opposite senses directed normal to their line of mutual intersection, and if neither set of faults consistently offsets the other. In this case σ_1 can be taken as parallel to the bisector of the "compressional angle" (usually the acute angle) between the two fault orientations. σ_2 is taken parallel to the line of intersection of the two faults.

The more common but less easily interpreted situation is that of a single fault in isotropic rock with a known sense and direction of initial displacement. Here the line lying in the fault plane normal to the displacement is tentatively taken as the direction of σ_2. If the angle between the fault plane and σ_1 is now arbitrarily taken as 30°, for instance, the directions of σ_1 and σ_3 may be determined as shown in Figure 7.36.

In most real rock bodies there may be sufficient anisotropy of one sort or another so that the above method yields erroneous results. About all that can be done with certainty in such cases is to identify certain orientations of the stress ellipsoid that could *not* give rise to the observed displacements, for example, orientations giving the wrong sense of displacement [e.g., see McKenzie (1969a).].

An unreliable kind of fault interpretation is that in which estimates of stress orientation are based entirely on knowledge of the attitude of a fault. Examples would be to say that a low-angle fault striking northeast indicates a NW-SE orientation for σ_1 or that a vertical fault striking NE indicates σ_1 or σ_3 directed E-W. For description of various methods used to measure the present-day orientations and magnitudes of the principal stresses in rocks, the reader is referred to Friedman (1972), Jaeger and Cook (1969, p. 213, 363), and Leeman (1964).

7.3.11 Fault History

In this section we discuss three aspects of fault history: the response to stress of brittle rocks before a fault propagates through them, the nature of sliding once a fault plane has been established, and the changes in the initial state of stress that may occur when movement starts.

Microfracturing and Dilatancy.   When a brittle rock is compressed in the laboratory and shortened at a constant rate until shear fracture occurs, it is found that permanent changes begin well before a through-going fracture forms. Figure 7.37 shows the results of such a test performed on granite at 1 kbar confining pressure and room temperature by Brace et al. (1966). Up to a stress difference of roughly half the fracture strength, the specimen shortens by a fraction of 1 percent and its volume decreases by an even smaller amount, governed by the compressibility of quartz and feldspar. Through this stage of the experiment the original dimensions of the specimen are entirely recoverable on unloading and the behavior is elastic. At stress differences greater than about half the fracture strength, inelastic effects begin to be noticeable. The volume of the specimen no longer decreases with continuing longitudinal shortening but instead *increases* slightly; the specimen is now *dilatant*. The volume increase is correlated with opening of numerous cracks or microfractures on a granular scale. Dilatancy appears to be a common feature of deformation of brittle rocks, but it is not restricted to them. Edmond and Paterson (1972), for example, have observed dilatancy associated with macroscopically ductile behavior in sandstone and marble.

In tests such as the one represented in Figure 7.37 the intensity of microfracturing activity is found to increase as the fracture strength is approached. Microfracturing events become more frequent and also more concentrated spatially near the eventual fault plane [Scholz (1968)]. Such results suggest that microfracturing and dilatancy may also precede fault displacements in

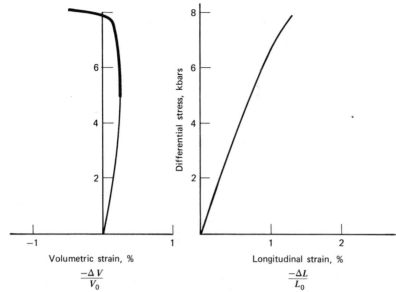

FIGURE 7.37   Volumetric strain and longitudinal strain of a granite
cylinder shortened at 1 kbar confining pressure, room temperature. The
specimen is dilatant along the part of the volumetric strain curve shown by
the heavy line. Positive ΔV corresponds to volume decrease. [Modified
from Brace et al. (1966, Fig. 1b), with permission of the American
Geophysical Union.]

nature, and recent attempts to establish a physical basis for earthquake prediction are based on this idea [Scholz et al. (1973); Whitcomb et al. (1973)].

The rate at which microfractures propagate is presumably an important
factor determining the rate or stress difference at which brittle yielding will
occur on larger scales. Martin (1972) and Scholz (1972) have shown that the
propagation of microfractures in quartz and other silicates is accelerated by
the presence of water, particularly at high water pressures. This effect is
attributed to weakening of the silicate framework at the highly stressed tips of
cracks by a *stress corrosion* process involving rapid hydrolysis of silicon-oxygen
bonds.

The manner in which propagating microfractures link up to give a macroscopic fault is still poorly understood, even in rocks deformed under laboratory conditions. In some tests, for example, extension microfractures are seen
to form early and to link up through suitably oriented grain boundary cracks
to form a fault [Dunn et al. (1973)]. In other tests, however, through-going
faults are thought to predate the associated extension microfractures [Fried-
man et al. (1970)]. For other observations and discussion of this problem, the

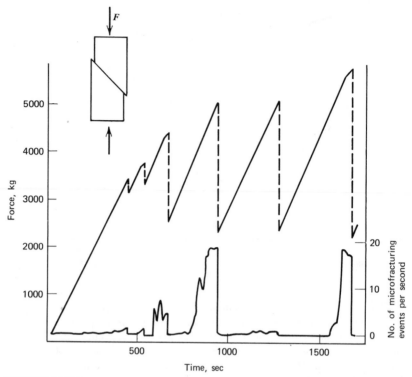

FIGURE 7.38   Force-time and microfracturing activity-time records for a
frictional sliding experiment on granite, at 1 kilobar confining pressure, room
temperature. [Modified from Scholz (1968, Fig. 7), with permission of the
American Geophysical Union.]

reader is referred to Brace and Bombolakis (1963), Hoek (1965), Bombolakis
(1968), Lajtai (1971), and Bombolakis (1972).

Displacement History.   Major faults show total displacements that have accumulated incrementally over active lifespans measured in tens of millions of
years or more. The instantaneous rate of slip at a point can range from zero to
rates comparable to the elastic wave velocity of the surrounding rock. Very
slow aseismic slip, at average rates typically of 1-2 cm/yr, is known as *fault creep*
[see Dickinson and Grantz (1968)]. At different points on a fault the instantaneous rates of slip may be quite different. Thus, an earthquake may emanate from one portion of a fault, where the slip rate is temporarily very high,
while simultaneously another segment of the same fault is not slipping at all or
is only exhibiting slow, aseismic creep.

The observation that slip on natural faults can occur with or without
generating detectable seismic waves is paralleled by the laboratory observation

of Brace and Byerlee (1966) and Byerlee and Brace (1968) that frictional sliding between rock surfaces can occur with or without detectable stress drops. Where sliding occurs in laboratory specimens without stress drops it is called *stable sliding*. Where it occurs with prominent, episodic stress drops the behavior is called *stick-slip*. In stable sliding the slip rate is more or less constant for a given stress difference. In stick-slip behavior the slip rate is alternately very high (during stress drops) and very low (between stress drops). Figure 7.38 shows a typical force-time record for a stick-slip experiment performed on dry granite at 1 kbar confining pressure and room temperature by Scholz (1968). Note that, as in the initial fracturing of granite, there is noticeable microfracturing activity associated with and premonitory of each of the stress-drop episodes during this frictional sliding experiment. Byerlee (1970) suggests that the cause of stick-slip behavior in dry laboratory specimens faulted at room temperature is sudden brittle fracture of locked regions on the fracture surface.

It is now widely believed that the most important cause of earthquakes at very shallow depths (e.g., less than 20 km) is large-scale stick-slip sliding on faults [Brace and Byerlee (1966); Griggs (1972)]. It is accordingly of great interest to determine what controls this kind of fault movement. Preliminary results suggest that temperature, effective confining pressure, pore fluid chemistry, and rock type are among the variables that determine whether stick-slip or stable sliding occurs. High temperature has been found to suppress stick-slip in granite and gabbro specimens [Brace and Byerlee (1970)], and this has been suggested as one possible reason why earthquake foci are restricted to shallow depths on transcurrent faults in California and elsewhere. Stick-slip is also absent at low temperatures in unaltered silicate rocks as long as the effective confining pressure is less than 1 or 2 kbar [Byerlee and Brace (1968, 1972)]. At higher effective confining pressures (where there is also higher effective normal stress on the fault plane) stick-slip behavior appears in these rocks and becomes increasingly violent with increasing pressure. The effects of pore fluid chemistry, as opposed to pore fluid pressure, are not yet clear, but it seems possible that important effects may exist if stress corrosion at crack tips is a factor governing the growth rate of microfractures. Rocks containing appreciable calcite or serpentine or pore space have been found to exhibit stable sliding only, at all confining pressures [Byerlee and Brace (1968)]. Other lithological variables of probable importance are the thickness, mineralogy, and fluid content of the gouge along an active fault [see Riecker (1972); Engelder (1973)].

An example of the detailed displacement history of part of a major fault is provided by recent studies of the San Andreas near Parkfield, California. A few details of activity on this segment of the fault are given here to underscore the idea that major faults move in a very heterogeneous manner and to show how seismological observations can be related to laboratory fracture studies.

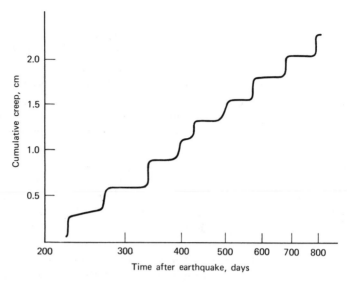

FIGURE 7.39 Cumulative creep at the Carr Ranch, Parkfield, California, following the Parkfield earthquake of June 27, 1966. Episodes of rapid creep (steep segments of curve) occur at predictable intervals. Notice that this creep record begins more than 200 days after the earthquake, so that most of the total creep has already occurred. [After Scholz et al. (1969, Fig. 10), with permission of American Geophysical Union.]

The data and interpretations are from Scholz, et al. (1969) and references cited therein.

An earthquake of magnitude 5.5 occurred near Parkfield on the San Andreas fault on June 27, 1966. The main shock is believed to have resulted from rapid slip of a region in the fault plane that was 40 km long and that extended in depth from a level of about 4 km down to a level of about 10 km. The average displacement over this area is believed to have been 30 cm. If the size of the whole San Andreas fault system is taken as 1000 km long by 20 km deep, then the part that moved during the Parkfield earthquake is roughly 1 percent of the whole fault area.

At the surface, where there was minimal slip during the main Parkfield shock, creep set in immediately following the earthquake, initially at a rapid rate that decayed logarithmically with time. Two years after the earthquake, total surface creep of 18-28 cm had occurred at various points above the initial break. Figure 7.39 shows the cummulative creep curve for one such station for the period 200–800 days after the earthquake. In this and other areas on the San Andreas fault [Tocher (1960)], the creep slip, though aseismic, is clearly episodic. The episodes of rapid creep are regular enough to be predictable (Fig. 7.39).

The main Parkfield shock was followed by a series of aftershocks in addition to the creep episodes mentioned above. Like the main shock, the aftershocks originated between depths of about 4 and 10 km in the fault plane. Because they tended to emanate from areas near the ends of the original break, the aftershocks are attributed to *delayed brittle fracture* in regions of stress concentration that must exist around the edges of a slipped segment. Delayed brittle fracture is familiar from *static fatigue tests* [Martin (1972); Scholz (1972)] in which rock specimens fracture at constant stress if the stress is applied long enough.

The general interpretation suggested for Parkfield by Scholz et al. is that this part of the San Andreas displays three different kinds of behavior at three different levels. Between the surface and a depth of about 4 km slip occurs by stable sliding. Between 4 and 10 km the behavior is stick-slip. Below 10 or 12 km the fault again moves in some stable manner, perhaps involving crystal slip. Laboratory studies suggest that the change in behavior around 4 km may result from suitable increase in effective normal stress on the fault plane and that the change in material properties at 10-12 km may be induced by increased temperature or by a change in rock type.

The general idea expressed above, that faults are seen to move in a very heterogeneous manner when examined in detail spatially or temporally, applies equally well to major faults of types other than transcurrent faults. Thus, for example, Sykes (1971) and Stauder (1972) present evidence to show that seismicity is highly localized in the fault or fault systems beneath the Aleutians arc, that aftershock sequences may migrate systematically along the arc, and that long-term mechanical decoupling may exist between adjacent segments of the structure.

The average rate of slip on major faults over long periods of time is of special interest in tectonics (Section 10.2). It may be estimated geologically by using the methods mentioned in Section 7.3.4 to obtain the displacement and dividing by some estimate of the elapsed time. Average slip rates may also be obtained on some faults (e.g., oceanic transform faults) by applying the theory of sea-floor spreading (Section 10.2.1). Two methods are available for making relatively short-term determinations of slip rate. The first is to carry out repeated surveys of networks of geodetic stations either side of a fault [e.g., Whitten (1956); Savage and Burford (1973)]. The second is a simple seismological method proposed by Brune (1968) [see also Davies and Brune (1970)] in which the surface wave magnitudes of all earthquakes occurring on a given fault in a given interval of time are used to estimate the total slip on the fault during that time and, hence, the average slip rate. The magnitude of each earthquake is correlated with a quantity called the *seismic moment* and this, when divided by the total area of a fault, leads to an estimate of the contribu-

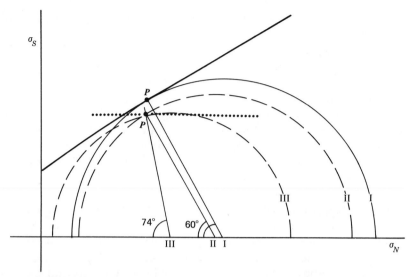

FIGURE 7.40 Mohr circles for states of stress before (I) and after (II, III) strike-slip faulting. Fault plane (*P*) assumed to be at 30° to σ_1 at the moment of failure. Circles II and III are drawn assuming a 10 percent drop in the shear stress on the fault plane and no change in normal stress. Circle II illustrates the possibility that the angle between the fault and σ_1 may remain 30° after faulting. Circle III is drawn so that this angle increases as a result of faulting.

tion made during each individual earthquake to the total slip on the fault. The method does not depend on knowledge of the actual slipped area during each event, but it does depend critically on correct estimation of total fault area. Calculated slip rates are in approximate agreement with geodetic measurements. In some cases the slip rates indicated by this method are lower than those indicated geodetically, suggesting either that some additional slip is occurring by creep or that strain is accumulating for an earthquake [Brune (1968)].

Stress History. If fault displacement history is complex, then it seems reasonable to expect that stress history may also be complex. Aspects of this problem have been discussed theoretically by Anderson (1951), McKinstry (1953), Chinnery (1966), and Lajtai (1969b). We consider here the question of when and how the principal stress directions may become reoriented by the onset of faulting. Some authors have proposed that such reorientation must be a general feature of active faulting and that the new state of stress gives rise to "second-order" fractures that are oblique to the primary fault [McKinstry (1953); Moody and Hill (1956)].

Consider a large segment of a fault plane across which a small and sudden shear displacement has just occurred. Such a segment may be a portion of an established fault on which an episode of stick-slip movement has just occurred or part of a newly generated brittle shear fracture on which movement has taken place for the first time. In either case the sudden shear displacement is accompanied by a sudden reduction in the shear stress on the fault plane. This is the *stress drop* or *stress release* of seismologists [see Chinnery (1964)] and it reflects release of stored elastic strain energy, much of which is radiated as seismic waves. We now look at a small rock volume adjacent to the fault but remote from the ends of the active segment. Figure 7.40 is a Mohr diagram on which the large circle (I) represents the two-dimensional state of stress in this rock volume just before movement occurred. The problem now is to find the Mohr circle representing the new state of stress. The dotted line represents the new shear stress in the fault plane if we assume a stress drop of 10 percent, for example. Note that there are an infinite number of Mohr circles that could be drawn representing the modified state of stress, each of them consistent with the specified stress drop of 10 percent. Two of these circles are shown in Figure 7.40. Circle II indicates no rotation of the σ_1 direction with respect to the fault plane. Circle III involves a rotation of σ_1 away from the trace of the fault plane. It is evidently necessary to know values for other stress components in order to specify a particular circle as the one correctly representing the new state of stress. While a case can be made for certain assumptions about these other components [see Lajtai (1969b)], it remains somewhat unclear what stress conditions will obtain under natural conditions.

If fault movement takes place by stable sliding, it is again difficult to predict whether displacement on the fault will lead to rotation of the principal stresses. The reason for this is illustrated in Figure 7.41, where we are again concerned with the stresses at a point *remote* from the ends of the fault (e.g., a point in the center of each specimen). In Figure 7.41a σ_1 is initially vertical and it remains in this orientation after a small displacement. In Figure 7.41b a similar displacement leads to counterclockwise rotation of σ_1. The difference in behavior results because the pistons in a are free to move sideways as well as up and down, so that force is applied vertically across the ends of the specimen whether or not sliding has occurred. In b this is not so and σ_1 near the center of the specimen becomes reoriented as shown. The degree to which natural loading systems will behave like the cases shown in a or b is unknown.

At points near the boundaries of a new fracture, or near the boundaries of a newly slipped region in an old fracture, major changes in the magnitudes and orientations of the principal stresses are very likely to accompany displacement. The patterns are complex and depend on assumptions about the original state of stress, the elastic properties of the rock, and the details of how displacement approaches zero as the end of the fracture is approached [see

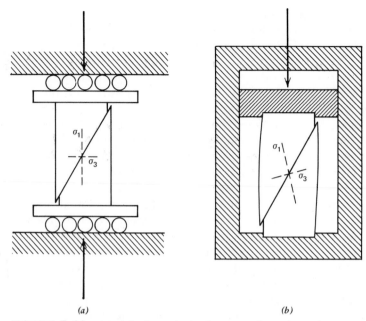

(a)   (b)

FIGURE 7.41   Models for principal stress orientations along an
active fault during stable sliding. In (a) σ_1 and σ_3, for a point in the
center of the specimen, remain near vertical and horizontal
because of sidewise movement of lubricated end plates. In (b) σ_1
and σ_3 become reoriented as shown from original vertical and
horizontal positions because ends of specimen are constrained to
stay in line.

Chinnery (1966)]. Figure 7.42 shows an example of principal stress directions
predicted for one set of assumptions. Since any part of a fault plane may have
been adjacent at one time or another to the edge of a fracture or of a slipped
region, it is possible that temporary perturbation of the regional stress field
may have occurred at any point and may have been responsible for certain
"secondary" features. An alternative is that some "secondary" features (e.g.,
feather joints) may have formed *before* the through-going fault developed, as
an early response to the same interval of loading [Lajtai (1969b); Gay (1970);
Dunn et al. (1973)].

7.3.12   Focal Mechanism Solutions

In addition to geological methods for determining the orientation of fault
planes and fault displacements, there is a powerful seismological method that

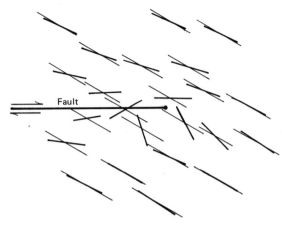

FIGURE 7.42   Predicted directions for σ_1 (short heavy lines) near the end of a fault, based on Chinnery (1966, Fig. 5). Prefaulting orientation of σ_1 (long light lines) assumed to be at 30° to the fault plane; prefaulting values of σ_2 and σ_3 are assumed to be zero. For further assumptions see Chinnery (1966).

yields similar information [Byerly (1926)]. The procedure in general is called *first-motion study* and the results, which can be interpreted in terms of fault movements, are known as *focal mechanism solutions.* This technique complements geological methods because it permits interpretations of present-day movements on deeply buried or otherwise concealed faults. First motion studies have played a central role in the development of plate tectonics and provide an important method for determining the relative motions of lithospheric plates (see Section 10.2). Summaries of the method are given by Honda (1962), Stauder (1962), and Khattri (1973).

Suppose an earthquake originates by sudden shear displacement over part of a fault plane as shown in Figure 7.43. Elastic waves are radiated in all directions from the source region or *focus*, but they will have different characteristics in different directions. Notice, for example, that the first compressional wave disturbance or *first motion of P* arriving at point 1 or point 3 will be a compression but that the first motion at point 2 or point 4 will be a dilation. The boundaries between regions close to the focus that receive compressional or dilational first motions of P are defined by the fault plane itself and by a second plane (the *auxiliary plane*) which is normal to the fault plane and to the slip direction. The fault plane and auxiliary plane divide the space near the focus into four quadrants. In alternate quadrants the first motion of P is compressional or dilational. This pattern is known as a *quadrantal* distribution,

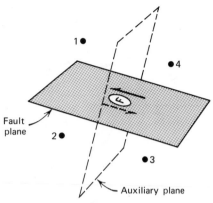

FIGURE 7.43   Fault plane and
auxiliary plane in the vicinity of the
focus of an earthquake (F). Directions of
relative displacement shown by arrows.
The first arrival of P is compressional at
points 1 and 3 and dilational at points 3
and 4, (the initial displacement of points
1 and 3 has a component *away* from the
focus, while the initial displacement of
points 2 and 4 has a component *toward*
the focus).

and it is known to be a characteristic feature of most natural earthquakes [see
Stauder (1962)]. A quadrantal distribution of the first motion of P does not,
however, prove that an earthquake originated by faulting because other *focal
mechanisms* can give the same pattern (for example, a sudden volume change
at the focus accompanied by a pure shear strain). Because the actual
mechanism is usually unknown, it is preferable to describe the result of a
first-motion study as a *focal mechanism solution* rather than a *fault plane solution*.
However, any pattern with quadrantal character can be analyzed to suggest
the orientations of possible fault planes and displacement directions as follows.

Figure 7.44 shows focal mechanism solutions obtained by Stauder (1972)
for several earthquakes that originated beneath the Aleutian arc in 1969. Each
solution is given in the form of a stereographic projection that shows two
planes at right angles to one another. Alternate quadrants between these
planes are shaded to indicate compressional first motions of P. Diagrams of
this type are prepared by determining the first motion of P at numerous
widely spaced seismograph stations and determining for each station the *direction in which the recorded compression or dilation left the focal region,* taking due
account of the way in which elastic waves follow curved paths through the

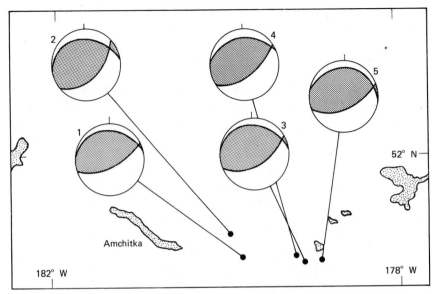

FIGURE 7.44   Focal mechanism solutions for five earthquakes in the vicinity of Amchitka, central Aleutian Islands. Epicenters are indicated by dots. Lower hemisphere stereographic projections are shown, with compressional quadrants shaded. [After Stauder (1972), with permission of the American Geophysical Union.]

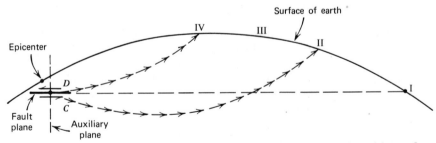

FIGURE 7.45   Schematic relations between fault orientation and positions of seismograph stations receiving compressional or dilational first motions of P. If the earth had uniform P wave velocity, point I would separate stations receiving compressional first motions (to the right of I) from stations receiving dilational first motions. Because the P wave velocity increases downward, however, P waves leaving the compressional quadrant (C) follow curved paths and can cross the trace of the fault plane, arriving at points such as II. The boundary between compressional and dilational stations is, therefore, at a point such as III. The first motion at point IV is dilational.

earth (Fig. 7.45). Stereographic projections such as those in Figure 7.44 are then prepared by plotting the orientations of the pair of orthogonal planes that best divides the space around the focus into quadrants in which the departing *P* waves had compressional or dilational character.

If an earthquake was caused by faulting, then one of the two planes indicated by the focal mechanism solution is the fault plane and the other is the auxiliary plane. The direction of displacement is perpendicular to their line of intersection, in a sense indicated by the arrangement of compressional and dilational quadrants. Thus, in Figure 7.44, the solution for earthquake number 4 indicates reverse faulting on a plane dipping steeply southeast *or* reverse faulting on a plane dipping gently northwest. This ambiguity is an inherent characteristic of focal mechanism solutions based solely on *P*-wave first motions. In principal it can be overcome by study of *S*-wave characteristics or by other means [see Khattri (1973)] but in practice this may be difficult. The solutions shown in Figure 7.44 were obtained using *P*- *and* *S*-wave data such that the ambiguity is removed and the northwest-dipping plane is indicated as the fault plane [Stauder (1972)]. Geological observations, especially the orientation of surface breaks associated with an earthquake, may also be helpful in distinguishing the fault plane from the auxiliary plane.

PROBLEMS*

1. Determine the shearing resistance of Potsdam sandstone at 0.55 kbar confining pressure, room temperature, using experimental data shown in Figure 7.21.

2. Construct diagrams like the ones in Figure 7.21c and d for the following room temperature test on olivine basalt.**

Strain (percent shortening)	Differential Stress (bars)
2	9650
5	15300

Ultimate strength   15400 bars
Confining pressure   5050 bars

3. Determine the angle between the fault plane and the direction of σ_1 for each of the six tests on Wombeyan marble represented in Figure 7.22.

*Answers are given in Appendix B.
**Data from Griggs, Turner, and Heard (1960) in Handin (1966, p. 245).

4. Tensleep sandstone† has cohesion (c) equal to 170 bars and coefficient of internal friction (μ) equal to 1.1. Use a Mohr diagram to predict the ultimate strength of a specimen of this rock at 50 bars confining pressure.

5. Predict the ultimate strength and the fault angle to be expected if a specimen of Frederick diabase is shortened at room temperature and 2 kbar confining pressure, using the data of Figure 7.24.

6. What is the cohesion and the coefficient of internal friction for Frederick diabase, according to the tests represented in Figure 7.24?

7. A block of Frederick diabase is subject to stress such that $\sigma_1 = 13$ kbar, $\sigma_2 = \sigma_3 = 3$ kbar, and the pore pressure is zero. In this state of stress the rock is stable. Now suppose that the pore pressure is raised until fracture occurs. Predict the pore pressure at which this will happen from the Mohr envelope of Figure 7.24. What is the effective normal stress and the shear stress on the fault plane at the moment faulting is initiated?

8. Figure 7.32 shows the shearing resistance and the shearing stress on various planes in Solenhofen limestone at the moment before failure. Trace this graph and add to it curves for the shear stress and shearing resistance (a) at the start of the deformation, when $\sigma_1 = \sigma_2 = \sigma_3 = 0.75$ kbar and (b) about halfway through the loading, when $\sigma_1 = 2.4$ kbar.

9. Determine the trend and plunge of each of the principal stress directions for the following conjugate faults in isotropic-looking rock.
   Fault 1:   strike 40°, dip 70° NW
   Fault 2:   strike 18°, dip 55° SE

10. Determine the trend and plunge of each of the principal stress directions from the following information on a fault in isotropic-looking rock.
    Strike   314°
    Dip   30° SW
    Striations
       Trend   250°
       Plunge   28°
    The dip-slip component of displacement is reverse. Assume that the slickensides are parallel to the direction of initial displacement and that the angle between the fault and σ_1 was 30°.

†Data from Handin (1966, p. 273).

8

GEOMETRICAL
ANALYSIS

8.1 INTRODUCTION

This chapter summarizes methods used in field studies to establish the form, extent, and arrangement of structures in a map area, together with the time sequence in which the structures developed. Investigations of this type are called *geometrical analyses* because spatial or geometric features of the rocks are studied and inferences are drawn about large-scale structures from observations made on a smaller scale; that is, a knowledge of the total structure is built up from its component parts.

A successful geometrical analysis is often an end in itself, in that it produces a three-dimensional picture of the structure and, possibly, an account of the stages through which the structure developed. Geometrical analysis is also a necessary first step in any attempts to determine the strain or stress history of a region and is thus a means to an end in these more specialized studies.

A basic problem in the study of folded areas is to establish the distorted shape of each folded layer or, in other words, to discover the geometry of all the folds and faults present. This is not a straightforward undertaking, even in the simplest areas, because the normal discontinuous nature of the outcrop makes direct observation impossible and interpretation, therefore, necessary. Furthermore since the amount of relief is generally very small compared to the area mapped, the observation is essentially two dimensional and thus, even in areas of almost continuous outcrop, to produce a three-dimensional picture, interpretation is still necessary.

We start by presenting a few basic concepts (Sections 8.2 and 8.3) that are essential in the study of complex areas; some of these are useful in the study of simple areas. We then discuss the methods employed in the interpretation of simple (Section 8.4) and of complex areas (Section 8.5).

8.2 BASIC CONCEPTS

8.2.1 Scale

It is convenient to distinguish three scales of investigation termed: macroscopic, mesoscopic, and microscopic. The most important difference between these is not simply the size of the area observed but the way in which the observations are made. *Microscopic* scale pertains to any structure that is sufficiently small or of such a character as to require observation with an optical or electron microscope. This includes deformation features of individual grains, crystallographic features such as c axes of quartz, and details of dislocation substructure (see Chapter 2). *Mesoscopic* pertains to any sample that is continuous and can be observed without the aid of the microscope. Thus, it is concerned with a hand specimen or a single outcrop regardless of the size of the latter. The important point here is that observations made on a mesoscopic scale are observations of continuous bodies and are, therefore, free of interpretation insofar as the shape of the structure is concerned. Thus, if we talk of the morphology of a mesoscopic fold it is implied that the shape of the fold can be observed in full, as opposed to being reconstructed from observations of a number of isolated outcrops. *Macroscopic* scale, on the other hand, pertains to bodies of rock that are not completely exposed and reference to the morphology of a macroscopic fold, therefore, implies the interpretive step of reconstructing the structure from data collected at a number of outcrops.

In understanding the structure of an area from the point of view of its geometry, the geologist is concerned principally with the mesoscopic and macroscopic scales. However, microscopic observations are invaluable to better establish the detailed characteristics of features, such as foliations, that are visible on a mesoscopic scale (e.g., it is commonly impossible to distinguish slaty cleavage from fine crenulation cleavage in hand specimen). Other aspects of microscopic observation are more concerned with understanding mechanisms of deformation than with geometrical interpretation (see Chapter 2).

8.2.2 Style

The phrase *tectonic style* is said to have been introduced by Lugeon [Turner and Weiss (1963, p. 79)] and is used in the description of tectonic structures in much the same way as the expression architectural style is used to describe buildings. Buildings can be classified on the basis of observable features into groups of a given style and the same is true of tectonic structures. Tectonic style is used mainly in reference to folds but can be used in reference to other structures such as foliations. Thus, style can embody all the morphological features of a given structure or group of structures. In describing *fold style* one is concerned with such features as the shape of the fold in profile, the presence or absence of an axial plane foliation, the type of axial plane foliation, and whether the fold is cylindrical or not. Other features such as the nature of related or deformed lineations and the related metamorphic assemblages and microstructures are also important aspects of style in some areas.

   In examining a deformed area, it is commonly found that the folds can be asigned to a small number of style groups. For example, in many areas there are folds that are tight to isoclinal and have an axial plane schistosity; these are accompanied by other folds that are much more open and which may have no axial plane foliation (e.g., Fig. 8.1) or alternatively may have a crenulation cleavage as axial plane foliation. In such an area, if all of the folds can be ascribed to one or other of the styles, it can be said that there are two distinct fold-style groups. In many areas, however, the picture is more complex. Style tends to vary with rock type so that, for example, a given fold in mica rich layers may have an angular hinge, when seen in profile, but if traced into a

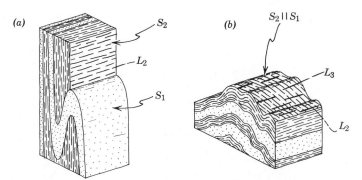

FIGURE 8.1   Contrasting fold styles, Broken Hill, Australia. (*a*) B_2 fold is tight, has an axial plane schistosity (S_2) and a lineation (L_2), parallel to the fold axis, defined by mineral alignment, especially alignment of sillimanite. (*b*) B_3 fold is open to gentle, has no axial plane foliation in this example but, in some other examples, has an axial plane crenulation cleavage, folds the earlier L_2 lineation, and has a lineation (L_3), parallel to the fold axis, defined principally by a second generation of sillimanite crystals that are coarser and scarcer than their B_2 counterpart.

more competent layer may become much more rounded. Similarly a given fold may have an axial plane foliation in one layer but not in another. Furthermore, in a given area folds that otherwise have the same style may vary greatly, but in a continuous manner, with respect to one prominent feature such as the sharpness of the hinge. Thus, in some areas it is impossible to divide the folds into a small number of style groups without resorting to arbitrary distinctions such as the magnitude of the angle between the fold limbs. The latter device would be futile as will become apparent from the discussion of fold generations later in this section. In areas where folds cannot be divided into style groups geometrical analyses may be impossible. Suffice it to say here that we are looking for natural, well-delineated groups when we divide structures on the basis of style.

The term style group is not synonymous with fold generation although they may be correlated as discussed below (Section 8.2.4).

8.2.3 Overprinting

Overprinting criteria form the principal basis for dating structures relative to one another in deformed rocks. For example, in Figure 8.2a, the axial surface of the B_1 fold, S_1, is itself folded by the B_2 folds (see also Figs. 8.3 and 8.4). In this situation B_2 is said to *overprint* B_1, that is, it postdates the formation of the B_1 fold, even though the interval of time between the formation of the

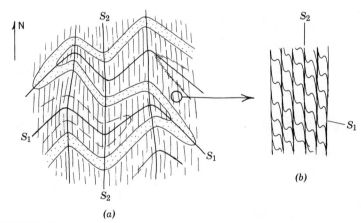

(a)

(b)

FIGURE 8.2   Overprinting relationships. (a) Tight folds with east-west axial surface traces (S_1) are overprinted by open folds with north-south axial surface traces (S_2). (b) Detail of cleavage in pelitic layer; S_1 is crenulated and overprinted by S_2.

two may have been very small. The B_1 fold may have changed shape during the development of the B_2 fold but it must, at least, have been initiated before the B_2 fold began to develop. By the same token if both folds have an axial plane foliation then S_2 will overprint S_1; an example of such overprinting is illustrated in Figure 8.2b. Overprinting of one structure by another may result in annihilation of the earlier structure and this is particularly true when one foliation overprints another [e.g., see retrograde zones at Broken Hill, Australia, Vernon and Ransom (1971)].

8.2.4 Generation

Ideally, in examining an area of rocks that have been deformed more than once, the geologist would like to be able to ascribe all the various structures and metamorphic assemblages to their correct position in the absolute time scale. In practice he has to be content with relative dating, and ascribing structures to a given fold generation is an attempt to achieve this end.

FIGURE 8.3   Overprinting of Hercynian folds by Alpine folds in Gotthard gneiss, Nufenen Pass, Switzerland.

FIGURE 8.4   Overprinting relationships between a group of folds with an axial plane slaty cleavage (running E/W) and a later group with an axial plane crenulation cleavage (running approximately NNW/SSE). Phyllite from Aston Massif, Pyrenees, from collection of H.J. Zwart. Negative print of a thin section.

A given *generation* of folds and related structures comprises a group of structures believed to occupy the same niche in a relative time scale. Thus, to establish a generation two steps are necessary; first the structures must be grouped with other structures believed to belong to the same phase of deformation, then the various groups must be placed in chronological order. The grouping is done on the basis of style. That is, folds in isolated outcrops are ascribed to style groups. Then the style groups are dated relative to one another by observing, wherever possible, the overprinting relationships between representatives of the different groups. If the initial grouping is valid, that is, the structures in one group do occupy the same niche in the relative time scale and occupy a different niche to all other structures, then the overprinting relationships should be consistent. However, a consistent overprinting does not prove that the grouping is valid. It has been shown, in some areas [Means (1963, 1966); Park (1969); Williams (1970); Olesen et al. (1973)], that folds having the same style can belong to more than one generation, and use of style as the basis for grouping structures is, therefore, not altogether satisfactory. However, there is no better basis in areas of discontinuous outcrop and this method of grouping does seem to work in many areas. Nevertheless, the geologist should be on guard against this problem and every opportunity should be taken to test the validity of using style for grouping folds into generations. If outcrop is continuous over areas that are large enough to contain many folds, it may be possible to group folds on the basis of overprinting relationships and separate these groups into generations without recourse to style groups. In such areas the validity of style as a criterion for grouping folds into generations can then be checked.

The need for using style in areas of discontinuous outcrop is demonstrated in Figure 8.5. In Figure 8.5a, outcrop is continuous and the folds can be grouped on the basis of overprinting. In Figure 8.5 the same structure is represented but outcrop is now discontinuous so that folds found in outcrops where there is no visible overprinting cannot be grouped on the basis of overprinting and style has to be used instead.

In some areas orientation of folds can assist in grouping but, in general, orientation is not a useful criterion for delineating groups that are to be ascribed to fold generations [see Weiss and McIntyre (1957)]. Certainly the fact that two folds have the same orientation of axial surface and of hinge line

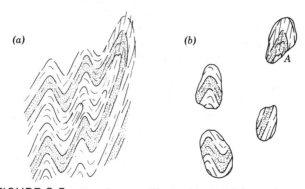

FIGURE 8.5   Sketch maps illustrating need for style grouping. (a) Outcrop is continuous and folds are readily grouped into generations on the basis of overprinting. (b) Outcrop is discontinuous and age relationships are no longer obvious. However, the folds can be grouped on the basis of style; as represented on the map the fold at A differs from the other two in that it has an axial plane foliation. In this example the style groups coincide with fold generations. For further discussion see text.

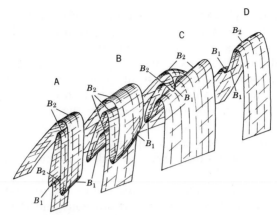

FIGURE 8.6 Sketch of a fold in a thin, biotite-rich, geneissic layer in marble from the Garwood Valley, Antarctica. This specimen illustrates a problem commonly encountered in multiply deformed areas. There are two generations of folds (B_1 and B_2) but these folds are recognizable only because the whole gneiss layer, as drawn, is exposed and the complete three-dimensional form of the structure and, therefore, the overprinting relationship, is visible. Both the B_1 and the B_2 folds fold an earlier foliation defined by the biotite and there are no style criteria for separating the two. Furthermore, although B_1 and B_2 folds have different patterns of orientation, if the specimen as a whole is considered, there are portions of the specimen in which B_1 and B_2 hinge lines and axial surfaces have almost identical orientations. Thus if exposure was restricted to a portion of the specimen (e.g., block D), so that the overprinting was no longer visible, it would be impossible to identify two generations of folds. Furthermore, even if the overprinting were visible it may be impossible to ascribe all of the folds to the correct generations. Consider, for example, block A; overprinting is obvious but it is impossible, without other evidence, to label the three folds in the center of the block.

does not necessarily mean that they belong to the same generation (see Fig. 8.6) and conversely folds of a given generation need not have the same orientation.

It is important to understand the chronological significance attached to the concept of fold generation. Two generations of folds may have formed dur-

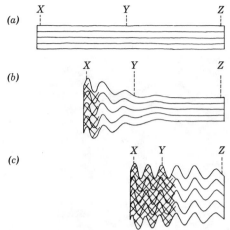

FIGURE 8.7 Diagram illustrating how
two generations of structures may overlap
in time of formation. Cross hatching
represents conjugate crenulation
cleavage. See text for explanation.

ing a continuous deformation or may have formed during two periods of
deformation separated by hundreds of millions of years; without techniques
such as absolute dating no distinction between these two possibilities can be
made. Furthermore, although B_2 folds, for example, may everywhere postdate B_1 folds in individual outcrops, it does not follow that all B_2 folds
postdate all B_1 folds. This point is demonstrated diagrammatically in Figure
8.7; a sequence of sediments is shortened in such a way that B_1 folds develop
first at X and then progressively toward Y. As the folds at X become tight,
further shortening is achieved by development of a conjugate crenulation
cleavage and its attendant B_2 microfolds (Fig. 8.7b). At this stage B_1 folds are
just developing at Y and have not even begun to form at Z but both areas can
be expected to go through the same sequence of deformational events if the
process continues (Fig. 8.7c).

8.3 FOLD SHAPE AND OUTCROP PATTERNS IN DEFORMED AREAS

There is infinite variety, in detail, in the geometry of the folded surface of
deformed areas. However, some of the structures can be grouped into broad
patterns that occur repeatedly and these basic patterns are described here.

In some areas where there is only one generation of folds, bedding or any
other marker surface may be folded into structures that approximate cylin-

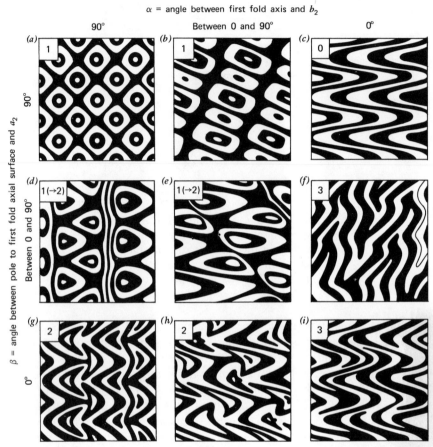

FIGURE 8.8 Two-dimensional interference patterns resulting from superposition of two generations of folds. The second-generation folds are assumed to be ideally similar so that the geometrical relationship between the two generations can be described, for B_2, in terms of kinematic axes a_2 and b_2 (see Section 4.7). The numbers in the top left-hand corner of each figure denote end members in the spectrum of interference patterns. Patterns 1, 2, and 3 result from overprinting relationships of the type depicted in Figures 8.9, 8.10, and 8.11, respectively. 1(−2) are transitional patterns resulting from overprinting relationships transitional between those depicted in Figure 8.9 and 8.10. O is a special case where B_1 and B_2 have parallel axes and axial surfaces and, therefore, do not produce any characteristic interference pattern. [From J. G. Ramsay (1967, *Folding and Fracturing of Rocks.* Copyright © 1967 by McGraw-Hill Book Co. Used with permission of McGraw-Hill Book Co.]

drical form (see Section 4.2). Elsewhere such folds may be doubly plunging and in some areas very large variation in plunge is believed to be the product of a single deformation [e.g., Hobbs and Hopwood (1969)].

FIGURE 8.9 Card model illustrating the type of interference pattern resulting from superposition of two groups of folds where B_1 and B_2 are orthogonal and a_2 lies in the B_1 axial surface (see Fig. 8.8a) and parallel to the faces of the cards. B_1 folds are represented by lines printed on the face of each card; the cards are all identical and thus in a stack of cards the lines define a series of curviplanar surfaces resembling a system of cylindrical folds. B_2 folding is achieved by displacing the cards relative to one another as in the slip-fold model (see Section 4.7.1) Note that although this interference pattern has been produced by superposition of two separate deformations it could equally well be produced by a single deformation. [From O'Driscoll (1962), with permission of the Alberta Society of Petroleum Geologists.]

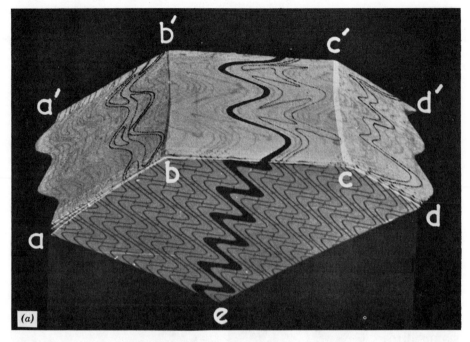

FIGURE 8.10 Card model illustrating possible outcrop patterns resulting from superposition of two groups of folds where B_1 and B_2 are orthogonal and a_2 is parallel to the card faces but inclined to the B_1 axial plane at a large angle [Fig. 8.8, intermediate to (d) and (g)]. B_1 folds are represented by lines printed on the face of each card; the cards are all identical and thus in a stack of cards the lines define a series of curviplanar surfaces resembling a system of cylindrical folds. B_2 folding is achieved by displacing the cards relative to one another as in the slip-fold model (see Section 4.7.1). (a) A photograph of the complete block. (b) A "close-up" of the three outcrop surfaces. [From O'Driscoll (1964), *Bull. Can. Pet. Geol.*, v. 12, p. 279–310.]

In areas of multiple deformation the structure may again be quite simple. For example in some areas a second deformation has simply tightened earlier folds [e.g., Williams (1971)]. In general, however, multiple deformation produces a form of interference pattern that is commonly more complex than structures produced by a single deformation. The appearance of such structures in outcrop depends on the number of generations of folds, on their relative orientations and sizes, and on the orientation of the interference pattern with respect to the topographic surface.

Ramsay (1967, p. 520) has pointed out that the interference patterns resulting from the intersection of two sets of folds can be divided into three basic groups which are represented by Figures 8.8, 8.9, 8.10, and 8.11. For the

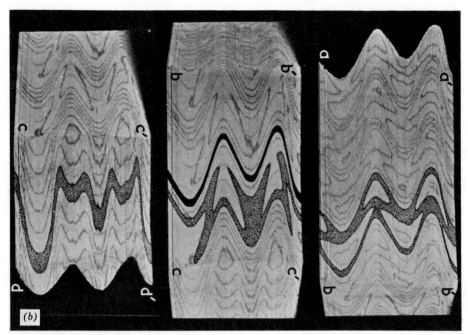

FIGURE 8.10 (Continued)

purpose of demonstration he has considered the slip fold model, but it is important to realize that the resultant patterns are not peculiar to this particular model but can result from any fold model. However having assumed the slip fold model the relative orientation of the two generations of folds can be

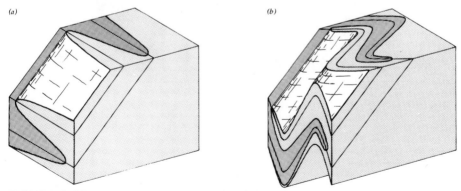

FIGURE 8.11   Interference pattern resulting from superposition of two groups of folds so that the two axis orientations are parallel and the two axial surface orientations are nonparallel. This particular situation is commonly referred to as coaxial folding (see Fig. 8.8f and i).

described in terms of the slip plane, which is parallel to the axial plane of related folds, and the slip direction a, which lies in the slip plane and is inclined to the related fold axes (see Section 4.7.1). The various patterns resulting from different relative orientations of the two generations of folds are summarized in Figure 8.8. Figure 8.8c represents the case of a first-generation fold that is simply tightened by the second deformation because a_2 is parallel to the B_1 axial plane. Figure 8.8f and i represents coaxial folding; the situation where B_1 is folded by B_2 but the B_1 and B_2 fold axes are parallel. This is a common type of refolding and may indicate that, as in material such as corrugated iron, it is easier to fold a folded sequence about an axis parallel to the existing axis than in any other direction.

For further discussion of this topic the interested reader is referred to Holmes and Reynolds (1954), O'Driscoll (1962, 1964), and Ramsay (1962, 1967).

8.4 ANALYSIS OF FOLDED AREAS WITH SIMPLE STRUCTURE

The simplest fold systems to analyse are sets of upward-facing folds with axes of constant trend and gentle plunge, in rocks with continuous bedding. We

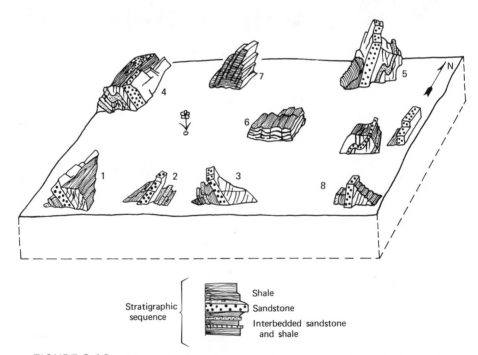

FIGURE 8.12 Schematic view of outcrops in an area of simply folded rocks. Outcrop numbers are referred to in the text.

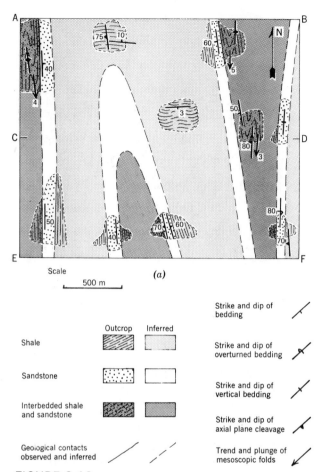

Scale

500 m

(a)

Strike and dip of
bedding

	Outcrop	Inferred	
Shale

Sandstone

Interbedded shale
and sandstone

Geological contacts
observed and inferred

Strike and dip of
overturned bedding

Strike and dip of
vertical bedding

Strike and dip of
axial plane cleavage

Trend and plunge of
mesoscopic folds

FIGURE 8.13   (*a*) Geological map of the area depicted
in Figure 8.12. Map shows factual data including
distribution of various lithologies, orientation data, and
distribution of outcrop. It also shows the geological
interpretation (inferred data). (*b*) Structure contour map
of the area represented in (*a*). Contours are drawn for the
base of the thick sandstone and give its height above a
reference plane. Solid lines represent contours below
ground level and broken lines represent those above
ground level. Overturning of the beds in the southeast
corner of the area results in crossing of contours along the
eastern edge of the map. For such a map to be reliable, in
a flat area like that depicted in Figure 8.12, would require
a lot of subsurface information. Such information can be
obtained, for example, from drill holes.

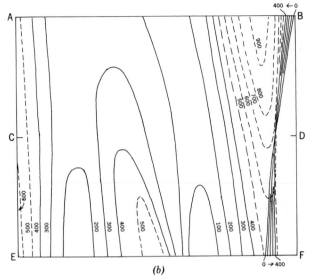

(b)

FIGURE 8.13 (Continued)

enumerate below some methods appropriate to the study of such areas; these methods formed the basis for structural mapping of folded areas prior to about 1950 and remain suitable for simple areas. It cannot be overemphasized, however, that many apparently simple areas are in fact complex, so that it is essential that the field-worker is alert to signs of such complexity.

1. The various rocks of the area are subdivided into recognizable stratigraphic units and the disposition of the units and the contacts between are mapped wherever possible. Thus Figure 8.13a is a map of the area represented schematically in Figure 8.12. Repetition of a stratigraphic sequence may suggest the positions of folds and in Figure 8.13a the repetition of the interbedded sandstone-shale unit at for example, outcrops 1 and 3, suggests at least two folds between these localities.

2. The orientation of bedding is measured and the data used to determine the shapes of the folds. The downward-converging dips at outcrops 1 and 2 (Fig. 8.13a), for example, indicate that the fold between these localities is a syncline and not an anticline. The angle between the dips suggests that the syncline is a fairly open fold. The northward-converging strikes at 2 and 3 suggest that the sandstone contacts connect with each other or "close" north rather than south of 2 and 3 and that the plunge of the anticline between these exposures is toward the north.

3. The vergence direction of mesoscopic folds (see Section 4.3) is used to determine the position of each exposure relative to nearby macroscopic folds.

At 4 (Fig. 8.13*a*), for example, the vergence of mesoscopic folds indicates a position on the east limb of an anticline or the west limb of a syncline. At 5, the vergence is opposite to that at 4, indicating the west limb of an anticline or east limb of a syncline. At 6, where the folds are symmetrical, the outcrop should be near the hinge of a fold.

4. The angular relationship between bedding and any axial plane cleavage that may be present is observed. It is assumed that the relationship between these surfaces, in macroscopic folds, is the same as that generally observed in mesoscopic folds (see Figs. 5.2 and 5.3) and the recorded data is then used in the following ways. If bedding and cleavage are perpendicular to one another, as at 7 (Fig. 8.13*a*), the outcrop is on the hinge of a fold. If bedding and cleavage make a small angle with one another the outcrop is on the limb of a fold, and which limb it is on can be determined. Thus, for example, at 3 the bedding-cleavage relationship is consistent with a position on the east limb of an anticline but it is not consistent with a position on the west limb of an anticline. If bedding and cleavage dip in the same direction but bedding is more steeply dipping than cleavage, as at 8, the beds are locally overturned. Otherwise, the beds are right side up. Finally, the intersection line of bedding and cleavage is taken to indicate the approximate orientation of macroscopic hinge lines. At 7, this direction plunges gently north.

5. In simple areas a reasonable idea of the trend of the fold hinge lines can usually be obtained by inspection, and the plunge can be obtained by observing the dip of bedding in the fold hinge (e.g., at outcrops 6 and 7 in Fig. 8.13*a*). A more accurate result, however, can be obtained by plotting the bedding orientation data on a stereographic projection as is done in complex areas (see Section 8.5). It is useful to remember that in areas of cylindrical folding the trend of the hinge line is parallel to the strike of any vertical beds that may be present and the plunge of the hinge line is equal to, or less than, the shallowest dip.

6. The final structural interpretation is represented in the form of vertical cross sections drawn perpendicular to the regional strike. The simplest interpretation of the area represented in Figures 8.12 and 8.13*a* is shown in Figure 8.14. The subsurface position of a fold hinge has to be interpreted and in areas of parallel folding good results can be obtained by the method of Busk (1929), in which the folded surfaces are represented by a system of circular arcs. If the folds are cylindrical good results are also obtained by projection down plunge (see Section 8.5.2); this method does not depend on the folds being parallel in form and is a preferable method for all cylindrical areas.

The accuracy of an interpretation obtained as above is much enhanced if subsurface information (from drill holes, mine openings, and so on) is available to augment surface observations. In some cases, the only information

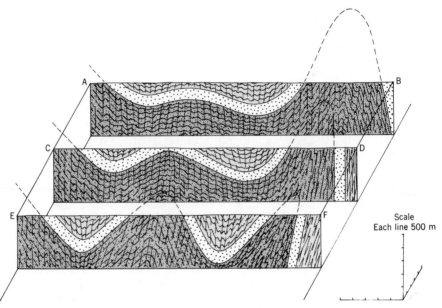

FIGURE 8.14   Vertical cross sections showing a possible interpretation of the macroscopic structure of the area represented in Figure 8.12 and 8.13. Outcrop topography is neglected.

available is drill-hole data. Here the form of each folded horizon must be pieced together primarily from knowledge of the depth below the surface at which various horizons are penetrated, the probable stratigraphic thicknesses between successive horizons, and the angle between small-scale bedding and the axis of recovered drill core. The results of such investigations are commonly presented in vertical cross sections (e.g. Fig. 8.14) and structure contour maps. The latter are maps in which the contours represent the height of some geological surface, above a given datum level, in exactly the same way that topographic contour maps represent the height of the ground. The surface represented may be a fault, the top of some stratigraphic unit, or any other geologically definable surface. For example, Figure 8.13*b* is a structure contour map of the bottom of the thick sandstone bed represented in Figure 8.13*a*.

8.5   ANALYSIS OF FOLDED AREAS WITH COMPLEX STRUCTURE

Folded areas with complex structure are characterized by one or more of the following features: fold shapes that are markedly noncylindrical, primary layering that has been made discontinuous by transposition, differentiated layering that is visible on the mesoscopic scale, folds of several sets that interfere with or overprint one another, folds that are downward facing, and folds that, in a given volume of rock, have different orientations in different surfaces (e.g., nonparallel folds in a transposition foliation and in its associated enveloping surface). In areas like these, application of the methods described in the previous section can lead to gross misinterpretations of macroscopic structure and to baffling contradictions in the results of an analysis.

During the 1950s workers in Britain, influenced by earlier European workers such as Sander (e.g., 1930, 1948 and 1950, see translation by Phillips and Windsor, Sander, 1970), Argand (e.g., 1911), Wegmann (e.g., 1929), and Kvale (1948) developed a method more appropriate to complex areas. This method embodies much of the earlier method but makes use of the previously neglected evidence such as the detailed spacial distribution and orientation of small-scale structures. This method never *assumes* that small-scale structures reflect the orientation and character of larger-scale structures but attempts to establish the relationship between the two. Early examples are provided by the work of McIntyre (1951), Weiss and McIntyre (1957), Ramsay (1958), Sutton and Watson (1959), and Weiss (1959b). This approach has become known as structural analysis [see Turner and Weiss (1963)] and here we are concerned with its geometrical aspects. It did not replace existing techniques but simply supplemented them. Thus, in areas where there is recognizable stratigraphy, for example, it is as important as ever to pay attention to the latter and to collect all the stratigraphic data available, including the disposition of gross lithological boundaries and as much information as possible concerning the direction of younging.

The underlying principle of structural analysis is that in a deformed group of rocks each phase of ductile deformation (there is commonly more than one) has left a penetrative imprint on the rocks such that small samples will ultimately yield a complete record of the deformation and, in principal, will provide all the information required to determine the sequence of events. This principle is readily substantiated by observation, although in practice it is generally essential to look at more than one sample to determine the complete history of the area. Nevertheless, in retrospect, one is able to see details of the complete history of an area in a well-chosen hand specimen. There are areas, however, in which earlier fabric is completely obscured by later deformation. Thus when considering a large area the technique relies very heavily on observation of small-scale structures in building up a large-scale interpretation.

8.5.1 Collection of Data

1. *Mapping.* Mapping is basically the same whether the rocks are simply folded sedimentary rocks or multiply deformed metamorphic rocks. It is important to record all data that may help in answering the questions asked.

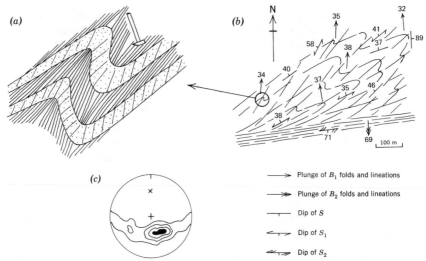

FIGURE 8.15   Form surface mapping: (*a*) A fold as seen in outcrop. (*b*) A
form surface map on which the fold depicted in (*a*) has been represented. In
total there are three foliations represented on this map. (*c*) A $\perp S$ figure for the
map area; the poles are contoured and the cross is the pole to the $\perp S$ girdle and,
therefore, the mean orientation of the axis of the macroscopic fold in S. Note
that in this area the macroscopic axis plunges parallel to the mesoscopic folds
represented on the map. Note that the direction of plunge of a fold need not
coincide with the trace of its axial surface [see (*b*)]. In fact the two are only
parallel in the special cases where the axial plane is vertical or the plunge is zero.
Note also that the size of the small fold [(*a*) and (*b*)] has been exaggerated so that
it can be represented on the map as an aid to interpretation. Exaggeration of this
order is permissible, provided that the fact that it has been done is
acknowledged.

Where the question is one of discovering the geometrical configuration of a
given group of rocks it is just as important to record stratigraphic data (including directions of younging) as it is to record fold data. However, there are
some techniques that assume greater importance in complex areas than in
simple areas. First, if a geological map is to be a representation of factual data
it should show the distribution of outcrop. In simple areas the interpretation
of the outcrop pattern may be so obvious that it excuses omission of outcrop
boundaries. In many complex areas, however, this is not true and a good map,
as a presentation of factual data, should show the approximate limits of outcrop (e.g., Fig. 8.13*a* and 8.16), Second, in many complex areas there is
important data preserved in layers that are too thin to record on the map. In
such areas *form surface mapping* (e.g., Fig. 8.15 and 8.16) is very valuable; it
consists of simply representing the trace of any penetrative surface on the

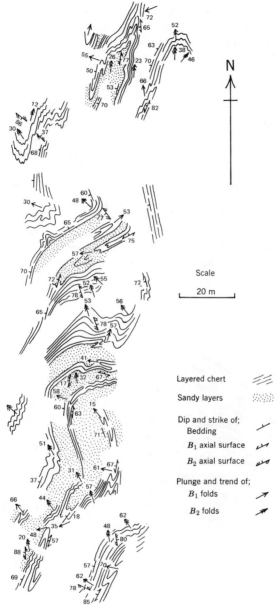

FIGURE 8.16   Form surface mapping in a multiply deformed sequence of cherts and sandstones. The form surfaces are drawn for bedding planes. [From Williams (1971), with permission of the Geological Society of Australia.]

map. Thus the fold in Figure 8.15*a* can be represented on the map (Fig. 8.15*b*) by a single line that does not necessarily represent any particular layer or surface but the overall structure. Such a structure need not be drawn exactly to scale (i.e., it may be treated more as a conventional symbol) and need not always represent the same surface (see Fig. 8.15*b*). This mapping technique enables the observer to see at a glance those areas in which the foliation is folded as opposed to those in which it is planar. It also reveals the sense of vergence of small folds and the traces of axial plane schistosities. An additional bonus of this method is that by judicious use of stippling and coloring a more realistic and aesthetically pleasing map, distinguishing various lithologies, can be made (contrast Fig. 8.17*a* and *b*).

2. *Observation of Mesoscopic Structures.* During the process of mapping it is necessary, in multiply deformed areas, to build up a picture of the interrelationship of the various structures. By this means the sequence of events is determined and an impression gained of the kind of macroscopic structures that may be encountered. For example, if interference of two groups of small folds produces dome and basin structures, visible in outcrop, then it is probable that such structures will be encountered on a macroscopic scale also [e.g., see Tobisch (1966)].

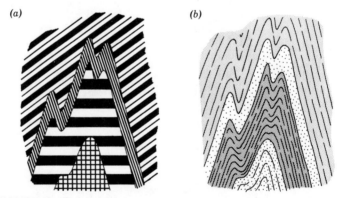

FIGURE 8.17   Sketches contrasting use of arbitrary ornamentation and ornamentation more closely related to the geology. The type of ornamentation used in (*b*) is generally aesthetically more pleasing and can immediately convey more geologically meaningful data than that used in (*a*). For example, use of coarse stippling for sandstone, fine stippling for siltstone, and lines for clay shale is so obvious that constant reference to the legend is unnecessary. Furthermore the lines can be used to represent details of the structure that cannot be conveyed by the arbitrary ornamentation.

The various structures are observed and wherever possible ascribed to style groups. All overprinting relationships are recorded and the various style groups ascribed to fold generations, wherever possible.

Fold generations are generally labeled B_1, B_2, etc., foliations S_1, S_2, etc. and lineations L_1, L_2, etc. The numbers are generally an indication of the chronological age of the structure but this is not necessarily true. F_1, F_2, etc. and D_1, D_2, etc., are used by some writers to denote folds and periods of deformation, respectively.

3. *Orientation Data.*   It is normal to plot dips and strikes and other orientation data on all geological maps but in complex areas this assumes even greater importance since there is generally far more variation in orientation. Furthermore, there are more fabric elements to be represented. The dip and strike of all foliations and plunge of all lineations should be measured extensively throughout the area. Figure 8.16 is a map of an area in which several fabric elements have been recognized and their orientations recorded. Mapping of this type is an important step in understanding the geometry of an area because the measurements are the basis of three-dimensional interpretation.

In addition to recording orientation data on the map it is useful to record them on an equal area projection (see Appendix A). It is stressed that these are complementary means of recording data — not alternatives — and as much data as possible should always be recorded on the map. Linear structures are plotted as points and surfaces are represented by their cyclographic trace or, more commonly, by their polar projection. Each fabric element is normally plotted on a separate diagram and points are commonly contoured in terms of the percentage of points per 1 percent area of the surface of the projection sphere. Poles to a given surface, for example, S_1, are referred to as poles to S_1 and represented by the shorthand $\perp S_1$ or πS_1. In an area in which S is cylindrically folded the poles to S spread in a girdle (see Fig. 8.15*c* and 8.18). Thus by plotting poles to a given surface the axis about which it is folded is readily determined (e.g., Fig. 8.15*c*), and the axial orientation of the large folds can then be compared with that of mesoscopic folds and lineations and a basis thereby established for interpretation of the relationship between mesoscopic and macroscopic structures.

Another method of determining and representing, the plunge of large scale folds was popular a few years ago and is still used by a few workers. It is based on what is known as a β diagram; the folded surface (S) is represented by its cyclographic traces on an equal area projection (see Appendix A) and points are plotted at the intersections of these traces. These points are then contoured and for a reasonably tight cylindrical fold, the maximum will be the same as that determined from a $\perp S$ diagram for the same data. For a less cylindrical fold, however, the method can be misleading, as was pointed out by Turner and Weiss (1963, p. 154) and Ramsay (1964). The problem is that the

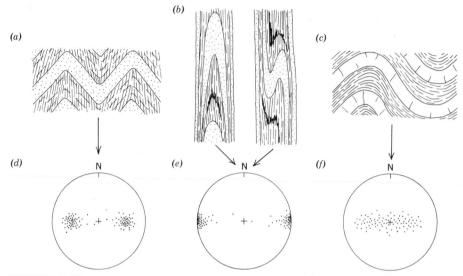

FIGURE 8.18   Illustration of the type of S pole figures [(d), (e) and (f)] that can be expected in areas where north-trending horizontal folds are cylindrical, have parallel axial surfaces, and have the types of profile depicted in (a), (b) and (c).

method produces many points that have no geological significance, thus making interpretation, at best, unnecessarily difficult. Furthermore, because the number of intersection points is very large, compared to the number of measurements made, it is an unnecessarily time-consuming technique. Beta diagrams have no advantages over ⊥ S diagrams and, therefore, in view of their limitations are of little value.

In measuring the orientation of a given fabric element the object is to obtain an accurate representation of its orientation pattern throughout the area being measured. For example, a foliation may have a constant orientation, may be folded cylindrically, or may be folded noncylindrically; whatever the structure its geometry should be indicated by the measurements. In order to do this it is necessary to make sufficient measurements. Without knowing the structure first it is impossible to know what the required number is, but the problem can be overcome reasonably by continuing to measure in an evenly distributed pattern, over the area of outcrop, until reproducibility is obtained. That is, measurements are made until the pattern does not change significantly with the addition of new measurements. A simple way of checking reproducibility is to plot alternate measurements on separate stereographic or equal area projections (see Appendix A) and see if they are reproducible. The advantage of this method is that it is very simple. The disadvantage is that the number of measurements made in order to establish reproducibility is twice as many as are required to obtain a representative sample.

8.5.2   Interpretation

Recognition of a Stratigraphy.   In areas where there is a recognizable stratigraphy the establishment of the sequence is an important step in interpreting
the large-scale structure and the history of an area. However, it is equally
important to realize that in many complex areas a sequence of juxtaposed
lithological units does not represent a stratigraphic sequence (see Section 5.4)
and in such areas an assumption to the contrary may lead to a misinterpreta-
tion of the geometrical aspects of the structure as well as the history. Furthermore, in deformed metamorphic rocks, layering as seen in outcrop, cannot be
assumed to be bedding (see Section 3.4). Thus it is important in deformed
metamorphic terrains to demonstrate — not assume — the existence of bedding and stratigraphy wherever possible. The existence of bedding can be
demonstrated by recognition of sedimentary structures (see Section 3.4).

The question as to whether or not a lithological sequence represents a
stratigraphic sequence, however, presents a greater problem. There can be no
positive solution to this question, only a lack of evidence to the contrary. If it
can be shown that layering within the sequence is bedding, that there are no
intrafolial folds (see Section 5.3.5) and that the direction of younging (see
Section 3.3) is constant throughout the sequence, then it is reasonable to claim
that the sequence is a stratigraphic one. The point, however, is still not proven
since there remains the possibility (even though in some areas it is improbable)
that portions of the sequence are missing or are repeated, for example, due to
faulting at a low angle to bedding.

Interpretation of Mesoscopic Structures.   This starts with the grouping of structures on the basis of style. Then, assuming that all the structures of a given
style group belong in the same niche in the deformational sequence of events,
an attempt is made to build up a picture of this sequence by ascribing the
groups to a series of generations. This is done by piecing together the fragments of the story found in individual outcrops. Thus if it is found that
members of style group A consistently overprint members of group B, other
members of which are seen in some outcrops to overprint structures of group
C, a sequence of events has been established with group C in any given area
being the oldest structures and group A the youngest. If, on the other hand, it
is found that, in some outcrops, group A structures overprint group B,
whereas in others B overprints A, then either the method of grouping the
folds is unreliable in that area or groups A and B are contemporary. Thus, in
summary, style is used to extend the conclusions drawn from overprinting
relationships of a few members of the various style groups, to all the members.

Interpretation of Macroscopic Structures.   The first step in producing a
geometrical interpretation of the macroscopic structure of an area is to produce a two-dimensional interpretation; this is necessary because the outcrop
map is an incomplete basis for the three-dimensional interpretation because

of the gaps between the outcrops. In areas of good exposure and varied lithology this may be a simple step that is not likely to introduce much error. Elsewhere, due to paucity of outcrop and monotony of lithology, the step may be highly interpretive. Basically it is a matter of joining up outcrops of like lithology in a manner that is internally consistent and consistent with any other constraints such as, for example, a known regional stratigraphy. As in simple areas (see Section 8.4) small-scale folds can be extremely useful in making this two-dimensional interpretation, especially in areas where there are few distinctive lithologies to use as markers. The symmetry or sense of asymmetry of the folds indicates whether they lie in the hinge or the left or the right limb of the larger fold of the same generation. In some areas, such as that of the Otago Schists of New Zealand [Grindley (1963); Means (1963, 1966); Wood (1963)], the only evidence for the existence of large-scale folds is in the form of mappable vergence zones; the symmetry of folds is constant within the zones but varies systematically from zone to zone. Axial plane foliations can be used in an analogous way to their use in areas of simple folding (see Section 8.4); in the hinge of large folds the folded surface and the axial plane foliation intersect at right angles. In the limbs the angle is generally smaller and the relationship between the two surfaces indicates on which limb a given outcrop lies.

Both the vergence and axial plane foliation methods work well in simple areas but in complex, multiply deformed areas they must be used with caution since, for example, the asymmetry of small-scale B_1 folds tells nothing of the disposition of the large-scale B_2 folds.

The shape of mesoscopic folds may also help in macroscopic interpretation in that folds of a given generation commonly have a very similar appearance in profile whatever their size. Thus if orientation diagrams for the folded surface indicate the presence of large folds with the same plunge as a group of small folds in the same area, it is reasonable to interpret the two as members of the same fold generation and it is then reasonable to seek a large-scale interpretation in terms of folds with shape similar to that of the small folds. The pole figures ($\perp S$ figure) for the folded surface may also throw some light on the appearance of the large-scale folds; in cylindrical areas, if sampling is representative, the figures indicate the shape of the fold as seen in profile. For example, in an area of isoclinal folds, limb orientations predominate and the $\perp S$ figure should therefore be a point maximum in a weak or partial girdle (see Fig. 8.18b and e). Note that this approach has to be used with considerable care since the pole figure may be ambiguous. For example, a simple point maximum may represent unfolded planar bedding or isoclinal folds (see also Fig. 8.18b and e).

In order to construct a three-dimensional structure from the two-dimensional interpretation, the area is divided into fold domains. That is, it is divided into subareas in which the bedding or principal foliation is planar or folded into structures that, within reasonable limits, may be described as cylindrical. This subdivision may be carried out at any stage up to the time when it is needed, but the experienced geologist can generally recognize subareas during the process of mapping. Having delineated the domains the three-dimensional interpretation is completed by projecting map data down the plunge of the fold axis and then by merging the structures at domain boundaries [e.g., see Weiss (1959b)]. The main limitation of this method is that the three-dimensional configuration of domain boundaries is generally something that has to be assumed. Otherwise the method is as accurate, within domains, as the folds are cylindrical. In general the most useful way of presenting the interpretation is in the form of a block diagram. Various projections can be used for this purpose and perspective drawings are perhaps the most aesthetically pleasing but are not so easy to prepare. Another projection that is easier to construct and very suitable for the purpose, is the orthographic projection; its construction is described by McIntyre and Weiss (1956).

In some areas it is not possible to divide the structure into domains of reasonable size [e.g., Williams et al. (1971)] because of a predominance of small noncylindrical folds. In such areas the interpretation cannot be extended beyond the surface of observation and must remain essentially two dimensional.

8.5.3 Limitations of the Method

The interpretation of multiply deformed areas, in practice, commonly rests very heavily on the assumption that the members of a style group all belong in the same niche in the deformational sequence of events. Unfortunately it can be shown that this is not always true. In some areas outcrop is sufficiently continuous to allow the overprinting relationships of all or most of the folds to be determined. In such areas the folds can thus be ascribed to generations, without recourse to style, so that style variation within a generation can then be studied. In some areas where this has been done [Means (1963, 1966); Williams (1970); Olesen et al. (1973)] it has been found that there is no simple one-to-one correlation between style groups and generations. For example Means recognized two style groups and three generations; the first and third generation are characterized by styles F_1 and F_2, respectively, but the second generation comprises folds belonging to both style groups.

Thus style is unsatisfactory as a means of grouping structures, at least in some areas. Unfortunately, except in areas of unusually good outcrop there is no better basis; overprinting relationships are generally too few and far between and orientation can be completely unreliable. However, in some areas, style grouping does seem to produce a reasonable interpretation and it may be

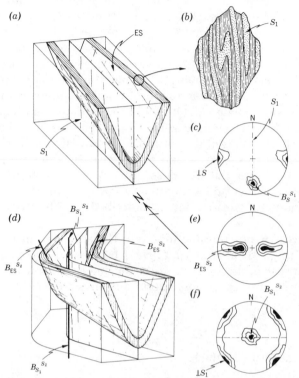

FIGURE 8.19   Transposition and the plunge of
small-scale folds. (a) Regional stratigraphy is folded into
a large, fairly tight structure and small-scale layering is
transposed parallel to the axial surface of the fold (S_1).
Thus the mapped distribution of various rock types gives
the position and orientation of the enveloping surface
(ES) but bedding as seen in outcrop (b) is parallel to S_1,
not to ES. (c) The orientation of the transposed S in an
area like that depicted in (a). In (d) the area has been
refolded and there are two types of B_2 folds. There are
folds in the enveloping surface $B_{ES}{}^{S_2}$ that have variable
orientations as shown in (d) and (e) but the orientations
cannot be measured and the presence of the folds is only
revealed by the mapped distribution of the various
lithologies. There are also folds in S_1 ($B_{S_1}{}^{S_2}$) and these
are visible and measurable in outcrop, but neither their
distribution nor the distribution of S_1(f) tell anything
about the orientation of $B_{ES}{}^{S_2}$ folds. See text for further
discussion.

that there are many areas in which it is a perfectly reasonable way of grouping folds. There is obviously an urgent need to check the validity of the method wherever extensive outcrop makes it possible to do so. Meanwhile in the absence of better criteria it is necessary to continue to use style as the principal basis for grouping structures and to look for other supporting evidence such as distinct orientation patterns or metamorphic assemblages associated with the style groups.

Another limitation of structural analysis is that it does not always provide a means of interpreting the macroscopic structure of areas in which the layering has undergone transposition. For example, in Figure 8.19a, layering has been transposed into a new orientation parallel to the axial surface of the large-scale fold. The latter structure is outlined by the enveloping surfaces (ES) enclosing various lithologies or, in other words, the fold outline is recorded in the distribution of the various lithologies. However, the enveloping surface is not a physical entity that can be seen and measured in the field, so that, in outcrop (e.g., Fig. 8.19b), the only surface for which the orientation can be determined is the transposition foliation (S_1), which in Figure 8.19a is vertical (see also Fig. 8.19c). Parasitic folds, however, indicate the plunge of the large fold so that there is no problem in interpreting the structure. If the area is now refolded about another axis as in Figure 8.19d there will be two types of B_2 folds; there will be B_2 folds in the enveloping surface and B_2 folds in the transposition surface. Because there are two types of B_2 folds a more explicit shorthand is required. The most informative notation is of the type $B_{S_1}{}^{S_2}$ which stands for a fold in S_1 with S_2 parallel to the axial plane [Turner and Weiss (1963, p. 133)]. This notation is cumbersome but very useful. The folds in the enveloping surface $B_{ES}{}^{S_2}$ will vary in plunge (Fig. 8.19d and e) but will not be visible in outcrop, whereas folds in the transposition foliation $B_{S_1}{}^{S_2}$ will all plunge parallel to the intersection of this foliation with the B_2 axial surface (i.e., plunge vertically, see Fig. 8.19f) and will be visible in outcrop. This means that if the lithological boundaries, recorded on the map, are projected down the plunge of the measurable B_2 axes ($B_{S_1}{}^{S_2}$) the resultant interpretation will be incorrect. In a simple, open structure like that depicted in Figure 8.19d, B_1 folds may be sufficiently well preserved to divide the area into B_1 domains and thus arrive at a reasonable three-dimensional structural interpretation. In extreme cases however, for example, if B_2 also results in transposition, it may even be impossible to determine whether the B_1 fold is antiformal or synformal.

STRUCTURAL ASSOCIATIONS

9.1 INTRODUCTION

Regional mapping has demonstrated that the earth's crust is made up of a wide variety of rock types in diverse structural configurations and various states of metamorphism. Despite this complexity, however, certain broad combinations of structures, or combinations of structures and lithologies, are found to occur repeatedly in many parts of the world and, to some extent, in rocks of different ages. We refer to these commonly occurring assemblages of structures as *structural associations*. Fifteen of them are described briefly in this chapter. No attempt is made to define each association precisely, because regional structures of various types grade into one another and preclude any neat classification. Some of the associations below overlap each other, and most of them are broad enough to include a number of different although related assemblages.

The aim of this chapter is to acquaint students with the main features of some prominent kinds of regional structure. A broad knowledge of structures from all over the world is a great asset for interpretation of any given map area, since interpretation tends to be based on past experience. The broader one's familiarity with structures elsewhere, the better the basis for interpretation locally. Besides this, of course, the study of structural associations is an important link between the geology of individual map areas and study of large-scale regional structure and tectonics.

In this chapter we emphasize the purely structural characteristics of structural associations, while recognizing that the stratigraphic, petrological, and geophysical characteristics may also be very significant.

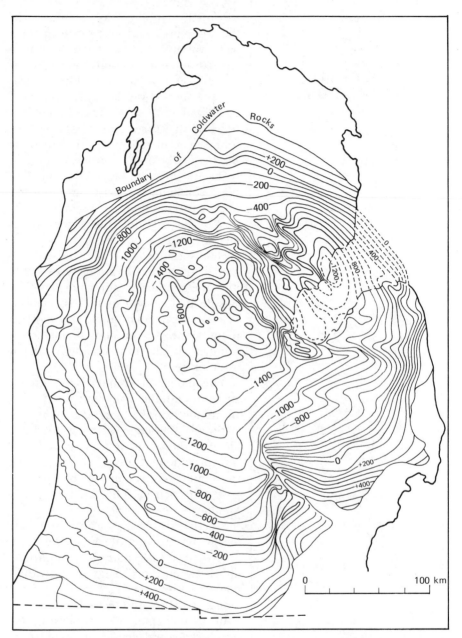

FIGURE 9.1 The Michigan Basin. Structure contours on the top of the Cold Water Formation. Contour values are in feet below sea level. [From de Sitter (1956), *Structural Geology*. Copyright © 1956 by McGraw-Hill Publishing Co. Used with permission of McGraw-Hill Book Co. After Brown-Monnet (1948).]

9.2   **FLAT-LYING SEDIMENTS**

In some areas of the earth's continental crust there are extensive deposits of sedimentary rocks that are essentially flat lying, or only gently tilted or warped, and that reach thicknesses of up to 5 or 10 km. An example is the Colorado Plateau region of the western United States, with about 3 km of sediments ranging in age from pre-Cambrian to Tertiary. The structure of bedding is generally very simple in such areas in that the dips are low; to a first approximation the layers are planar and horizontal. Seen in more detail, however, the surfaces of stratigraphic units typically display an irregular and complex form, as shown by *structural contour maps* (e.g., Fig. 9.1). Such departures from planarity may be significant commercially even if they are not very marked morphologically.

The most prominent structures in flat-lying sediments, apart from bedding, are systems of steeply dipping joints. These are generally found to have complex patterns, sometimes related geometrically to slight uplifts or depressions in the stratification. Although in many single exposures there may be as few as

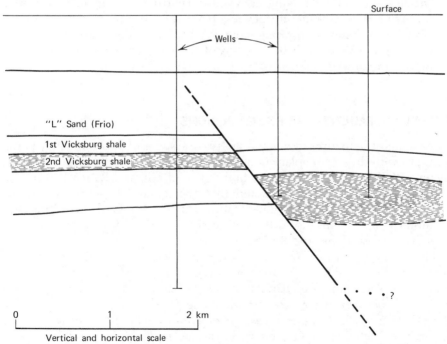

FIGURE 9.2   Growth fault in the Yturria oil field, Starr County, Texas. [From Bornhauser (1958). Published with permission of the author and of the American Association of Petroleum Geologists.]

two prominent joint sets at right angles to one another, it is usual to discover much more complexity than this when an inventory of joints is made over a wide area, and in rocks at different stratigraphic levels [e.g., Kelley and Clinton (1960); Nickelsen and Hough (1967)].

Regions of flat-lying sediments commonly display a peculiar kind of fault known as a *growth fault* [Ocamb (1961)]. These are typically normal faults across which there is abrupt thickening of stratigraphic units, from the hanging wall side to the footwall side, and along which there is a downward increase in the displacement of stratigraphic horizons (Fig. 9.2). Growth faults are interpreted as normal faults that are moving contemporaneously with deposition of sedimentary or volcanic rocks. They are well known in the Gulf Coast region of the United States [see Bornhauser (1958)] and are familiar in other deltaic environments, in association with salt domes, and around the margins of intracontinental fault-bounded depositional basins.

Areas of generally flat-lying sedimentary rocks may display zones termed *monoclines* where the dips are locally steeper. Where a monocline in cover rocks occurs over a high angle fault in basement rocks, the structure is sometimes called a *drape fold*.

The largest areas of flat-lying sediments on earth are sediments of the deep ocean basins. These range in thickness from very thin veneers near oceanic ridges to thicknesses of approximately 0.5-1.0 km in the main ocean basins [see Ewing and Ewing (1970)]. Details of the structure of deep ocean sediments are unknown.

9.3 BLOCK-FAULTED AND RIFTED REGIONS

The youngest structures in many regions are systems of approximately parallel faults with dip-slip displacements measured in hundreds or thousands of meters. Well-known examples are faults of the Basin and Range Province in the western United States, the system of rift valleys in East Africa, the Rhine graben in Germany, and the Baikal rift in Russia. Others of special interest, because of their continuity with oceanic rifts (Section 10.2.1), are the fault troughs of Iceland and the Red Sea.

The major faults may be concentrated in narrow zones either side of a central downthrown block, or they may occur quite evenly spaced over a wide area and isolate numerous parallel blocks (Fig. 9.3). All transitional patterns are known between these two extremes. The physiographic trough over the downthrown block in the first case is referred to as a *rift* or *rift valley*. The structure in both cases can be referred to as *block faulting*, a general descriptive term for all situations involving differential vertical movements of large, fault-bounded blocks. Individual fault blocks range in width from several tens of kilometers to upwards of 100 km. In regions like the Basin and Range

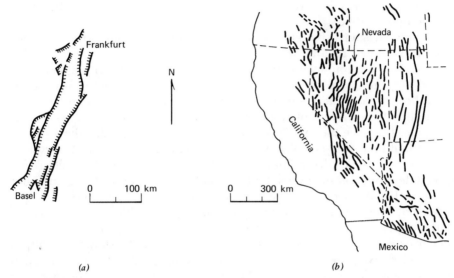

FIGURE 9.3   Block-faulted regions. (*a*) The Rhine graben between Basel and Frankfurt; faults hatched on downthrown side. [From Illies (1970), with permission of E. Schweizerbart'sche Verlagsbuchhandlung (Nägele u. Obermiller).] Basin and Range province and adjacent areas with similar structure, southwestern United States. Faults shown by heavy lines. [From Gilluly (1963), with permission of the Geological Society of London.]

Province the blocks may be displaced with or without tilting of initially horizontal surfaces. In rifted regions, the rift valleys tend to be situated on the crests of regional arches, and originally horizontal surfaces dip gently away from the central block [see Illies (1970)].

The classic interpretation of block-faulted regions is that they have been extended perpendicular to the strike of the faults [e.g., Gregory (1921)]. This interpretation is sometimes carelessly expressed by referring to such regions as "tensional." Note, however, that tension describes stress whereas extension refers to strain. Even if the evidence for extension is compelling, this does not necessarily mean that any of the principal stresses were tensile.

The evidence for extension of block-faulted regions is strong, particularly if we include the evidence for spreading across oceanic ridges (Section 10.2.1). Present-day extension across block-faulted regions has been measured using geodetic methods in the Basin and Range Province [Tompson (1966)] and in Iceland [Decker and Einarsson (1971)]. The rates of continental extension appear lower by one or two orders of magnitude than typical sea floor spreading rates. Oceans spread at rates of centimeters per year whereas the Rhine graben has extended 4.8 km in 45 m.y. [according to Illies (1970)] and the

Dixie Valley graben in Nevada has extended 3 km in 10 or 15 m.y. [according to Thompson (1966)].

Two qualifications should be attached however to any tentative conclusion that block-faulted regions are commonly extensional.

First, there is often uncertainty about the direction of dip of the faults. The classic interpretation takes them to be normal faults dipping toward the downthrown block, and this has been confirmed locally in many instances, especially where drilling or tunneling have permitted three-dimensional observations [e.g., see Illies (1970)]. It remains true, however, that the dip of the master faults has not been observed directly in most cases, so that the possibility exists that some of the faults are vertical faults or reverse faults. Widespread development of reverse faults would be associated with regional shortening rather than regional extension perpendicular to the faults.

Second, extension indicated by continental block faulting is only extension of the uppermost 5 or 10 km of crust. It may or may not be associated with similar extension of the rock at deeper levels. Figure 9.4 shows several proposed relationships between block-faulting and regional displacements. All of these models involve extension of the earth's surface, but only one (Fig. 9.4a) necessarily involves extension throughout the crust.

The strains associated with block faulting are incompletely known but the stresses are even less well known. One reason for this is that block faulting

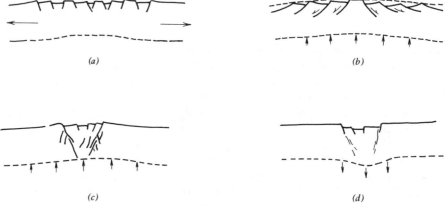

(a)

(b)

(c)

(d)

FIGURE 9.4   Some proposed relationships between near-surface block faulting and regional displacements at deeper levels. (a) Block faulting in response to crustal extension, with ductile yielding and necking of the lower crust [Artemjev and Artyushov (1971)]. (b) Normal faulting in response to crustal uplift, by large-scale "landsliding" [Moore (1960)]. (c) Rift valley formation along crest of a regional upwarp [Cloos (1939)], perhaps due to intrusion of a deep flat layer of igneous rock [Illies (1970)]. (d) Rift valley formation in response to downward vertical movement at depth [Cloos (1936)].

seems commonly to have its orientation controlled, not by stress alone, but by preexisting structures. Ratcliffe (1971) has presented evidence to show that Mesozoic block faulting in the Newark Basin of the eastern United States was controlled in orientation by preexisting strike-slip faults of possible pre-Cambrian age. McConnell (1972) has emphasized the importance of similar control upon the East African Rift system and suggests moreover that the Cenozoic fault system is geometrically and *genetically* related to pre-Cambrian structures.

Burke and Dewey (1973) and Hoffman et al. (1974) suggest that the break-up of continents that generates new oceans is initiated at mantle-generated uplifts, upon which three-armed rift systems commonly form. It is proposed that two of the rifts often spread and become filled with oceanic crust. The third rift or "failed arm" never opens far and is preserved as a sediment-filled fault trough that intersects a coastline, or perhaps ultimately a mountain belt (Section 10.2.6). A proposed example [see also Wright (1968)] is the northeast-trending Benue trough, which intersects the present western coast of Africa, at the reentrant between its North-South and East-West trending segments.

9.4 WRENCH FAULTS AND ASSOCIATED STRUCTURES

Major transcurrent faults in continental rocks are typically hundreds of kilometers long, have dominantly strike-slip displacements, and dip steeply. Examples include the San Andreas fault in California, the Alpine fault in New Zealand, the Anatolian fault in Turkey, and the Great Glen fault in Scotland (Fig. 9.5). The first three of these fault systems are seismically active at present; the earthquake foci are shallow, at depths of 20 km or less. At the surface such faults are represented by zones hundreds or thousands of meters wide, containing numerous individual shear surfaces in an intricate anastomosing arrangement. Cataclastic rocks such as breccia and mylonite may be well developed [e.g., see Waters and Campbell (1935)]. Recent average rates of displacement, obtained by geodetic and seismic measurements, are of the order of 1-10 cm/yr. Associated regional strain rates in the rocks adjacent to major wrench faults have been measured in a few places. Adjacent to parts of the San Andreas fault, the strain rate is 10^{-14} sec^{-1} [see Whitten (1956); Burford (1968); Scholz and Fitch (1969)]. The average displacement rate on wrench faults is maintained, in large part, by numerous sudden individual jumps of the order of 3 meters or less, which occur at any given time along a restricted segment of the fault trace (Section 7.3.11).

The most conspicuous major structures associated with wrench faults are regional systems of fractures approximately parallel with, and commonly branching off, the main present-day fault (Fig. 9.5). In some cases, one or

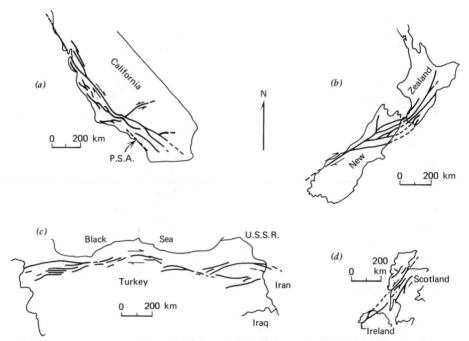

FIGURE 9.5   Four major wrench fault systems. (*a*) San Andreas fault system. [From Moody and Hill (1956), with permission of the Geological Society of America.] Dashed segment labeled P.S.A. is "Proto San Andreas" of Suppe (1970), see text. (*b*) Alpine fault system. [From Lensen (1961), with permission of the Earth Physics Branch, Department of Energy, Mines and Resources, Canada.] (*c*) Anatolian fault system. [From Ilhan (1971), with permission of Petroleum Exploration Society of Libya.] (*d*) Great Glen fault system. [From Pitcher (1969). Published with permission of the author and of the American Association of Petroleum Geologists.]

more of these currently less-active faults may have been the main locus of displacements at some earlier stage in the history of the whole fault system. Thus, Suppe (1970) proposes that the most active part of the San Andreas system south of the Transverse ranges shifted eastward in middle Tertiary time as indicated in Figure 9.5*a*. Wrench faults are also commonly associated with shorter strike-slip faults at a high angle to the main faults; these may be conjugate faults of approximately the same age as the main fault.

A second kind of fracture pattern associated with wrench faulting occurs in poorly consolidated materials overlying active faults at depth. An example of several stages in the development of such a pattern is shown in Figure 9.6. Another example of cover structures believed to be caused by strike-slip faulting at depth are lines of *en-échelon* folds shown in Figure 9.6*d*. These patterns of *en-échelon* fractures and folds associated with transcurrent movements are

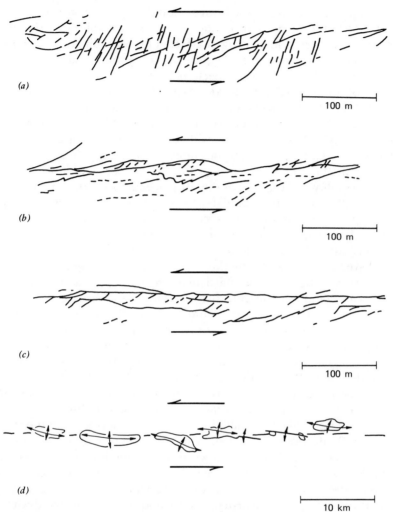

(a)

100 m

(b)

100 m

(c)

100 m

(d)

10 km

FIGURE 9.6 Surface structures associated with underlying wrench faults. (a), (b), and (c) are fractures in ground above a wrench fault in Iran. [From Tchalenko (1970), with permission of the Geological Society of America.] (d) are *en-échelon* anticlinal oilfields above the trace of the Inglewood fault, California. [From Moody and Hill (1956), with permission of the Geological Society of America.] (a) and (b) are believed to represent successive stages in the development of (c). (a) Development of earliest fractures in two orientations, neither parallel to the underlying fault. (b) Intermediate stage; prominent fractures more continuous and at lower angle to underlying fault. (c) Through-going fractures developed parallel to underlying fault. (d) has been reversed so that the sense of fault displacement corresponds to (a), (b), and (c).

of interest not only because of the information they yield on concealed faults, but also because they may help to suggest the structural and stratigraphic responses to be expected in an orogen affected by a combination of transcurrent and normal displacements (e.g., an orogen evolving at a trench with mixed trench-transform character, Section 10.2.5).

Most geologists now accept the idea that displacements of the order of several hundred kilometers can occur on major wrench faults. This was not the case about 20 years ago when Kennedy (1946), Hill and Dibblee (1953), and Wellman (1955) first set out the evidence for large displacements on the Great Glen, San Andreas, and Alpine faults, respectively. These and other contributions in the 1950s played a major role in persuading geologists to admit the possibility of large horizontal displacements on the earth. The acceptance of such movements stimulated renewed interest in continental drift and helped pave the way for plate tectonics. The total displacement on the San Andreas fault, the most studied of the major wrench faults, is now considered by many current workers to be of the order of 300 km in the area south of the Transverse ranges and something in excess of 500 km in the area north of the Transverse ranges. On this particular wrench fault, and others, current geological investigations are largely directed toward establishing the history of displacements by which this total accumulated [see Dickinson et al. (1972)].

A major unsolved problem is concerned with the genetic relations suggested by the repeated occurrence of wrench faulting in association with fold and thrust belts [e.g., see Dickinson (1966)]. Examples like the San Andreas and Alpine faults seem to have been long-lived structures that were active during at least the last stages of folding of rocks now exposed nearby, and which may be associated even now with continuing folding not far below the surface. Examples like the Phillipines fault and others around the margin of the western Pacific occur behind and subparallel to oceanic trenches [Allen (1962)] in an association that again suggests some genetic connection between wrench faulting and folding. There are probably a variety of possible relationships between these two processes. One of them, which has been suggested for the San Andreas system in particular, is that folding was associated with underthrusting at a trench bordering Mesozoic California [Hamilton (1969); Hsu (1971)] and that part of this trench was subsequently changed into a transform fault by collision of the trench with a part of the oceanic ridge system [McKenzie and Morgan (1969); Atwater (1970)]. A schematic diagram explaining this evolution is given in Figure 10.26.

The role and importance of wrench faults in the early history of orogens is almost completely unknown. One difficulty here is that it may be difficult to recognize whether or not a given mapped discontinuity originated as a wrench fault or, for example, as a thrust fault. It seems possible that some of the longitudinal boundaries that separate different structural-stratigraphic zones in orogens were at one time zones of active strike-slip faulting. A possible example from New Zealand has been suggested by Landis and Bishop (1972).

9.5 SHALLOW FOLD AND THRUST BELTS

In some areas it has been argued from deep drilling, three-dimensional exposure, and seismic investigations, that an internally folded and thrusted sequence of shallow rocks passes downward abruptly, across an extensive undeformed surface of discontinuity, into rocks that have not been deformed, at least not during the same interval of time. The surface or zone of discontinuity is called a *décollement* zone. Examples of this kind of structure are found in the Canadian Rocky Mountains, the Jura Mountains, the Zagros Mountains of Iran, the foothills region of western Papua, and the westernmost part of the Appalachian Mountains. In fact, shallow fold and thrust belts are found on the continental or "foreland" margins of virtually all present-day fold mountain belts, and they are sometimes referred to as *foreland thrust belts* [see Coney (1973)]. The rocks of such belts are typically un-

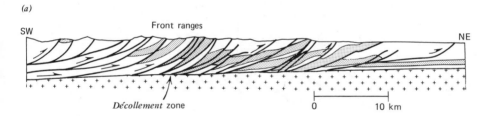

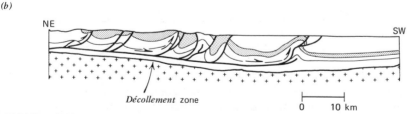

FIGURE 9.7   Shallow fold and thrust belts. (*a*) A portion of the Rocky Mountains thrust belt in southwestern Alberta, as interpreted by Price and Mountjoy (1970). Shaded rock is Devonian through Permian in age. [From R. A. Price and E. W. Mountjoy (1970). Reprinted with permission of the Geological Association of Canada.] (*b*) Foreland folded belt of Papua, as interpreted by Smith (1965). Shaded rock is Paleogene to Upper Miocene in age. [From Smith (1965), with permission of Elsevier Scientific Publishing Co.] Crystalline basement shown by crosses in both sections.

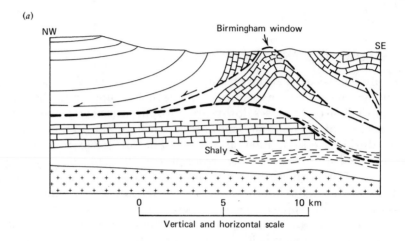

(a)

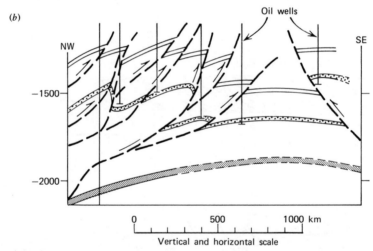

(b)

FIGURE 9.8   Two common relationships between thrusts and folds in shallow fold and thrust belts. (a) Anticlinal folding above a step in an underlying thrust fault (heavy dashed line). In the case illustrated, from central Pennsylvania, an earlier thrust is folded along with bedding in the underlying limestone, giving rise to a window. (b) Thrusts on limbs of an anticline, dipping away from the hinge line of the fold, South Summit oil field, Pennsylvania. Dotted unit is Devonian sandstone; shaded unit at base is Silurian salt. Figures to the left of the section give depth below surface. [From Gwinn (1964), with permission of the Geological Society of America.]

metamorphosed sedimentary sequences of miogeosynclinal or exogeosynclinal type [Kay 1951)]. These rocks, deformed at a shallow level in the crust, are typical of what is referred to as the *suprastructure* of mountain belts, in distinction from the once-deeper, more metamorphosed, and differently deformed rocks of the interior parts, which comprise the *infrastructure*. This distinction is not always easy to make.

The structure of shallow fold and thrust belts may be dominated by thrusts or by folds. A transition along strike, from dominant folding to dominant thrusting, is reported in the Appalachians [Gwinn (1964)] and in the Canadian Rocky Mountains [see Wheeler et al. (1972)]. Figure 9.7a shows a typical example of structure dominated by nested or *imbricate* thrust faults. The faults are concave upward and dip southwestward, toward the metamorphic parts of the mountain belt. The *stratal shortening* accomplished by telescoping the sedimentary sequence is estimated in Figure 9.7a as about 50 percent [Price and Mountjoy (1970)]. The accompanying *crustal shortening* of the underlying crystalline basement may have been zero.

Thrust faults within these belts, as well as the basal zone of décollement commonly follow incompetent rock units containing shale, salt, or gypsum. A given thrust often follows an incompetent horizon for some distance and then "steps up" through a competent unit to the next incompetent layer. An anticlinal fold is normally associated with each step in the fault surface, as shown in Figure 9.8a. Another important feature is the occurrence of subsidiary thrusts up both limbs of anticlines as shown in Figure 9.8b. A final characteristic of the fault pattern, well developed in the Jura, are strike-slip *tear faults* oriented at a high angle to the length of the belt, sometimes in conjugate sets [see Laubscher (1972)].

Figure 9.7b shows an example of a belt with better development of folds and less prominent faulting than Figure 9.7a. The folds in such regions are characteristically rather large open structures with wavelengths measured in kilometers. Their profiles are typically parallel in style and are often somewhat angular (kink or boxlike folds), but they may also be rounded. Anticlines tend to have narrower hinges than synclines, which may have very broad flat troughs. Axial plane cleavage is not conspicuously developed.

Major problems in the interpretation of these belts include the question of whether the folds and thrusts really were generated entirely independently of any deformation in rocks below the décollement [see Rodgers (1964)], and the related question of the source of the forces responsible for the thrusting and folding.

9.6 TECTONIC MÉLANGES

In strongly deformed rocks, originally continuous stratigraphic units are usually found to be disrupted in some degree by faulting, boudinage, or transposition. In certain deformed rock bodies, however, the destruction of stratigraphic continuity is a very prominent and pervasive feature and blocks or fragments of more brittle rock types are found *dispersed* and *mixed* in a more ductile matrix. The mixing process is sufficiently large-scale so that *exotic blocks* (originating remote from their present lithic surroundings) are found juxtaposed with blocks and matrix of entirely different lithology or age. Such bodies, where they originate by deformation, are called *tectonic mélanges*. The term mélange was used first by Greenly (1919) for rocks on Anglesey and has since been applied especially to very big areas characterized by this kind of structure. Two examples, both involving areas measured in thousands of square kilometers, are the Ankara mélange in Turkey [Bailey and McCallien (1950)] and mélanges in the Franciscan formation of California [Bailey et al. (1964); Hsu (1971)].

Mélange structure seems to be characteristic of low-grade eugeosynclinal terrains made up dominantly of greywacke and siltstone and containing members of the ophiolite suite (cherty sediments, pillow lavas, gabbros, and ultramafic rocks). Blocks of high-pressure metamorphic rocks [see Miyashiro (1973, p. 308)] may also be present. The matrix between blocks of all these lithologies is typically fine-grained pelitic rock or serpentine. The blocks themselves range in size from microscopic fragments to flat slabs several kilometers in their maximum dimensions. In the Franciscan formation, there is a weak regional structural grain parallel to the outcrop belt of Franciscan rocks, defined by preferred orientation of the long axes of the larger blocks and by preferred orientation of bedding and shear surfaces. The blocks may or may not be internally deformed. Where internal deformation is prominent, as for example in glaucophane schist blocks in the Franciscan formation, it is abruptly truncated by the faults separating the block from the surrounding matrix. The matrix itself is pervasively sheared and often intricately folded. The shears are commonly approximately parallel to bedding and are seen in practically every exposure [Bailey et al. (1964)]. Detailed mapping of mélange structure is made difficult by the relatively chaotic nature of the structure itself and by the fact that it occurs in rocks that may *never* have contained good continuous stratigraphic units and marker horizons. Detailed maps for parts of the Franciscan formation are given by Ernst et al. (1970).

Mélange structure was attributed by Greenly (1919) to thrust faulting and was seen to be associated locally with imbricate thrust faulting by Bailey and McCallien (1950). Hsu and Ohrbom (1969) compared mélange development to very large-scale land-sliding, such as occurred at Brientz, Switzerland, where a section of rock about 200 meters thick by 1.2 km long slid down a 20° slope, with much internal shearing and deformation [Heim (1932)]. Hamilton (1969), Ernst (1970), and Hsu (1971) have subsequently interpreted the Franciscan mélanges as a product of large-scale underthrusting of the American lithospheric plate (Section 10.2) by rocks underlying the adjacent Pacific Ocean. In this view, the mélange was built up by mechanical mixing of flysch turbidites of relatively local origin with fragments of oceanic crust and mantle, the ophiolite suite, emplaced from the West by large-scale horizontal movement of the sea floor. For Hsu (1971) the essence of a mélange of Franciscan type is that it contains, tectonically juxtaposed, rocks from widely separated realms of deposition. To explain how mélange structure develops over belts many tens of kilometers wide in the Franciscan formation, Hsu proposes that the zone of maximum deformation migrated progressively westward.

If the foregoing interpretation applies to a given mélange, then its deformational history is likely to be long and complex. The ophiolitic components may show structures connected with faulting or other deformation predating their emplacement in the trench. The trench deposits proper may show early structures due to relatively shallow slumping or tectonic deformation. Superposed on such earlier structures, both components of the mélange will show whatever structures result from their interaction at the trench. This is the site at which most of the breaking apart and dispersal of the ophiolitic components in the clastic matrix must occur. As Hsu (1971) points out, some of the blocks, particularly of the glaucophane schists, may be derived by subaerial erosion and enter the mélange as very large sedimentary clasts. A tectonic mélange may accordingly contain some large fragments that are not tectonic in origin. A body consisting mainly of such fragments would properly be classed as a very coarse sedimentary breccia, or as a *sedimentary mélange* or *olistostrome*, if there is marked mixing of rock types and chaotic structure due to submarine slumping [see Elter and Trevisan (1973)]. Tectonically sheared olistostromes may closely resemble ordinary tectonic mélanges [see Hsu (1974)].

The study of mélange structure is still in its infancy. The details of the mixing process remain very obscure, it is hard to distinguish sedimentary from tectonic features, and serious questions can be raised about the origin of the ophiolitic components. Thus, Blake and Jones (1974) have seen evidence that serpentinite blocks in some Franciscan mélanges are not pieces of far-traveled Pacific lithosphere, but are instead derived more locally, perhaps from beneath a marginal basin (Section 10.2.2) at the western edge of the American plate.

9.7 DOME AND BASIN STRUCTURE

Although ideal domes and basins are highly symmetrical structures (Fig. 4.4a, b) with approximately circular outcrop patterns, the terms dome and basin are also used more loosely for doubly plunging anticlines and synclines (see Section 4.2) with slightly more elongate closed outcrop patterns that characterize the structural association discussed here. Areas with these structures widely developed are commonly weakly metamorphosed, if at all, and tend to lack widespread development of slaty cleavage parallel to the axial planes of folds.

Dome and basin structures are common in foreland areas of fold belts [Lees (1952)]. Some classical examples from oil-producing areas are the Kettleman Hills area in California [Woodring and Stewart (1934)] and many fields in Iran [Lees (1952)].

The folding in dome and basin areas tends to be concentric or parallel in style, so that the folded surfaces can commonly be represented quite accurately by arcs of circles at each point. This is a common method of construction of cross sections in dome and basin areas and is referred to as the *Busk method* [Busk (1929)]. A section constructed using the Busk method is illustrated in Figure 9.9.

In many areas of dome and basin structure there is very little internal deformation of beds so that, as noted above, the pelitic beds lack a slaty cleavage whereas the sandstones tend to be strongly jointed rather than foliated and lineated; clastic grains show little sign of internal deformation although fracturing or weak development of deformation lamellae may be characteristic of some grains. Slickensides are commonly developed on bed-

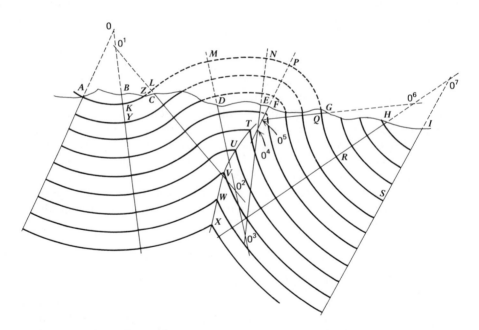

FIGURE 9.9   Diagram illustrating the Busk method. At each point in the structure the orientation of bedding planes has been approximated by the arc of a circle [from Busk (1929)]. At H and I, for instance, lines HR and IS are drawn normal to the dip at H and I, respectively. These lines meet at 0^7, which then forms the center of a family of concentric circles that represent the beds in that interval. The other points 0 through 0^6 have the same significance within their respective intervals.

ding planes as well as on joints and faults indicating that bedding plane slip is an important part of the deformation mechanism.

The lack of deformation within beds on the granular scale means that most of the overall shortening in these areas can be estimated by comparing the

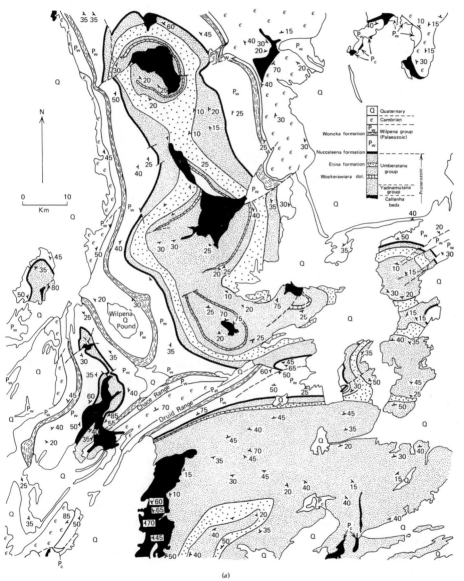

(a)

FIGURE 9.10 (a) Map of the Wilpena Pound area in South Australia. [After Parachilna 1:250,000 sheet of S.A. Geological Atlas Series. Sheet H54-13 Zones 5 and 6 (1966).] (b) Stereographic block diagram of the area shown in (a) looking due south.

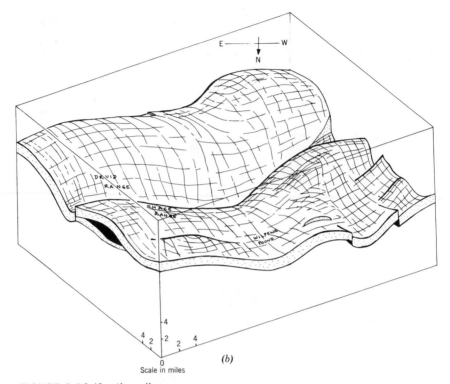

FIGURE 9.10 (Continued)

width of the fold belt with the arc length measured along a folded bed that
crosses the area. De Sitter (1939) has pointed out that the maximum shortening that can result from pure concentric folding without internal deformation
of beds is 36 percent. Such a number probably represents quite closely the
order of shortening in many dome and basin areas. Dome and basin areas,
therefore, correspond to regions of relatively low overall shortening. This is
illustrated in Figure 9.10 where in (a) a map of part of some dome and basin
structures in Cambrian rocks of South Australia is shown. Although the structure in the map appears to be rather complicated it actually consists of a series
of gentle warps in an otherwise flat dipping sheet as illustrated in Figure
9.10b. This is true of many dome and basin areas. This, together with the lack
of internal deformation within these beds (except for development of deformation lamellae in quartz grains) means that the overall shortening in
this South Australian example is relatively small — perhaps of the order of
30 percent.

The fact that folding is essentially parallel in style means that a particular
fold may meet volume problems upward and downward along its axial sur-

face. This is admirably brought out by use of the Busk construction as shown in Figure 9.9. Figure 9.11 shows the various regions within a particular fold that result from these volume problems. A particular layer B may, in fact, be simply folded in a parallel manner. However, another layer A above B may need to be extended in order to conform to the fold shape imposed by B. In so doing normal faults may develop along the hinge of the fold in layer A. Similarly, another layer, C, in the core of the fold may develop thrust faults in order to conform to the shape change prescribed by layer B. A map of an area that appears to show this kind of behavior is illustrated in Figure 9.12.

Another kind of fault pattern is common in dome and basin areas and is associated with changes in plunge along the hinge of an individual fold. This is known as *cross faulting* and is common in such situations; examples are shown in Figure 9.13.

The origin of dome and basin structure is imperfectly understood. In many areas these structures are associated with salt domes (e.g., in many of the southern states of the United States) or with other diapiric structures. For example, the dome and basin structures illustrated in Figure 9.10 are associated with shale diapirs (see Fig. 9.10*a*). The widespread development of faulting together with the overall lack of metamorphism implies that these

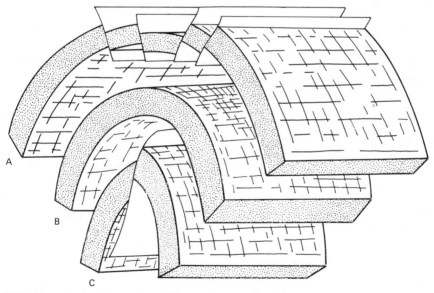

FIGURE 9.11   Block diagram showing the various regions within a concentrically folded sequence. Layer A has been folded so that there is normal faulting in the hinge. Layer B has been folded with no faulting. Layer C in the core of the structure has been folded and thrust faulted.

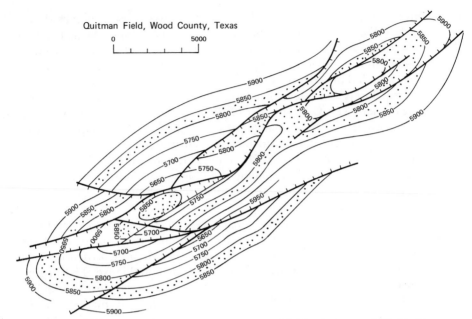

FIGURE 9.12   Faults parallel to hinge region of folds. Quitman Oilfield, Texas. [After Smith (1951) in de Sitter (1956)].

areas have been deformed at relatively shallow depths in the earth's crust. It is not clear, however, whether dome and basin structure is necessarily associated with diapiric behavior or whether the domes and basins represent embryonic folds that are just starting to be amplified in a more-or-less undeformed sheet.

9.8 SALT DOMES

Salt is a sedimentary rock that, on account of its low density (2.2 g/cm³) and low equivalent viscosity $[10^{18}\text{-}10^{21}\text{P}$; Heard (1971)], plays a number of special roles in structural geology (see also Section 9.5). In this section we review briefly the nature of thick, columnar, stock or pluglike bodies of salt known as *salt domes*. Such bodies occur by the hundreds in the flat-lying sediments of the Gulf Coast region of the United States. Other well-known examples occur in the Persian Gulf and in Germany.

Salt domes are one example of rock bodies that move upward and pierce overlying strata. Such bodies in general are called *diapirs* [see O'Brien (1968)]. Diapiric structures may or may not be associated with regional deformation. In the case of the Gulf Coast domes, it is considered that the salt is driven upward entirely by buoyant forces resulting from the density contrast between the salt and the heavier rocks above it [Nettleton (1934)].

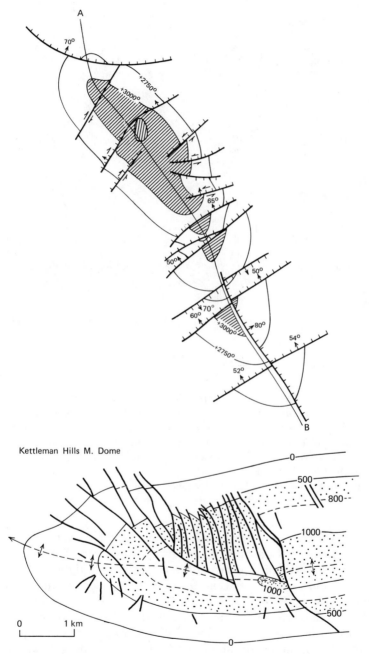

FIGURE 9.13 Faults at a high angle to fold hinge. (*a*) Elk Basin
Oilfield. [After Bartram, (1929)]. (*b*) Kettleman Hills, Middle Dome,
California. [After Woodring et al. (1941).] [Both in de Sitter (1956).]

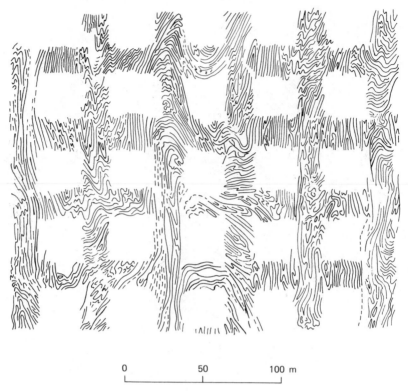

0   50   100 m

FIGURE 9.14   Map showing typical internal structure of a Gulf Coast salt dome. Map section at 200 meters below sea level, Morton Salt Co. mine, Weeks Island dome, Louisiana. Layering defined by alternation of pure halite layers and halite layers with several percent anhydrite. Dips are very steep, typically greater than 80°. Note evidence for refolding in central part of map. [From Kupfer (1968). Published with permission of the author and of the American Association of Petroleum Geologists.]

Salt domes are typically 2 or 3 km in diameter and have very steeply dipping sides that may extend downward for several kilometers. The internal structure of the salt is dominated at middle levels by very tight folds with near-vertical axes and axial planes (Fig. 9.14). Refolding or multiple refolding is common [Kupfer (1968)]. This may be connected in part with episodic upward movement of the salt, which is believed to rise in a complex fashion by the advance of local spines and lobes [Muehlberger and Clabaugh (1968)].

Structures associated with salt domes in the surrounding rocks are systems of normal faults over the tops of the domes and upturned strata along their flanks (Fig. 9.15).

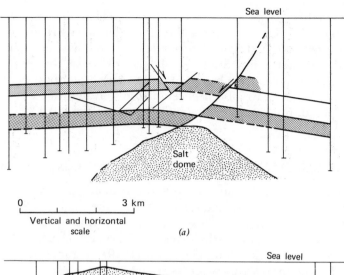

Sea level

0   3 km

Vertical and horizontal
scale   *(a)*

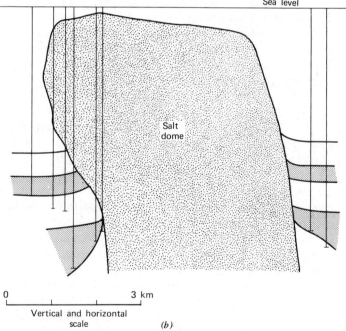

Sea level

Salt
dome

0   3 km

Vertical and horizontal
scale   *(b)*

FIGURE 9.15   Structures in rocks surrounding salt domes. (*a*)
Normal faulting and arching of strata above salt, Iowa salt dome,
Louisiana. (*b*) Upturned edges of strata on flanks of salt dome, Cote
Blanche salt dome, Louisiana. Vertical lines in both sections are
drill holes. [From Atwater and Forman (1959). Published with
permission of the author and of the American Association of
Petroleum Geologists.]

Salt dome salt, like glacier ice, is a metamorphic rock in the sense that it has undergone extensive flow and recrystallization. Salt domes in areas like the Gulf Coast are examples of highly deformed bodies that have arisen independently of regional shortening. Study of salt structures may therefore help to establish whether or not certain bodies of deformed silicate rocks (such as the cores of gneiss domes) have also been emplaced and deformed by buoyant forces as proposed, for example, by Ramberg (see Section 10.3.3).

9.9   SIMPLE FOLD SYSTEMS WITH STEEPLY DIPPING AXIAL SURFACES

Many deformed areas in the earth's crust consist of regular trains of gently plunging folds all more or less the same size with the same orientation and steeply dipping axial surfaces. Classical localities include the Appalachian fold belt in Pennsylvania and nearby states [Rodgers (1950, 1953, 1970)] and the front ranges of the Rocky Mountains [Link (1949)]. Figure 9.16 shows part of the Rocky Mountain structure whereas Figure 9.17 shows similar structures in the Central Western fold belt of New South Wales [Hobbs and Hopwood (1969)]. These areas are characterized by extreme regularity in fold shape and disposition, and individual fold hinges may be traced for perhaps up to 20 km with very little change in plunge or trend.

Areas deformed in this manner are unmetamorphosed or weakly metamorphosed in the low-to-medium greenschist facies and may possess well-developed axial plane slaty cleavage in suitable rock types. In fact many slate belts (see Section 9.10) contain folds of this type. The fold style tends to be close to parallel although there is some thickening on the hinges of folds leading to class IC folds in competent beds and class III folds in incompetent beds. Faults tend to be relatively rare in these regions and are restricted either to the margins of the zone or to zones of relatively high deformation where the faults tend to be reverse faults approximately parallel to the axial planes of the folds. Quartz veining is a prominent characteristic of these belts, the veins tending to occur in joints approximately normal to the fold axes or as prominent elongate quartz masses parallel to the axial plane cleavage.

Surprisingly, relatively little modern structural work has been done in these areas and little is known as to what governs the wavelength of these structures or why they tend to be so uniform and continuous. It is probable that the wave trains produced in these rocks are controlled by particular layers in the sequence that are thick and strong. These layers are the first to start to fold and folds nucleated in these layers are amplified to ultimately dominate the structure of all surrounding rocks. In the sequence shown in Figure 9.17, for instance, the dominant members of the sequence that may control the wavelengths of folds are probably the Merrions Tuff near the top of the sequence and the Sofala Volcanics near the base. Both units are thick, weakly

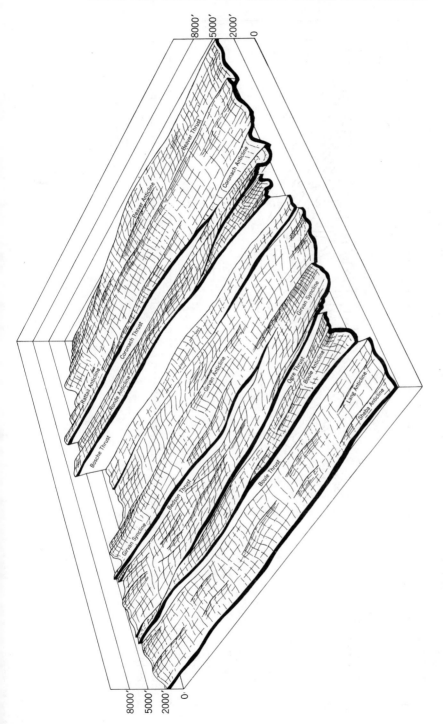

FIGURE 9.16   Block diagram of part of the front ranges of the Rocky Mountains, U.S.A.

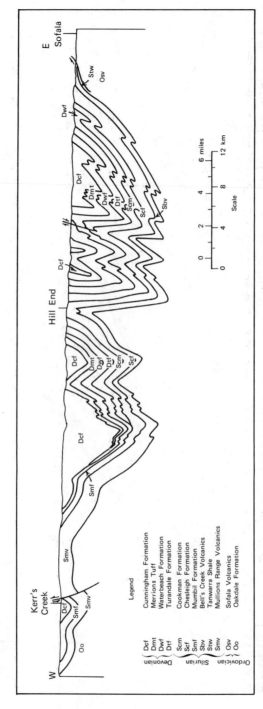

FIGURE 9.17 Cross section through part of the central western fold belt of New South Wales [from Hobbs and Hopwood (1969)].

deformed, but highly competent rocks. Detailed discussion of this matter is presented by Johnson (1970, pp. 75–126).

Areas deformed in the manner described above are completely gradational to slate belts (Section 9.10), the main difference being one of lithology.

9.10 SLATE BELTS

Slates are a common rock type in many of the worlds orogenic belts. Typically they outcrop in narrow zones that are elongate parallel to the orogenic belt, but there are exceptions. The area of North Wales occupied by slates, for example, is almost equidimensional [Smith and George (1961)]. However, if this area is subdivided on the basis of age, the Cambrian slates are found to occur in a narrow belt along its northern edge. In view of its general applicability, the term *slate belt* is used to describe this association despite the fact that not all slate belts have a linear outcrop pattern.

The definition of the slate belt association is, of course, as dependent on lithology as on structure. Slates are low-grade metamorphic rocks of restricted parentage. Metamorphically they belong in the greenschist facies and are commonly found below the biotite isograd. Typically they are metasediments of silt or finer grain size but they are sometimes derived from igneous rocks; for example, some of the slates of the English Lake District are derived from lapilli tuffs [Taylor et al. (1971, p. 14)] and some of the Mariposa slate of California is believed to be derived from igneous rock of gabbroic composition [Dale (1914, p. 68)]. Generally slates are interbedded with other rocks of sedimentary origin, such as sand and gritstones, conglomerates, novaculites, and limestones. They are also very commonly pyritic and/or carbonaceous. In some belts, such as the commercially producing slate areas, the slate may be the most abundant rock type, but elsewhere, it may be a minor constituent of the stratigraphic column.

The most obvious structural feature of slate is the cleavage (see Section 5.2.4) along which it is readily split. Less obvious and less perfect is another surface parallel to which the rock may be split: the *grain*. This surface is exploited by quarrymen in the dressing of slate into workable pieces or into the final product. It is perpendicular to the cleavage and parallel to a lineation that is generally visible on the cleavage plane. This particular lineation is usually inclined to the trace of bedding on the cleavage surface, at an angle close to 90° (see Fig. 6.3). It is referred to as the *down dip lineation* because of its common orientation parallel to the direction of dip of the cleavage (see Fig. 6.3).

Folding in slate belts is generally simple, although the folds are commonly tight to isoclinal. In most areas the folds are upward facing, hinge lines are doubly plunging with shallow plunges predominating, and axial planes are steep to vertical with the axial planes of parasitic folds commonly fanning across the major anticlinoria and synclinoria [e.g., the Arkansas slate belt, see Purdue (1909)]. In commercially producing slate areas the slate beds are generally thick (several meters or more) and folds tend to be large with amplitudes measurable in meters or tens of meters. In areas where thin (a few centimeters) beds of different lithology are intercalated, small parasitic folds, with amplitudes measurable in centimeters, generally abound. If the slates are interbedded with other lithologies, the cleavage generally changes orientation as it passes from one rock type to another (see Fig. 5.3). In the slate the fans are almost invariably divergent though the amount of divergence may be very small and in the other rock types the cleavage is convergent (see Fig. 5.2).

Various joint patterns are found in slate belts but two systems occur recurrently: a system of joints perpendicular to the down dip lineation and a system of *ac* joints, which are often parallel to the grain. Both systems, but particularly the first, are commonly occupied by quartz and/or carbonate veins. Various patterns of faulting may also be found in such regions, but again there is one type that is common in this association, and that is faulting parallel to the cleavage. Such faults are very well developed in some areas [e.g., North Wales, see Morris and Fearnsides (1926)] and vary in magnitude of displacement from a few millimeters to tens of meters. They may or may not be genetically related to the development of the cleavage (see Section 5.2.2).

It is common to find kinks and/or crenulation cleavage overprinting slaty cleavage in slate belts. The kinks tend to occur in "swarms" that, in some areas, are spatially related to large faults. They quite commonly occur as conjugate pairs (see Fig. 4.16a) but even where they are abundant, their contribution to the large-scale structure is generally small. The crenulation cleavage may be related to a second generation of folds and, if such folds are large, the structure becomes more complex and there is gradation into the multiply deformed belt association (Section 9.11). In other areas however the crenulation cleavage is not accompanied by much folding [e.g., see Hoeppener (1956)] and the structure remains comparatively simple.

Various attempts have been made to analyze the strain in slate belts using such objects as deformed fossils, lapilli, or reduction spots. They indicate that the cleavage is now perpendicular, or approximately perpendicular, to the direction of maximum shortening. The most reliable of these analyses appear to be ones based on reduction spots and they indicate shortening perpendicular to cleavage generally in excess of 55 percent [see Wood (1973)]. Another interesting point has emerged from Wood's work (1974) on reduction spots in the slates of North Wales. He has shown that the amount of shortening perpendicular to the vertically oriented cleavage varies from point to point within the belt and that, where the amount of shortening is large, the amount of vertical extension is also large, and conversely, where shortening is small, vertical extension is small. Furthermore, there is a spatial correlation between large vertical extension and fold culminations and small vertical extension and fold depressions, indicating that, in the Cambrian slate belt of North Wales, the variation in fold plunge is due to heterogeneity of strain.

9.11 MULTIPLY DEFORMED BELTS OF LOW AND MEDIUM METAMORPHIC GRADE

Multiple or "polyphase" deformation, indicated by the presence of folds or other structures of more than one generation (Section 8.2.4), is a normal feature of regionally metamorphosed rocks. It is especially obvious in rocks of low and medium metamorphic grade, where very abundant layer silicates are present to define foliations, but only if late-stage deformation and recrystallization have been moderate, so that early structures are not obscured. In high-grade rocks, on the other hand, multiple deformation may be equally common, but it may leave a less clear record in the rocks (Section 9.16). Although many multiply deformed areas are characterized by complex outcrop patterns (see Fig. 9.18) and complex three-dimensional structures, there are also areas with complicated histories that appear simple at first sight because of the intensity of late deformation.

Major portions of virtually all the world's fold-mountain belts are made up of multiply deformed rocks of low and medium metamorphic grade. Such regions have no consistent shape. They are "belts" only in the general sense that they tend to be elongate parallel to the length of the mountain chain.

The following geometrical features are characteristic of multiply deformed belts:

1. Noncylindrical folds are common [Reynolds and Holmes (1954); King and Rast (1955); Ramsay and Sturt (1973a)] so that equal area projections of poles to the first observable layering are generally complicated [see Weiss and McIntyre (1957)].
2. Lineations tend not to be straight but have complicated orientation patterns depending on whether they have initially formed in a curved foliation plane [Weiss (1959a)] or they have been distorted after they developed [Weiss (1959a); Ramsay (1960); Ramsay (1967); Hobbs (1971)].
3. Interference of different generations of folds of about the same size leads to complicated outcrop patterns (see Section 8.2.4); large-scale, very tight dome and basin structures appear to be very common [Ramsay (1962b); Tobisch (1966); Hansen (1971)].
4. Lineations of an extensional type are commonly parallel to fold axes [Hossack (1968); Hooper (1968)]. There remains a problem in determining if this always implies large-scale extension parallel to the fold axes [see Section 6.4 but also see Ramsay and Sturt (1973b)].

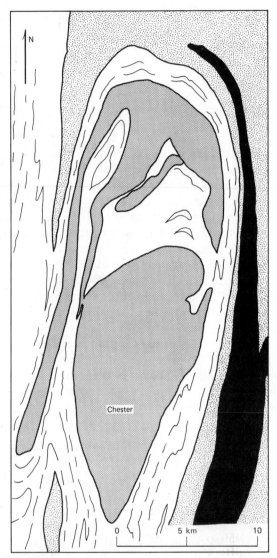

FIGURE 9.18 Outcrop pattern in the vicinity of the
Chester Dome, Vermont. Rock units are as follows:
pre-Cambrian Mount Holly Complex (gray); Cambrian
and Ordovician rocks (white); Devonian rocks
stratigraphically below the base of the Standing Pond
volcanic member (stippled); Devonian rocks above the
base of the Standing Pond member (black). [After Doll et
al. (1961).]

A certain time sequence of fold styles (Section 8.2.2) is now very familiar from detailed mapping in many mountain belts. First-generation folds are commonly tight to isoclinal and may have slaty cleavage or a schistosity as axial plane foliation. These folds are usually followed, at any given place, by one or more generations of folds having a crenulation cleavage as axial plane foliation. Such folds are commonly more open than the first-generation folds, but they may also become very tight, and their axial plane foliation may pass from an obvious crenulation cleavage to a schistosity that can mimic the first-generation structure. Finally, there may be one or more generations of folds that lack an axial plane foliation. These late folds are commonly kink or chevron folds in rocks with a well-developed prior foliation. If such foliation is absent, the late folds may be open warps with rounded hinges and curved limbs. Common variants on this style sequence are (1) very early, tight folds lacking axial plane foliation, and (2) more than one generation of folds with axial plane slaty cleavage or schistosity.

As examples, parts of this style sequence, or the whole of it, have been described from the Scottish Highlands by Rast (1963), Weiss and McIntyre (1957), Ramsay (1958a and b), from Scandinavia by Ramsay and Sturt (1973a), from the Appalachians by Fyson (1964), Freedman et al. (1964), and from Australian, New Zealand, and Indian areas by Hobbs (1965), Means (1963), and Naha and Halyburton (1974). In some cases, as many as five or six generations of folds are postulated. Recognition of such large numbers of generations is generally quite speculative, however, because it is usually based on assuming that members of different style groups necessarily belong to different generations, and because it usually relies on the correctness of long-range correlation of style groups (see Section 8.5.3).

Less regularly observed than the style sequence described above, but also quite familiar, is an orientation sequence as follows: the early folds with schistosity parallel to their axial planes are commonly recumbent. The later folds, with crenulation cleavages parallel to their axial planes, have more steeply dipping axial surfaces. The latest folds, without axial plane foliations, may have various orientations, but their axial planes are often moderately to steeply dipping.

What has been said so far pertains specifically to small folds seen in individual outcrops, but it also applies in some degree to folds on larger scales, although these are always subject to more interpretation. Large first-generation folds are often said to be tight to isoclinal, particularly in medium-grade terrains. There is usually maximum uncertainty about the shape of these folds, or even of their existence, however. Large folds of the second group have more open profiles and more steeply dipping axial surfaces. Late folds of large size tend to be very open and to have steeply dipping axial surfaces.

There are no simple generalizations about the relation between the relative ages and sizes of large folds. In many areas, however, it appears that the prominent features of the outcrop pattern are reflections of second- or third-generation folds. The earlier folding, which may have been very tight and associated with extensive transposition, may be obscurely expressed in the outcrop pattern. This natural tendency for early folds to be less obvious is often compounded by an accepted, but incorrect, stratigraphy that has unknowingly assumed the same rock units on opposite sides of early isoclinal folds to be independent stratigraphic units. (An error of this type was made in early interpretations of the area shown in Fig. 5.30.) It is possible that there are many other cases extant of stratigraphy that is incorrectly understood because of failure, for lack of detailed structural mapping, to recognize early folds or stacks of thrust sheets.

While some areas appear to contain very large early folds, associated with the slaty cleavage or schistosity, there are others where no such folds have been demonstrated and where they may not occur. In such areas, there may be an extensively developed first-generation foliation and numerous first-generation mesoscopic folds but no large folds. Such areas may include cases where there has been early development of a series of thrust nappes (see Section 9.12).

A major geometric problem in multiply deformed regions is to determine the extent to which any of the fold hinge lines formed oblique to, or at a high angle to, the general trend of the fold belt. Such "cross folds," particularly early cross folds, or very late cross folds, have been proposed in many places. It is still too early to know how common these are, or what they reveal about the tectonic history of individual belts. Field-workers should be well aware, however, that some of the folds they are mapping may have hinge lines that trend at a high angle to the regional strike of the fold belt [e.g., reclined folds (Fig. 4.25)].

Questions concerning the ages of deformational phases in multiply deformed belts are difficult to answer. One wants to know the ages of each generation of structures, and whether different generations represent entirely separate episodes of deformation, or successive responses during a single episode (see Section 8.2.4). There has been some progress on such questions locally [Harper (1967)].

Another group of difficult questions concern the timing of deformational events (marked by fold generations) relative to thermal events (marked by changes in metamorphic mineralogy). This is a complex and poorly understood subject; the reader is referred to Johnson (1963), Spry (1969), Sturt and Taylor (1971), and Trouw (1973).

Some of the ways in which the tectonic setting could change at a point in an evolving mountain belt, and thus give rise to new generations of structures, are discussed in Section 10.2.6.

9.12 NAPPE STRUCTURES

Prior to about 1880 most of the rocks in the European Alps were considered to be *autochthonous,* that is, they were more or less in the same place that they had been deposited originally. Few people have doubted, since the work of

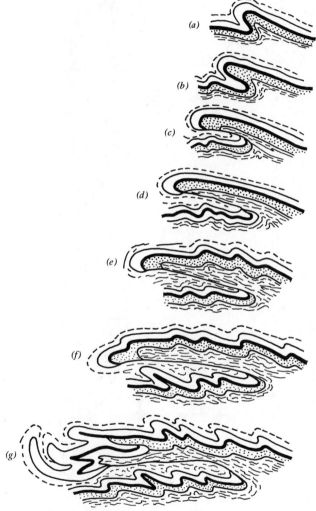

FIGURE 9.19   Progressive development of a fold nappe.
(*a*) Fold with overturned limb. (*b*) Overturned limb thinned.
(*c*) Extreme thinning of overturned limb. (*d*) Thrust fault
replaces overturned limb. (*e*) Commencement of folding on
upper limb. (*f*) Overthrust fold (nappe). (*g*) "Plunging"
nappe with brow digitations. [From Heim (1919).]

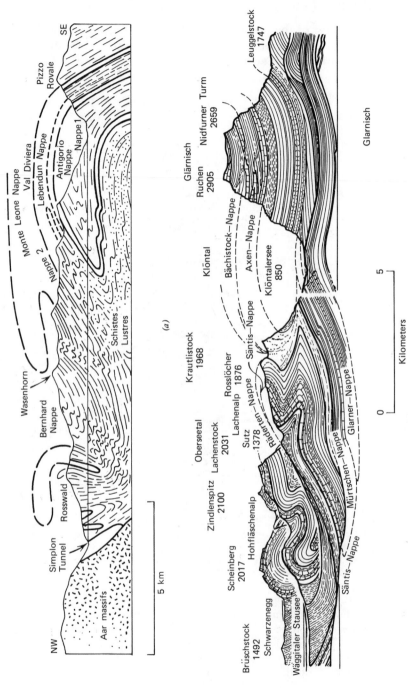

FIGURE 9.20 Cross sections through parts of the European Alps. (a) Cross section through the Lower Pennine Nappes from the Aar Massif to the Root Zone. [After Schardt (1904) in Badgley (1965).] (b) Cross section through the Glarner Alps. [From Heim (1919).]

Heim (1878), that these rocks were involved in very large recumbent folds but the idea was widely held that the alpine rocks had not traveled very far from their original place of deposition. However, during the last 20 years of the nineteenth century [see Bertrand (1884); Lugeon (1901, 1902)], the idea evolved that many of the rocks of the European Alps had traveled a considerable distance from the place where they were first deposited, that is, they were *allochthonous*. The displacement was seen to involve large-scale thrusting and recumbent folding [see Heim (1919)], as illustrated in Figure 9.19. The word *nappe* once meant simply a large, more or less horizontal, undisturbed sheet of material. In alpine usage the word quickly became an abbreviated form of *nappe de recouvrement* and the term is now applied to any large, more-or-less horizontal sheet of rocks that has been displaced a considerable distance from its original place of origin. Examples of these horizontal sheets are illustrated in Figure 9.20 [see Heritsch (1929) for a discussion of the development of the nappe theory in the European Alps. A good English review of the Western Alps is given by Ramsay (1963.)]

Early workers in the European Alps, such as Bertrand (1884) and Heim (1919) considered that a true nappe developed by the shearing out of the lower limb of a recumbent fold as illustrated in Figure 9.19. However, the idea that large horizontal displacements can take place on thrust zones without large recumbent folding is now well established, and the concept of a simple *thrust nappe* as opposed to a *fold nappe* has long been discussed in the literature. In a thrust nappe, more-or-less undisturbed material is simply carried along on the thrust. Figure 9.21 shows an example of one of these thrust nappes developed in the Arltunga metamorphic complex in Central Australia. The thrusts shown in this figure are pervasively developed in the basement and cover rocks and tend to duplicate the sequence up to 10 times at any one place. At the same time, there is no, or little, inversion of the stratigraphic sequence so that no recumbent folding is involved on a large scale. There is a strong schistosity developed parallel to the thrust slices but very little in the way of large-scale folding associated with the schistosity development. Most of the large-scale folds developed in this area fold the schistosity and sometimes

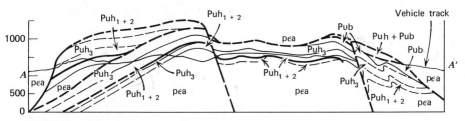

FIGURE 9.21   Cross sections through thrust nappe complex Atnarpa Range, Central Australia. [From Yar Khan (1972).]

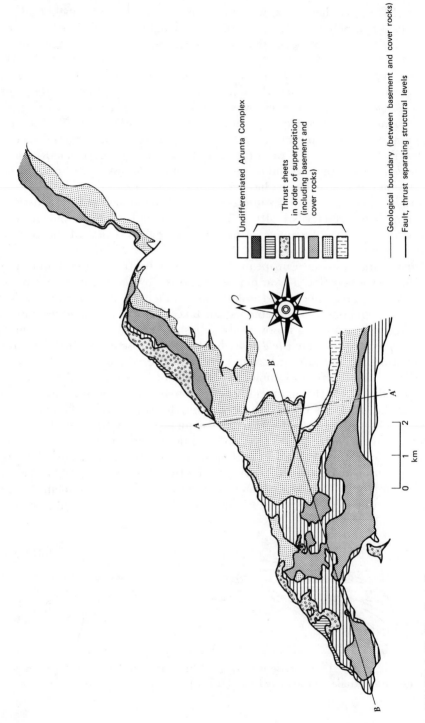

FIGURE 9.21 (Continued)

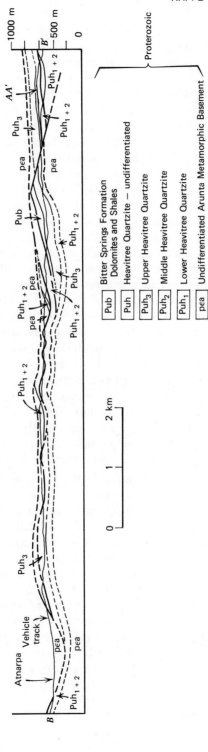

FIGURE 9.21 (Continued)

the thrusts as well and hence are late-stage structures. This kind of structural history appears to become more widespread as more detailed structural work is done in complicated metamorphic terrains.

Nappe structures are now recognized in most of the large fold belts of the earth. Some examples are Kennedy (1955), Sutton and Watson (1962b), and Johnson (1970a, b) from Scotland; Strand (1961), Henley (1970), Strand and Kulling (1972), and Trouw (1973) from Scandinavia; Lombard (1948), King (1950), and Thompson et al. (1968) from the Appalachians; Lugeon (1901, 1902), Argand (1916), Heim (1919), Bailey (1935), and Collet (1936) from the Alps; Haller (1956, 1971) from Greenland; Gansser (1964) and Naha and Ray (1972) from the Himalayas; Forman (1971) from Central Australia; and Price and Mountjoy (1970) from the Canadian Cordillera.

Although there has been now a hundred years of detailed geological work in the European Alps, most of this has concentrated on elucidating the large-scale structural geology and, in particular, on unraveling the detailed stratigraphy and paleotectonic evolution [see Trumpy (1960)]. The idea of large horizontal displacements has persisted, and an involved sequence of nappe formation has been postulated on the basis of essentially stratigraphic arguments [see Trumpy (1973)]. There has been very little modern structural work published from the European Alps [see, however, Milnes (1965), Kobe (1966), and Wenk (1973)], but a considerable amount of unpublished work exists in the form of Ph.D. theses at Imperial College, London, and the Institute of Geology at Leiden, as well as at a number of other European universities. The general picture that seems to emerge from this work is one of great complexity in alpine structure and metamorphic history.* The nappes are not the product of one generation of folding, and there appears to be a general sequence of development in the Pennine Nappes, which is outlined below (Fig. 9.20*a* is a section across the lower Pennine Nappes).

1. The first stage of deformation in the Pennine Nappes is postulated on the basis of the large-scale distribution of rock units. Thus, the Antigorio Gneiss and the Bosco Group (Fig. 9.22) consist of Hercynian gneisses and schists but are separated from each other by a relatively thin layer of Mesozoic calcareous schists and marbles. The precise structural significance of this layer of Mesozoic rocks is uncertain; some stratigraphic work suggests that it could be synformal in character. However, even though the precise nature of this early deformational event is uncertain, it is clear that some quite significant event is required to intercalate this thin layer of Mesozoic rocks between quite distinct older rocks prior to the formation of the obvious large folds.

2. The "stratigraphy" generated by early thrusting (and perhaps folding) is folded by a group of large folds that have a strong schistosity as axial plane. These large folds are commonly recumbent. In Figure 9.22 some folds of this type are marked F_2.

*We thank W. D. M. Hall and H. J. Zwart for informal discussions on this matter.

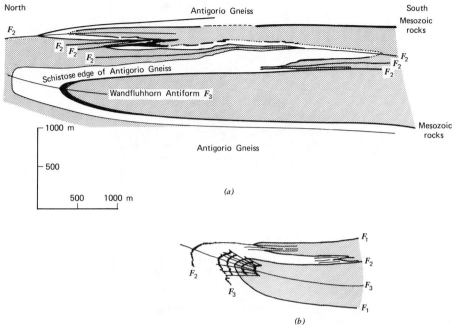

FIGURE 9.22   Profile of folds in the Lower Pennine Nappes. (*a*) Detail of folding. (*b*) Interpretation of structures in terms of folding episodes F_1, F_2, and F_3. [From Hall (1972).]

3. The schistosity is deformed by a predominantly horizontal crenulation cleavage. Again, this cleavage is axial plane for very large folds (e.g., the Wandfluhhorn Antiform of Figure 9.22).

4. In some areas there is an even later, predominantly steeply dipping crenulation cleavage which deforms all earlier structures. The classical root zone of the Alps (Fig. 9.20), which supposedly is the area from which most of the nappes originated, appears to be a very late structure associated with this steeply dipping crenulation cleavage.

Thus, the large alpine nappes that were so clearly recognized by Lugeon (1901) and Argand (1916), and which form the classical pictures that writers present in many textbooks (see Fig. 9.20), appear to postdate an earlier period of very extensive thrusting and are commonly later-generation structures belonging perhaps to a second or third phase of deformation.

There does not appear to be as great complexity in the Helvetid Nappes of the European Alps. Here the rocks within the nappes are commonly only mildly deformed, if at all, and metamorphism is weak or absent. The large recumbent folds at times have their lower limbs intact and are not removed by large thrusts (see Fig. 9.23). Evidence of multiple deformation is not common.

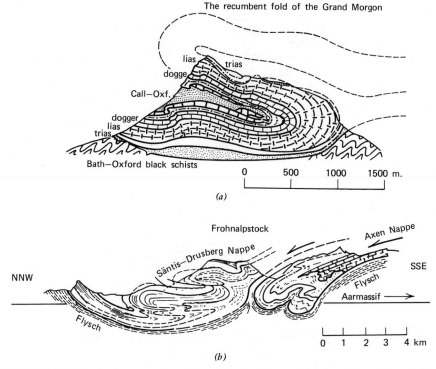

FIGURE 9.23   (a) Preservation of lower limb of recumbent syncline. Nappe of Ubaye-Embrunais, French Alps. [After Schneegans (1938) in de Sitter (1956)] (b) Nappes of the Helvetid Alps, north of the Aar Massif. [After Heim (1919) in de Sitter (1956).]

In the past there has been a widespread opinion [see Trumpy (1973)] that these nappes of the Helvetid Alps developed by gliding downhill under the influence of gravity.

9.13   AREAS WITH GENTLY DIPPING SCHISTOSITY

Several metamorphic regions are known in which the prominent schistosity has a gentle dip over areas measured in thousands of square kilometers. Examples occur in the Otago schist terrain of New Zealand and the San-bagawa metamorphic terrain of Japan. Further details of the Otago example follow.

The Otago schist terrain occupies a wide, arcuate belt that extends south and east from the Alpine fault (Section 9.4) and is bounded on the northeast and southwest by folded greywacke sequences with more steeply dipping axial planes. The schists themselves are dominantly quartzofeldspathic, commonly displaying a differentiated layering parallel or approximately parallel to the schistosity, and are thought to have been derived from a thick sequence of greywackes containing occasional basic igneous rocks (now greenstones). Chlorite-bearing assemblages are widespread in the quartzofeldspathic schists as well as in the greenstones. The prominent schistosity dips less than 30° throughout most of the central and eastern parts of the belt, an area of roughly 5000 square kilometers. Toward the northwest the schistosity steepens, the metamorphic grade increases, and large folds with steeply dipping axial planes can be mapped [Grindley (1963); Wood (1963)].

Detailed structural studies of the Otago schist were initiated by F. J. Turner in the middle 1930s. He recognized the general significance of multiple lineations and foliations in the rocks and suggested that they had suffered a long and complex history, an interpretation borne out by subsequent mapping of Robinson (1958), Wood (1963), Means (1966), and Brown (1968). Most parts of the schist belt are now considered to show evidence for at least three phases of deformation. The major structure is dominated by large recumbent folds (Fig. 9.24a). The evidence for such folds is considered strong because three somewhat independent lines of evidence have been used to delineate them. Wood (1963) suggested their presence using mappable vergence boundaries; Means (1966) made use of enveloping surface attitudes observed on the outcrop scale; and Brown (1968) was able to discern the attitude of very large stratigraphic units, as seen in oblique aerial photographs, for example, and could use these to establish the existence of major fold hinges.

The major recumbent folds in the Otago schists are difficult to recognize for two main reasons. First, the parent greywacke sequence is notably poor in good marker horizons; and second, strong transposition of schistosity and layering on the outcrop scale [Bishop (1972)] gives rise in many places to a single dominant flat-lying foliation, even when other evidence shows that stratigraphic units on a larger scale are steeply dipping.

Many metamorphic terrains contain belts with early recumbent folds that have later been refolded around steeply dipping axial planes. The distinctive feature of regions like the Otago schist seems to be that the later folding, while it occurred on a small scale, did not succeed in appreciably reorienting or replacing the flat schistosities developed during the earlier recumbent phases.

Daly (1915) suggested that flat-lying schistosity could be developed independently of folding, by "load metamorphism" in deeply buried rocks. In this interpretation, little or no strain is required for, or indicated by, a subhorizontal schistosity. We know of no well-documented examples of areas for which this process has led to regional development of a flat-lying schistosity. It is clearly unacceptable for regions where the schistosity is associated with recumbent folds, as in Otago and in the Sanbagawa terrain [see Kawachi (1968); Ernst et al. (1970)].

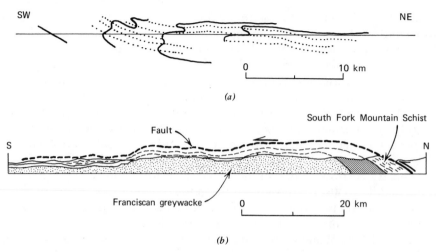

(a)

(b)

FIGURE 9.24   Large-scale structures associated with regional development of flat-laying schistosity. (a) Recumbent fold profiles as interpreted by Brown (1968), in the schists of east-central Otago, New Zealand. [From Brown (1968). Reproduced from the New Zealand Journal of Geology and Geophysics.] (b) Flat-lying thrust fault as interpreted by Blake, et al. (1967) in northwest California. Franciscan greywacke without schistosity (stippled) is believed to grade upward through a transitional zone (shaded) into the completely recrystallized South Fork Mountain Schist. [From Blake et al. (1967), with permission of the U.S. Geological Survey.]

Blake et al. (1967) have described an extensive belt of flat- to gently dipping schistose Franciscan rock in California and Oregon that is associated with the sole of a major thrust fault (Fig. 9.24b). Recumbent folds are seen on a small scale but may or may not exist on a large scale. The striking feature is the progressive development upward of a strong schistosity, so that nonschistose greywacke some distance below the fault passes into schist immediately below the fault. This transition occurs through a thickness of as little as 500 meters in Figure 9.24b. It is not yet clear what connection there may be, if any, between flat schistosity of the Otago type and flat schistosity associated with thrust faulting.

9.14 MYLONITE ZONES

Mylonite zones are narrow planar regions in which deformation is intense relative to that of the adjacent rocks. The rocks in such zones are generally fine-grained equivalents of the adjacent rocks, the reduction in grain size being due to cataclasis, recrystallization, or a combination of both. Typically,

mylonites have a strongly layered appearance, the layering being due to variations in the amount of deformation or to variations in composition. A lineation is commonly developed within the foliation and is defined by elongate
rods of mineral aggregates or ovoid patches of weakly deformed material
surrounded by strongly deformed rock. Folding is invariably developed in the
mylonite layering and commonly the axes of these folds are parallel to the
lineation [for some examples see Hooper (1968); Trouw (1973); Hossack
(1968), (Scandinavia); Christie (1963), (Scotland); Zwart (1958), (Pyrenees);
Brown (1967), (Grenville Front, Canada); Stirewalt and Dunn (1973), (North
Carolina, United States); Forman (1971), (Central Australia)]. Although this
parallelism of fold axes and lineation is common it is important to point out
that the relationship is not always developed [for examples see Kvale (1941,
1953)]. The folds vary from open through tight to isoclinal and are commonly
intrafolial. Invariably where such folds are developed there appears to be a
secondary mylonite layering developed parallel to their axial planes. The
geometrical relationships between layering, lineation, and folding within
mylonite zones have been poorly studied for the most part and there is a need
for more detailed study.

Mylonite zones range in scale from the size of thin sections (Fig. 9.25) up to
zones that are hundreds of kilometers long and perhaps 30 km or so in width

FIGURE 9.25   Thin-section scale mylonite zone consisting of fine-grained quartz
and mica. Arltunga Nappe Complex, Central Australia. Scale bar is 0.3 mm.

(see Fig. 9.26). The classical mylonite zone is associated with the outcrop of the Moine Thrust in the northwest highlands of Scotland (Fig. 9.26a). This zone is gently dipping and approximately 200 km in length. It ranges in thickness from 10 to 100 meters [Lapworth (1885); McIntyre (1954); Christie (1963); Johnson (1967); McLeish (1971)]. The Moine Thrust itself though apparently postdates the formation of the mylonites [see Johnson (1957, 1960 and 1970a)]. Mylonite zones occur in rocks of all metamorphic grades ranging from greenschist to granulite facies [Waters and Cambell (1935); Scott and Drever (1953); Hsu (1955); Sutton and Watson (1959); Hobbs (1966a); Eisbacher (1970); Vernon and Ransom (1971); Ross (1973); Bell and Etheridge (1973, 1974)]. Where mylonite zones occur in high grade metamorphic rocks they are commonly associated with retrogression and appear as zones of retrograde schists. Particularly in such situations these zones are loosely referred to as shear zones. Recently the microstructure of quartz mylonites has been systematically studied experimentally by Tullis et al. (1973).

The term mylonite was introduced by Lapworth in 1885 to describe rocks that appeared along the Moine Thrust in Scotland. Lapworth envisaged that these rocks had been produced by strong grinding or milling of the Moine schists: hence the term mylonite from the Greek *mylon* — a mill. He considered these rocks to have been formed solely by crushing with no concurrent recrystallization. In the formation of these rocks a strong layering or banding had developed. To Lapworth then the important features of a mylonite were that it had been produced solely by cataclasis without any associated recrystallization of the constituent particles and that a well-developed layering had formed. The terminology of mylonites is discussed at length by Christie (1960) who points out that recrystallization is commonly very widespread in the Moine mylonites. However, he considered that recrystallization postdated the deformation. It is important to note that most of the classic microstructures described by older workers and interpreted by them as cataclastic structures are due to recrystallization. Such microstructures particularly include the classical *mortar structure* in which an aggregate of fine-grained, new recrystallized material is developed around the margins of older deformed grains (see Fig. 2.20b). The subject is discussed at length by Bell and Etheridge (1973). Christie employs the term *blastomylonite* to describe those rocks that have been strongly recrystallized. The term was introduced by Sander in 1912. In nature there are presumably all gradations from strongly deformed rocks where the deformation is solely of a cataclastic nature through to these where recrystallization has accompanied or postdated a ductile mechanism of deformation. Moreover, there are presumably many rocks that have been deformed in a cataclastic manner but that have been sintered or recrystallized at a later date. The lack of any systematic descriptive material in the literature on mylonites from different environments prohibits any definite statements.

The situation, however, is somewhat unsatisfactory if workers insist on using Lapworth's original term to signify that the deformation in mylonites has

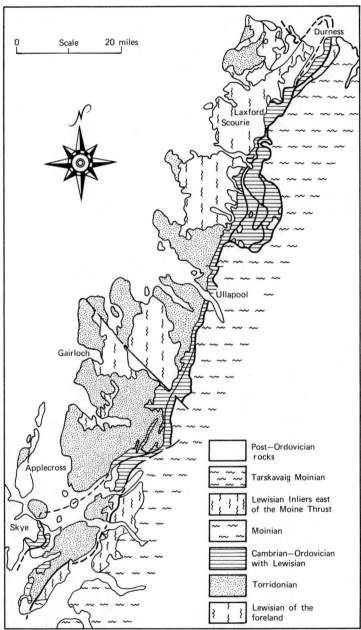

FIGURE 9.26   (a) Map of Moine Thrust Zone, Scotland [from George (1970).
From Fig. 1.10, *The Geology of Scotland* edited by C.Y. Craig with permission of the
editor and publishers, Oliver and Boyd, Edinburgh).] (b) Map of mylonite zones in
Central Australia. These particular zones are up to 500 km in length [after
Marjoribanks (1974).]

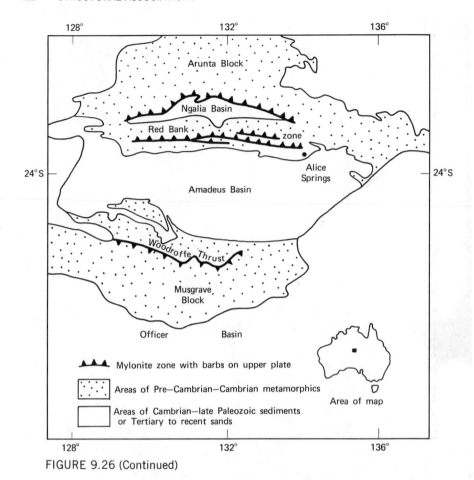

FIGURE 9.26 (Continued)

been solely of a cataclastic nature with no accompanying recrystallization. This arises because within many mylonite zones there are rocks in which the shapes of original grains have been strongly distorted [see Christie (1963, plate VIII); and Fig. 9.27b)] but where the deformation throughout has been of a

FIGURE 9.27   Microstructural sequence in a mylonite zone: Scale bar in each case is 0.3 mm. (a) Original microstructure of country rocks. Detrital grain shapes outlined by inclusions. (b) Flattening of initial quartz grains. Notice that the dusty trains of inclusions outlining the original grains are clearly distinguishable. (c) The beginnings of recrystallization. (d) Recrystallization with some relicts of detrital quartz grains remaining. (e) Complete recrystallization to form a blastomylonite. (f) Pseudotachylite. All from Arltunga Nappe Complex, Central Australia. (Specimens courtesy of M. Yar Khan.)

(a)

FIGURE 9.27 (Continued)

FIGURE 9.27 (Continued)

ductile nature produced by the propagation of dislocations through the crystalline structure of the grains [see McLaren and Hobbs (1972)]. For these rocks the strict Lapworth term, mylonite, is not applicable nor is the term blastomylonite. We prefer to use the term mylonite in a general sense covering rocks that occur in these zones of relatively high deformations no matter if the deformation has been cataclastic or ductile on the scale of grains, and to use the term blastomylonite to refer specifically to rocks in these zones that have completely recrystallized, either during or after the deformation. This terminology is compatible with Bell and Etheridge (1973).

In many mylonite zones there is evidence that the rocks have melted, and extremely fine-grained material comprises dikes and sills that run out from the mylonite zone and intrude the neighboring rocks. These fine-grained rocks are called *pseudotachylites* [Shand (1916)]. The term is also discussed by Christie (1960).

An example of one type of mylonite zone is illustrated in Figure 9.21 [Yar Khan (1972)], where zones of high strain are associated with thrust nappes in a sequence of orthoquartzites where the primary sedimentary grains are easily recognizable and outlined by dusty inclusions. These grains are cemented together by optically continuous quartz overgrowths (Fig. 9.27). One hundred meters structurally below the thrusts the detrital grains begin to be deformed

FIGURE 9.27 (Continued)

FIGURE 9.27 (Continued)

FIGURE 9.27 (Continued)

and deformation lamellae with undulatory extinction appear. As the thrust is approached the grain shapes become progressively distorted until a well-developed foliation appears in the rock defined by these flat grains of quartz. Ultimately, at high strains in excess of 40 percent shortening, the rocks begin to recrystallize around the margins of the grains to produce what is classically a mortar structure. Finally, a polygonal mosaic of grains is developed. The sequence is illustrated in Figure 9.27. The final product close to the thrust is a pseudotachylite that, in places, crosses adjacent rocks as dikes. The thrust plane itself is a stratigraphic discontinuity across which there has been considerable movement. Microstructural progressions similar to this have been recorded by Christie (1963) and Bell and Etheridge (1974). They are typical of many mylonite zones.

Another type of mylonite zone is the discrete shear zone described by Ramsay and Graham (1970) and Hara et al. (1973). These shear zones are widely developed in many metamorphic terrains [e.g., see Vernon and Ransom (1971)].

The strain associated with mylonite zones has been a matter of some discussion. A widespread interpretation is that such zones are the product of large shear strains. Where the displacement of the rocks either side of the zone can be established, this is commonly the case [see Ramsay and Graham (1970); Vernon and Ransom (1971); and Hara et al. (1973)]. However, in most very large mylonite zones such displacements cannot be demonstrated and then interpretation is not so clear. Johnson (1967) has proposed that in many mylonite zones and, particularly, the Moine Thrust, the strain is essentially a flattening normal to the foliation and the mylonite layering has an origin much the same as slaty cleavage. Johnson bases this interpretation on the symmetrical patterns of preferred orientation of quartz that are developed in the Moine mylonites [see Fig. 2.22 and Christie (1963)]. This may, in fact, be the case for the Moine mylonites but for many other zones such patterns of preferred orientation are strongly asymmetric [see, for instance, Hobbs (1966a); Hara et al. (1973); and Bell and Etheridge (1974)]. It is important to also note that typical mylonite microstructures commonly thought of as the result of shear strain have been produced in axially symmetric straining by Tullis et al. (1973).

Another problem associated with the strain in mylonite zones is the common parallelism between lineation and fold axes. Many workers hold that the lineation is an axis of principal extension whereas others claim it to be a direction of shearing within the foliation plane. Generally, however, there is very little evidence to define the strain very clearly, but a problem exists in establishing the relationship of folds and lineation in mylonite zones to the distribution of strain within those zones and in correlating these features with the overall displacement associated with the zone.

9.15 MANTLED GNEISS DOMES

In 1949 Eskola described structures that he called *mantled gneiss domes* from
the Caledonides of Finland and pointed to the common occurrence of similar
structures in other orogenic regions. As described by Eskola, these structures
comprise a core of granitic migmatites or gneisses overlain by a layered
metasedimentary and metavolcanic cover or "mantle." The core rocks in the
central portions of some domes have an igneous microstructure, but near to
their contact with the mantle they are foliated and this foliation, the contact,
and the layering in the mantle are all parallel to one another and generally dip
outward, away from the gneiss core, in such a way that they define a dome.
Thus in outcrop the mantle rocks form a ring of generally outward dipping
strata (where the domes have steep walls the simple pattern is sometimes
complicated by locally overturned dips) surrounding an area of granitic
gneiss. The obvious contrast between the granitic gneisses of the core and the
overlying layered metasediments led an earlier worker, Trüstedt (1907), to
interpret the contact, in the area in which he worked, as an unconformity.
Eskola points out that this interpretation is supported by the way in which the
contact faithfully follows the same stratigraphic horizon in the mantle rocks
within a given area. He also points out that such a relationship can be proven
for some domes where the basal mantle rock is a conglomerate containing
pebbles of the granitic rocks of the core. Not withstanding this, the core rocks
are locally seen to intrude the mantle as dikes so that the situation is complex.
   Similar structures are known from other orogenic regions. Excellent examples are found in the Appalachians in the Maryland-Pennsylvania region
[Cloos and Broedel (1940); Higgins et al. (1973)] and in New England [e.g.,
Balk (1946); Thompson et al. (1968)] where more than 25 domes occur in two
linear belts that parallel the local Appalachian trend. The easternmost of the
latter two belts stretches from northeastern New Hampshire to Long Island
Sound (*ca.* 420 km) and includes approximately 20 domes [see Naylor (1960);
Thompson et al. (1968)]. The core rocks of these particular domes are again
predominantly granitic in composition and mainly of igneous, plutonic origin
but there are also probable metasediments and metavolcanics present. Along
most of the length of the belt the core rocks are in contact with the same
stratigraphic layer of the mantle, the Ammonoosuc Volcanics.
   Another example of gneiss domes is found in the Pine Creek Geosyncline of
northern Australia. For example the Rum Jungle granite is a complicated
body including schists, gneisses, and intrusive granites [Heier and Rhodes
(1966); Richards, Berry and Rhodes (1966)]. It is essentially dome shaped and
is mantled by a sequence of low-grade metasediments, the lowest member of
which contains clastic material derived from the core. The core differs from
those of the Finnish domes in that it is not foliated parallel to its contact with
the mantle.

Haller (1962) has described similar structures from Greenland and other examples cited by Eskola include the Crystalline Massives of the European Alps and the Vredefort structure of South Africa [see Hall and Molengraaff (1925)]. Ramsay (1967, p. 384) illustrates yet another example from Uganda and Ruanda.

Various origins have been postulated for the various examples of mantled gneiss domes. Eskola (1949) envisaged the development of the Finnish domes and possibly the development of gneiss domes elsewhere as a process requiring two "orogenic revolutions." During the first orogeny, granite plutons were emplaced in metasediments and metavolcanics that were then lowered by erosion to expose the plutons. The plutons and country rock were then covered by a new sequence of sediments. Finally, during the second orogeny, the old plutons were reactivated by injection of new granitic magma causing them to expand upwards and thereby deform the overlying strata into a dome. The new magma and deformation converted the old granite into migmatites and gneisses and gave rise to the small igneous bodies seen to intrude the mantle in the Finnish gneiss domes. As explanation of the orientation of the foliation in the core rocks, parallel to the core/mantle contact, he claimed (1949, p. 467) that ". . . . the granitic materials, when the dome was swelling up, would naturally be transported along the paths of differential shear." This explanation of the foliation seems inadequate and we suggest that in general, where there is a foliation parallel to the contact in the rocks of the core and mantle, it is more likely that the foliation predates doming. In some cases the core/mantle contact may be a thrust plane or thrusted unconformity and the foliation and thrust may be contemporaneous. In the Chester Dome of New England we have seen two generations of tight to isoclinal folds predating development of the dome and there, the folded foliation and possibly the contact are related to the earlier deformations.

Chapman (1939) believed that the Bronson Hill gneiss domes of New England were intruded as a series of laccoliths. He explained the foliations as being partly of igneous flow origin, in the core, but mainly due to extension parallel to the core/mantle contact as the contact, overlying strata, and early solidified core rock were deformed from a planar configuration to the present dome shape. A more commonly accepted explanation of these domes was first suggested by Thompson [ms. (1950) cited by Fletcher (1972)]. He postulated that the granite core rose like the salt of a salt dome in response to the gravity instability resulting from the low density of the core relative to the mantle. Fletcher (1972) has modeled this hypothesis mathematically and is able to simulate the Bronson Hill domes very successfully. The process has also been simulated experimentally by Ramberg (1967).

As in the case of the New England domes the Rum Jungle granite was initially interpreted [Sullivan and Matheson (1952); Malone (1962)] as an intrusive body. However subsequent work showed that the core was older than the mantle [Richards, Berry, and Rhodes (1966)] and unpublished work by one of the writers indicates that both the Rum Jungle and neighboring Waterhouse granites represent culminations (see Section 4.3), occurring where anticlines, of two separate generations and different trends, combine to produce a structural high (see Fig. 8.8a and b).

This mechanism has also been suggested by Ramsay (1967, p. 384) as a possible mechanism for the domes of Uganda and Ruanda. In addition he points out that the complex form need not be due to multiple deformation but could result from a single deformation in which "compressive strain was *acting in all directions* within the surface layers" (our italics). It should be noted that this mechanism is not necessarily incompatible with the gravity instability mechamism described above. Both processes could operate synchronously.

It has been postulated that the Vredefort structure is an astrobleme, that is, the impact feature formed by collision of a meteorite with the earth [see Daly (1947); Dietz (1961b)]. While this explanation may be correct for this particular structure it cannot be considered a generally applicable explanation for mantled gneiss domes.

9.16 GNEISSIC REGIONS

In areas of upper amphibolite and granulite facies metamorphism the geometry of the structures developed is commonly similar to that developed in multiply deformed rocks of lower metamorphic grade (see Chapter 8 and Section 9.11). However, a number of new problems arise:

1. An increase in the grain size of micas, quartzes, and feldspars obliterates or obscures many important details, such as sedimentary structures, so that facing criteria, although present, may be difficult to establish.
2. At high grades of metamorphism, even rocks of pelitic composition may lack micas and consist of minerals such as quartz, K feldspar, sillimanite, and garnet. In such rocks, many foliations that presumably appear at low grades as slaty cleavage or crenulation cleavage now appear simply as a gneissosity and it may be very difficult to establish any kind of overprinting criteria for the various foliations developed.
3. Many areas appear to be structurally simple in that the conspicuous layering in the rocks may superficially appear to have the same orientation over large areas. Evidence for multiple deformation may be rare because of lack of structural detail due to coarseness of grain size; the temptation is to interpret the structure as a sequence of uniformly dipping metamorphosed rocks. However, in rocks where micas are present there is invariably a preferred orientation of micas parallel to these layers and this is deformed around the hinges of many but not all folds. In addition the layering is quite commonly lenticular, probably due to

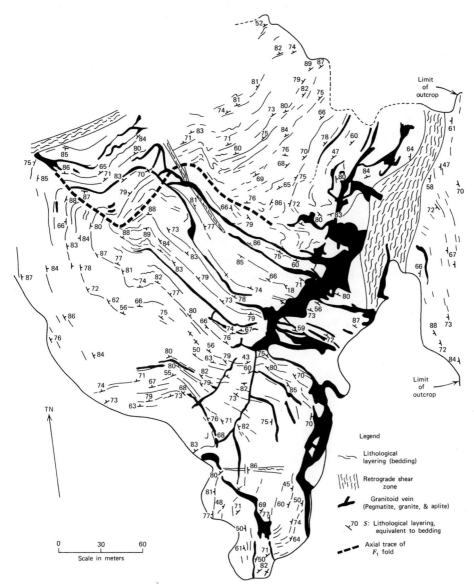

FIGURE 9.28   Map of structure in biotite-sillimanite gneiss, Broken Hill, Australia. (Map courtesy of W. Laing, Adelaide University.)

widespread transposition. Detailed examination produces evidence of multiple deformation in many of these rocks.

4. Partial or wholesale melting, if present, obscures structural detail and is sometimes associated with considerable inhomogeneity in the orientation of structural features such as fold axes and axial surfaces.

Few structural analyses exist in very high-grade rocks but some examples are those of Hobbs (1966b), Williams (1967), Williams et al. (1971), Naha and Chaudhuri (1968), and Ray (1974). These problems of course are well shown in pre-Cambrian metamorphic terrains. An example of the types of complexities developed in high-grade metamorphic areas is illustrated in Figure 9.28.

The area shown in Figure 9.28 is from Broken Hill, Australia, and consists predominantly of coarse-grained sillimanite gneisses in which the sillimanite needles may reach 2-3 cm. in length. The gneisses are regularly interlayered with quartzite. Two generations of sillimanite growth can be recognized in the area, both of them associated with the formation of an axial plane schistosity and lineation.

A superficial walk across the area reveals many folds but the impression generally is of a uniformly dipping sequence that perhaps is fairly tightly folded; there is an obvious swing in strike of the foliation. Detailed form surface mapping and very close attention to the relationships between bedding and schistosity reveal the map shown in Figure 9.28. The area clearly is not simple and contains large structures belonging to at least two generations. This is an example of the detail required to establish the structural geometry in many gneissic terrains.

10
TECTONICS

10.1 INTRODUCTION

Tectonics in the broad sense includes study of earth structures on every scale. In the narrower sense of this chapter, tectonics deals specifically with very big structures — like oceanic ridges, geosynclines, and mountain belts — and in this sense it is sometimes called "geotectonics" to emphasize its global aspect.

Tectonics and structural geology are both concerned with *relative movements of parts of the earth,* for such movements give rise to observed structures like foliations or faults or mountain belts. There is a difference, however, in the kinds of evidence used to work out the movements on different scales. In structural geology the scale of the moving parts is often small and the relative displacements across a structure are deduced mainly from direct observation of the structure itself. As the scale increases, for example to the scale of regional mapping, the structures become more difficult to examine directly, and the evidence for their geometry and the displacements across them comes increasingly from stratigraphy or other lines of evidence. In tectonics, where the parts in motion are very large and the structures between them are often quite inaccessible, the evidence from other earth sciences becomes very important. Tectonic theories are, therefore, largely based on and tested against evidence from geophysics, physiography, stratigraphy, paleontology, petrology, and geodesy. A chapter on tectonics nevertheless belongs in a book on structural geology because the nature and sequence of structures in an evolving rock body depends, like the depositional and thermal history of the rocks, on the

433

tectonic setting of the body and the way in which this setting changes with time. By *tectonic setting* we mean the position of the body in the pattern of large-scale movements.

Much of this chapter is devoted to *plate tectonics,* a new unifying theory of present-day earth movements. We give plate tectonics prominent treatment because it provides an internally consistent framework for most of the data presently available, because it is a quantitative and testable theory, and because it offers plausible explanations for many geological phenomena of the past. While none of the new interpretations for continental structures need yet be regarded as compelling, such interpretations do provide a welcome stimulus for rigorous regional studies. Good mapping of key areas, combined with improved basic understanding of the processes of rock fracture and flow, will ultimately show whether plate tectonics as presently conceived played a major or a minor role in the development of individual bodies of deformed rock.

A topic that is given only slight attention in this chapter is "mountain building," that is the internal history of mountain belts. This subject requires an approach that goes well beyond structural geology, to consider in chronological detail, the stratigraphic and petrological development of individual mountain chains. We limit outselves in this book to describing some of the main *components* of mountain belts, the structural associations of Chapter 9.

10.2 PLATE TECTONICS

Plate theory says, in brief, that the crust and uppermost mantle of the Earth are divided into a small number of thin and rigid spherical caps or *plates* that are moving tangentially with respect to one another (Fig. 10.1). The material of the plates themselves (which extend down to a depth of the order of 100 km) is the *lithosphere,* and it rests on a part of the upper mantle known as the *asthenosphere.* The dividing zone between two plates is a *plate boundary.* The region where three plates meet is a *triple junction.*

The boundaries of six major plates as delineated by LePichon (1968) are shown in Figure 10.2. A comparison of Figure 10.2 with Figure 10.3 (showing epicenters of recent shallow earthquakes) indicates that the plate boundaries follow and, in fact, are *defined* by zones of localized seismic activity. Notice that several smaller plates, not shown in Figure 10.2, are discernible in Figure 10.3. Maps showing how the major plates of Figure 10.2 can be subdivided into smaller plates are given by Morgan (1968) and Dewey (1972).

There are three basic types of plate boundaries in oceanic regions, each associated with a distinctive kind of seismicity. They are known as *ridges, trenches,* and *transform faults.* Although the most revealing differences between them are seismological, the three classes of plate boundary are named for other characteristics. *Ridges* are named for their tendency to follow the crestal

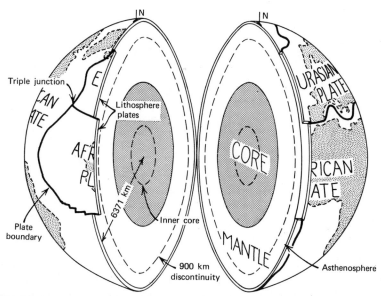

FIGURE 10.1   A sketch of the Earth, showing parts of the American, Eurasian, and African lithosphere plates and various features of the interior. Many plate boundaries are omitted for simplicity.

parts of submarine mountain ranges. *Trenches* are named for their association with furrows or trenches on the deep sea floor. *Transform faults,* which have more variable physiographic expression, are named for their structure.

No attempt is made here to comprehensively trace the development of plate theory. For more complete historical accounts and additional discussion of the evidence behind the theory, the reader is referred to recent books by Takeuchi et al. (1970), Wyllie (1971), and Cox (1973), and to the pioneering papers in plate tectonics by McKenzie and Parker (1967), Morgan (1968), LePichon (1968), and Isacks, Oliver, and Sykes (1968).

10.2.1 Ridges

With the major exception of the plate boundary that follows the East Pacific Rise, ridges tend to occupy medial positions in each of the main ocean basins (Fig. 10.2). The physiographic high with which they are associated may be a very broad swell on the sea floor, such as the East Pacific Rise, or a sharper feature such as the Mid-Atlantic Ridge. Note, however, that even the Mid-Atlantic Ridge appears very subdued, similar to continental mountain systems, when drawn without vertical exaggeration (Fig. 10.4).

Ridges may be associated with local physiographic features attributable to block faulting (Section 9.3). This is most conspicuously true of the ridge

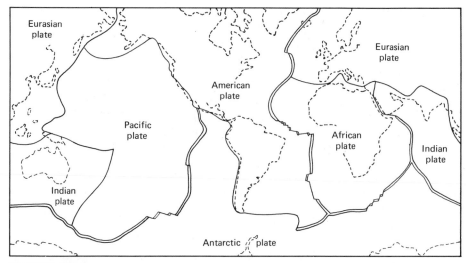

FIGURE 10.2   Six major lithospheric plates as delineated by LePichon (1968).
Ridges shown by double solid lines. Trenches, transform faults, and boundaries of
uncertain character show by single solid lines. [After LePichon (1968), with permission
of the American Geophysical Union.]

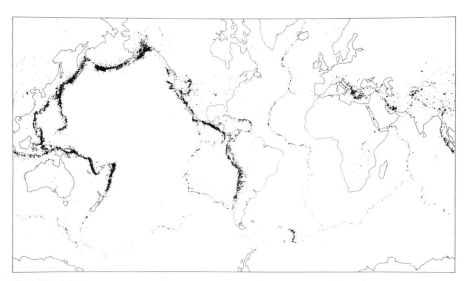

FIGURE 10.3   Epicenters of earthquakes with foci shallower than 100 km, between
1961 and 1967. [From Barazangi and Dorman (1969), with permission of the
Seismological Society of America.]

through the Atlantic Ocean, which is followed along most of its length by a well-defined elongate valley in the sea floor [Heezen and Ewing (1963)]. This valley, or *rift,* is similar in width and depth to parts of the continental rift system in East Africa [Heezen (1969)]. The ridge along the East Pacific Rise is not associated with a single large rift, but in places there is sea-floor topography that suggests longitudinal block faulting [Menard (1964, p. 119); Atwater and Mudie (1968)]. Many active volcanoes are located along ridges. The principal extrusive rock is tholeiitic basalt [Engel et al. (1965)]. In places, ridges appear to be zones of generally high heat flow, but the heat flow pattern remains largely unknown and is not always simple [see Langseth and Von Herzen (1970); McKenzie and Sclater (1969); Sclater et al. (1971)].

The earthquakes that occur beneath ridges originate at shallow depths (<30 km), and those analyzed yield focal mechanism solutions (Section 7.3.12) consistent with normal faulting on planes striking parallel to the trend of the ridge [Sykes (1968)]. While the seismicity of ridges defines their positions and suggests the nature of short-term displacements across them, magnetic anomaly patterns provide the most revealing guide to their long-term behavior. A magnetic anomaly profile published by Pitman and Heirtzler (1966) across part of the Pacific-Antarctic Ridge is shown in Figure 10.5. Over the plate boundary itself, which follows the crest of the physiographic ridge, there is a broad positive anomaly. This is flanked on either side by a series of narrower positive anomalies. The surprising property of this and many other profiles

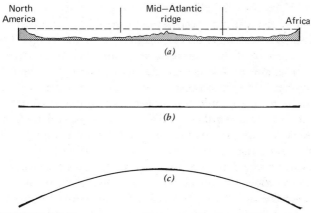

FIGURE 10.4   Three sections across the North Atlantic. (*a*) After Heezen and Ewing (1963), with permission of John Wiley and Sons, vertical exaggeration 40:1. (*b*) Same section redrawn without vertical exaggeration. (*c*) Same section without vertical exaggeration and showing curvature of the earth. Length of each section is approximately 5850 km.

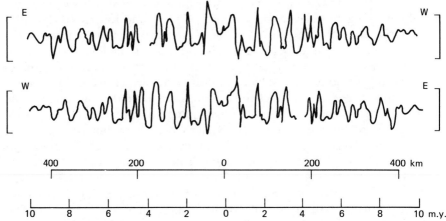

FIGURE 10.5   Magnetic anomaly profile across part of the Pacific-Antarctic ridge. Top trace shows profile as seen from the north. Bottom trace shows same profile as seen from the south. Correspondence of features in the two traces shows high degree of symmetry in the profile. Distances from center of ridge and inferred ages shown by scales. Length of vertical scale with each profile is 1000 γ. (From Pitman and Hiertzler, Science, 154, p. 1164–71. Copyright 1968 by the American Association for the Advancement of Science.)

across ridges [Vacquier (1972)] is that there is a reasonably good one-to-one correspondence between any anomaly on one side of the ridge and another anomaly at the same distance on the other side of the ridge. This bilaterally *symmetrical arrangement* of the magnetic anomalies can be confirmed in Figure 10.5 by comparing the profile as seen from the north with the same profile as seen from the south. Where sufficient coverage is available, the anomalies can be correlated from one profile to another. In most areas near ridges such maps show more or less regular parallel "stripes" of normally and reversely magnetized rock that are symmetrically disposed either side of the ridge. An example is shown in Figure 10.6.

The cornerstone of plate tectonics is the theory of *sea-floor spreading,* an interpretation of ridges that has grown out of suggestions by Heezen (1960), Hess (1962), and Dietz (1961a). According to this theory the floor of the ocean is moving or spreading away from ridges so that two points on opposite sides of a ridge become progressively more distant from one another. Simultaneously with spreading, and to fill the gap that might otherwise develop along the ridge, new rock is emplaced in this region by intrusion and extrusion from below. When first proposed, the idea of sea-floor spreading provided an elegant but unconvincing explanation for aspects of ridge vulcanism, seismicity, and physiography, and for the apparent youthfulness of sea-floor rocks com-

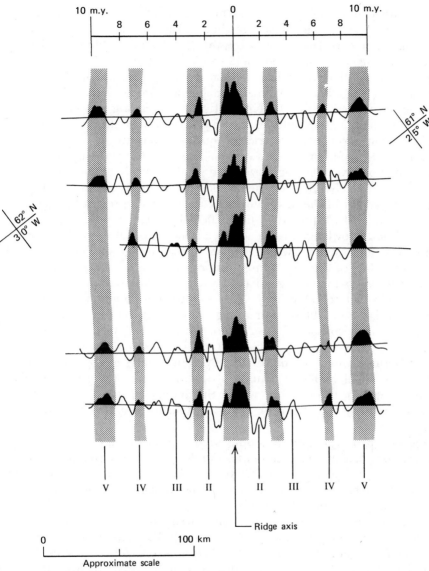

FIGURE 10.6   Magnetic anomalies plotted along ship's tracks for several crossings of the Reykjanes Ridge, south of Iceland. Some of the more prominent positive anomalies are correlated as shown by the stippled stripes running parallel to the ridge axis. [The correlations in this area are based on twice as many profiles as those shown here, in the original figure of Talwani et al. (1971).] For explanation of anomaly numbers (Roman numerals) and inferred ages, see Heirtzler et al. (1968), and Coulomb (1972, p. 57). [After Talwani et al. (1971), with permission of the American Geophysical Union.]

pared with continental rocks. Hess himself called the idea an exercise in "geopoetry." What transformed the idea to a serious hypothesis was the confirmation by Cox, Doell, and Dalrymple (1963) and McDougall and Tarling (1963) that the earth's magnetic field had reversed itself at least once during the past three million years, and the realization by Vine and Matthews (1963) and Morley (1963) that the symmetric magnetic striping associated with ridges could be explained as the combined effect of spreading and field reversals. Thus, for example, the central belt of positive anomaly shown in Figure 10.6 could be split into two symmetrically equivalent positive stripes if spreading continues and a polarity change occurs. The youngest rocks along the ridge would then be reversely magnetized and would give rise to a new central belt of negative anomaly. This belt then could be split in two by continued spreading and the onset of the next interval of normal (i.e., present day) polarity.

If sea-floor rocks acquire their magnetic polarity when they are emplaced at a ridge, then knowledge of the age of an anomaly and its present distance from the parent ridge allows an estimate of the average spreading rate to be made. Estimates of this sort, using distances measured normal to the nearest ridge axis and anomaly ages based on the ages of supposed equivalent anomalies in continental rocks, yield spreading rates ranging from about 1-15 cm/yr [Isacks et al. (1968)]. Spreading rates and directions are regarded as being reasonably constant for periods of the order of 1 m.y. but may change significantly over longer periods of time [e.g., see Menard and Atwater (1968); Pitman and Talwani (1972)]. Spreading rates are generally considered to be equal or nearly equal on opposite sides of a given ridge, but instances have been described where the anomaly pattern, and hence the overall spreading rates, appear to be asymmetric [e.g., Weissel and Hayes (1971)].

The theory of sea-floor spreading predicts that the age of the basement igneous rock beneath an ocean basin should increase progressively with distance from a ridge, provided the material has come from the ridge. This prediction can be checked in two ways: by dating samples of igneous rock and the overlying sediments obtained by drilling into the sea floor and by dating the oldest volcanic rocks exposed in islands well removed from present ridges. A deep-sea drilling program, known as the JOIDES program, has been in progress for several years and had drilled approximately 250 sites through 1972. The results are reasonably consistent with the spreading hypothesis [see Maxwell et al. (1970); Fischer et al. (1970); and continuing reports in *Geotimes*)]. Dating of volcanic islands has also provided some support for the theory, particularly where groups of progressively older islands are found aligned along directions perpendicular to ridges, as in the case of some island chains in the south Pacific [Wilson (1963a); Morgan (1972a); Jackson et al. (1972)].

Plate tectonics incorporates sea-floor spreading as one of its essential parts and makes explicit the idea that the sea floor *and* the lithosphere beneath it are simultaneously being created at ridges. Seismic observations suggest that typical lithosphere beneath oceans is 70-80 km thick, and that lithosphere beneath continents is up to 150 km thick [see Kanamori and Press (1970); Walcott (1970); Wickens (1971); Press (1973)].

10.2.2 Trenches

In contrast with ridges, trenches tend to occur around the *edges* of major ocean basins, either adjacent to continental margins (e.g., the west coast of South America) or along the ocean-facing side of island arcs (e.g., in the western Pacific). Trenches thus tend to have islands or continents on *one side only,* and this is one aspect of a very general *asymmetry* in the properties of trenches.

Trench vulcanism is characteristically andesitic. In many cases there is a lateral variation in the composition of the volcanic rocks from tholeiitic basalt on the oceanic side of an island arc system to increasingly alkaline and andesitic lavas on the continental side [see Kuno (1959); Dickinson (1970); Miyashiro (1974)]. A zone of high heat flow is known on the continental side of some arcs, for example, those of the western Pacific [Sclater (1972)]. A more prominent geophysical feature of trenches is a very marked negative gravity anomaly, which occurs beneath the physiographic trench or between the trench and the arc or continental margin [see Hayes and Ewing (1968)].

Figure 10.7 shows the physiographic profile and distribution of earthquake foci associated with several trenches. Note that trench earthquakes, in distinction from ridge earthquakes, may come from depths as great as 700 km. The foci associated with any given trench are found to lie in a narrow zone that intersects the surface of the earth in the general vicinity of the physiographic trench and dips beneath the adjacent arc or continental margin. Such zones are known as *Benioff zones*. They commonly dip at about 45° but much steeper or flatter dips are well known (Fig. 10.7).

Focal mechanism studies (Section 7.3.12) of trench earthquakes give a complex picture summarized by Isacks, Oliver, and Sykes (1968) and Isacks and Molnar (1971). At every shallow depths (<5-10 km) in the neighborhood of the physiographic trench, the results are consistent with normal faulting on planes striking approximately parallel to the trench. At slightly greater depths (10-70 km), beneath the area between the arc and trench, thrust solutions are obtained. One of the two possible fault planes tends to parallel the Benioff zone itself. For earthquakes originating at depths greater than about 70 km the most common result indicates two planes striking parallel to the Benioff

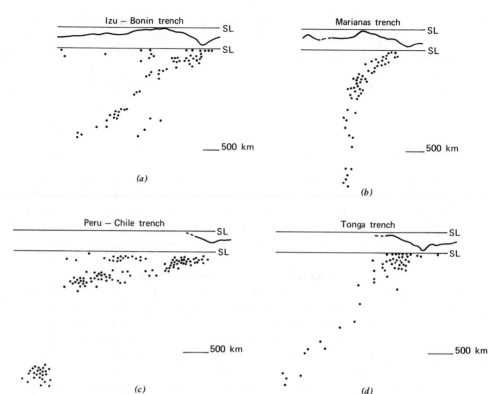

FIGURE 10.7 Physiographic profiles and Benioff zones defined by earthquake foci (dots) for portions of several trenches. Vertical exaggeration of physiographic profiles about 10:1. No vertical exaggeration in part of each figure below the lower sea level (SL) line. (*a*) and (*b*) From Katsumata and Sykes (1969), with permission of the American Geophysical Union. Earthquake foci in (*c*) and (*d*) from Isacks and Molnar (1971) and Sykes (1966), respectively. Physiographic profiles in (*c*) and (*d*) are from Hayes and Ewing (1968). (With permission of American Geophysical Union.)

zone and inclined to it at angles of about 45°. The indicated senses of movement suggest that intermediate (70-300 km) and deep focus (300-700 km) earthquakes commonly reflect lengthening or shortening parallel to the dip of the Benioff zone. Shortening is predominant for deep-focus earthquakes.

The plate tectonics interpretation of trenches is that they are the sites of underthrusting on a grand scale, along which the lithosphere of one plate disappears or is "consumed" as it passes beneath the edge of another plate (Fig. 10.8). This process is called "subduction" and trenches are sometimes referred to as "consuming plate margins" or "subduction zones." Gutenberg and Richter (1951), Benioff (1954) and others had previously suggested that trenches were somehow associated with thrust faulting, but it was not until

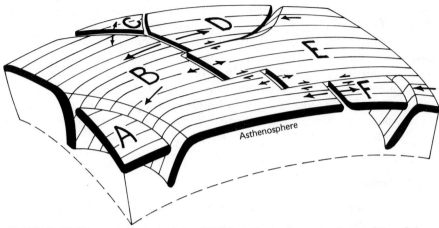

FIGURE 10.8   Schematic drawing of lithosphere plates on the surface of the
earth, showing various types of plate boundaries and triple junctions. Arrow
lengths proportional to displacement rates (velocities) *relative to local plate margins*
(the center lines of ridges and transform faults, the hanging wall edge of trenches).
Character of boundaries as follows: A-B, trench; B-C, ridge; B-D, ridge; B-E, ridge
with dextral transform segments locally; B-F, ridge; D-E, dextral transform at near
end, transitional into dextral trench at far end; E-F, sinistral transform and trench.
Triple junctions at B-C-D, B-D-E, B-E-F. The nature of the boundary C-D can be
worked out as an exercise using the relation given by Equation 10.6. Inspired by
Fig. 1 of Isacks, Oliver, and Sykes (1968).

ridges had been interpreted as spreading centers (capable of generating millions of square kilometers of new sea floor) that trenches were seen as the
probable sites for simultaneous consumption of similar amounts of old sea
floor, along with the whole of the underlying lithosphere [McKenzie and
Parker (1967); Morgan (1968)].

The dip of the Benioff zone is seen to reflect, in a general way, the dip of
the underthrust lithosphere, but the earthquake foci may or may not lie on
the upper surface of the descending slab. The earthquakes giving thrust
solutions at shallow depths may, in some cases, originate near the upper
surface of the slab, but those occurring at greater depths are believed to
develop *within* the slab [see Isacks and Molnar (1971)]. These intermediate
and deep-focus earthquakes may be due to faulting brought about by dehydration embrittlement or sudden movements due to some kind of hot creep
instability or rapid phase changes [Griggs (1972)]. The vulcanicity and high
heat flow associated with trenches are thought to be associated with partial
melting of the underthrust slab, possibly including entrained sediments at its
upper surface [see Dickinson (1970)].

An important but little understood tectonic element associated with island arcs, and therefore with trenches, are the *small ocean basins* or *marginal basins* that occur behind all currently active arcs (e.g., the Sea of Japan, the Aleutian basin). These basins are underlain by crust of normal oceanic thickness and seismic properties. They lack well-defined medial ridges and simple magnetic anomaly patterns, but they are tentatively interpreted as regions across which some kind of slow extension or spreading is taking place. The rate of extension may be lower than the rates of spreading across active ridges in the major ocean basins [Karig (1971)].

10.2.3 Transform Faults

Transform faults by definition [Wilson (1965); Atwater (1972)] are large transcurrent faults that terminate in some other kind of structure, for example, ridges, trenches, or triple junctions. Certain segments of present-day plate boundaries appear to have this character. The best-known oceanic

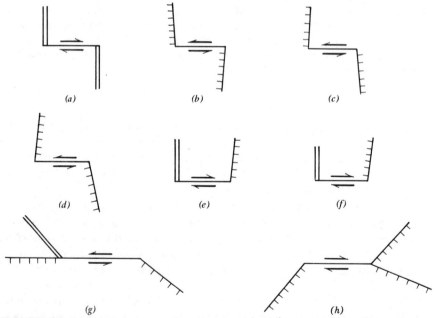

FIGURE 10.9   Various types of transform faults. Ridges shown by double lines, trenches by single lines with hatching on hanging-wall side. (*a*) Ridge-ridge type: (*b*), (*c*), (*d*) Trench-trench types: (*e*), (*f*) Ridge-trench types: (*g*), (*h*) Two examples of transform faults that terminate in triple junctions. The length of the transform fault stays constant with the passage of time in (*a*), (*b*); increases in (*d*), (*f*), and (*g*); and decreases in (*c*), (*e*), and (*h*).

examples are the seismically active zones, often hundreds of kilometers long, that truncate oceanic ridges and appear to displace the ridge segments in a faultlike manner (e.g., in the equatorial Atlantic, Fig. 10.2; Fig. 10.8).

Earthquake foci on transform faults, like those beneath ridges, seem to be restricted to shallow depths (<30 km). The focal mechanisms are, however, entirely different [Sykes (1968)]. The solutions obtained from transform fault earthquakes suggest transcurrent displacement on planes approximately parallel or perpendicular to the trace of the seismic zone. Of these two orientations, the planes parallel to the overall trace of the seismic zone seem more likely to be the real fault planes. The surprising feature about these solutions is that the sense of displacement indicated by the first motions is opposite to the apparent sense of displacement of the ridge segments. Thus in the equatorial Atlantic, the first motions are consistent with *dextral* displacements but the ridge offsets are *sinistral* [Sykes (1968)]. This discrepancy was predicted and explained by Wilson's (1965) concept of the transform fault. In this interpretation the offset of the ridge may be an original feature of a spreading system and spreading can occur without any necessary increase or decrease in the distance between ridge segments. Figure 10.9 shows this situation and some other kinds of transform faults. Note that the length of a transform fault may in some cases increase or decrease with time.

Figure 10.10 shows the development of an idealized ridge-ridge transform fault. There are "extensions" of the fault that extend outward from the ends

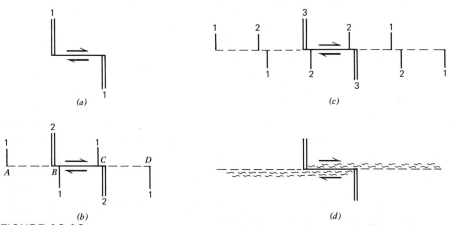

FIGURE 10.10 Development of transform fault "extensions" and offsets of magnetic anomalies across them. (*a*) Time 1, inception of spreading. (*b*) Time 2, anomaly 1 has split and moved off the ridge in both directions, generating "extensions" *AB* and *CD*. The transform fault remains segment *BC* only. Ideally there is no slip and no seismicity across any parts of *AB* and *CD*. (*c*) Time 3, "extensions" have grown longer. (*d*) Same as (*c*) but showing the predicted occurrence of mylonites on one side only (the older side) of each "extension."

of the active segment. These are lines across which there is an offset in the magnetic anomaly pattern equal in sense and magnitude to the offset of ridge segments. Such extensions are sometimes referred to as the seismically "inactive segments" of the transform faults, although ideally they are not and never have been transcurrent faults (Fig. 10.10). They are simply lines across which sea floor of different ages has been "welded" together. Such transform fault extensions may be expressed physiographically by lines of scarps or other features [Menard (1964)]. The largest known examples are the great "fracture zones" of the eastern Pacific.

Transform faults play an invaluable role in the interpretation of plate motions. For while ridges and trenches indicate sites of spreading and consumption respectively, the associated transform faults reveal the *directions* of spreading or consumption. Transform faults associated with ridges tend, with a very few exceptions noted, for example, by McKenzie (1972, p. 328), to trend at right angles to ridges. This is taken to indicate that the direction of spreading is typically perpendicular to ridges [LePichon (1968); Morgan (1968)]. Transform faults terminating at trenches are less common, but many examples are known, and the conclusions drawn from their *variable angle of intersection with trenches* is that subduction may be, but commonly is not, by displacement parallel to the dip direction of the Benioff zone. Trenches appear free to consume lithosphere in such a way that the displacements involve large strike-slip components in addition to the dip-slip component. This observation has important implications for orogenic history and is mentioned further in Section 10.2.5 and Section 10.2.6 below.

10.2.4 Strain Within Plates

In the plate *model,* the current idealized conception of plate behavior that is useful for making calculations, lithospheric plates are said to be *rigid.* This quite obviously cannot be true of real plates on the earth. A rigid body is one in which all strains are zero for a given stress distribution. The rocks of a lithospheric plate, on the other hand, must certainly undergo small elastic strains; and they almost certainly undergo time-dependent deformations, or creep, even at very low levels of differential stress. At trenches, large strains are assumed in all pictures that show one plate bent beneath the other. Therefore, when the behavior of real plates is described as "rigid," all that is really implied is that there has been negligible distortion of near surface horizontal lines on the plates (e.g., magnetic anomaly stripes). Little is known of the strain of vertical lines through plates, or of horizonal lines deep within plates. There is some evidence that the interiors of plates are in fact nonhydrostatically stressed and locally seismic [Sykes and Sbar (1973)] although the long-term strains are unknown. Evidence consistent with minimal strains within plates during the past 200 million years includes the excellent fit of the opposing margins of Africa and South America [Bullard et al. (1965)], the fit of magnetic anomalies to the Mid-Atlantic Ridge and to corresponding anomalies across the ridge [Pitman and Talwani (1972)], the relative aseismicity of plate interiors (Fig. 10.3), and the undeformed-looking sediments on the deep sea floor. There is also considerable agreement between displacement at trenches deduced from focal mechanism solutions and displacements predicted from spreading data by application of the theory of rigid plates [LePichon (1968); Isacks et al. (1968)]. We emphasize again, however, that the evidence for small strains applies mainly to the uppermost part of the lithosphere. The magnetic stripes beneath the sea floor, the sediments, and the continental margins that appear undeformed are all features of the uppermost 10 or 20 km. They reveal nothing directly about the state of strain farther down.

Whether strains appear small or large within present-day plates depends in part on how plate boundaries are defined. This chapter has been concerned so far primarily with plate boundaries in oceanic regions that are marked out by narrow zones of seismic activity. There are also, however, several major plate boundaries that are associated with very wide belts of recent deformation and seismic activity. These typically involve continental rocks. The most obvious examples in Figure 10.3 are the southern boundary of the Eurasian plate, which extends from the Mediterranean region through the Himalayas, and the western boundary of the American plate at the latitude of the United States. To maintain the view that plate interiors are essentially unstrained in cases like these, one can either regard the plate boundaries as coextensive with such "wide soft zones" or one can resolve such zones into a number of smaller plates each of which is supposed to be internally undeformed. The latter approach has been used, for example, in study of the Mediterranean region by McKenzie (1970) and Dewey et al. (1973).

10.2.5 Geometry of Plate Motion

In this section we discuss the purely geometric features of plate motion for the simple model of rigid spherical caps. Some basic concepts are introduced in detail for students with little experience of vector methods. Additional background information will be found in books such as Den Hartog (1948) and Spiegel (1959). All stereographic projections in this section are plotted on the *upper hemisphere,* to make them look somewhat like the corresponding views of a globe.

Angular Displacement of a Plate. A plate in motion over the asthenosphere occupies different positions at different times. Its change in position over some interval of time is described by an *angular displacement.* Note that an angular displacement only specifies a *change* in position, and not the actual history of movement by which the displacement occurred.

An angular displacement of a plate is described as a rotation about some line through the center of the earth. That there *is* such a line for any two

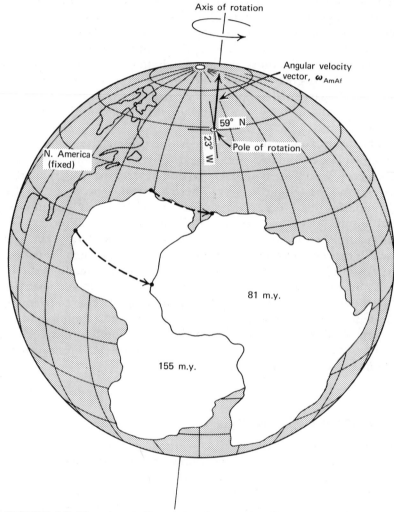

FIGURE 10.11   Sketch illustrating the angular displacement of Africa, in a reference frame fixed to North America, between 155 and 81 million years ago. The angular displacement is described by a rotation of +34.5° about a pole at 23° W 59° N. The average angular speed is about 0.5° per million years. Oceanic parts of plates are omitted for convenience. Data from Pitman and Talwani (1972).

positions of a plate follows from a theorem of Euler [see Thomson and Tait (1912, p. 69); Synge (1960, p. 13)].

> If two successive positions of a rigid body have point O in common, there must be a line through O that is common to the body in both positions. Rotation about this line as axis will always be capable of bringing the body from one position to the other.

Euler's theorem applies to spherical plates on a spherical earth because there is a point O at the center of the earth that is fixed or "common" to a plate in any two positions. The line through the center of the earth, about which a plate can be rotated from one position to the other, is called the *axis of rotation.* This axis intersects the surface of the earth in two antipodal points, either of which can define the *pole of rotation.* A complete description of the angular displacement of Africa shown in Figure 10.11 is rotation of +34.5° about a pole situated at 23° W 59° N relative to North America. The positive sign of the angular displacement indicates a rotation that is clockwise when viewed along the axis of rotation, looking away from the center of the earth.

Any motion must be described with respect to a designated reference frame. In Figure 10.11 the frame is fixed to North America and the pole of rotation is described in this frame. The latitude and longitude of the pole would be different if described relative to the African plate.

A finite angular displacement has a direction and magnitude associated with it, but it is *not* a vector quantity. Vector quantities obey laws of vector addition, one of which says that the sum of several vectors is independent of the order in which they are combined [Spiegel (p. 2)]. Figure 10.12 demonstrates that finite angular displacements do not obey this law.

Angular Velocity of a Plate.   The *angular velocity* of a plate specifies the instantaneous *angular speed* at which the plate is rotating about its instantaneous axis of rotation *and* the orientation of this axis. The sign of an angular velocity specifies the sense of rotation, positive for clockwise rotation as seen from the center of the earth. Units of angular speed are radians per million years and degrees per million years.

The exact or instantaneous angular velocity of a plate cannot be measured and is never known. What can be measured is the so-called *average angular velocity,* specified by the pole of rotation for a finite angular displacement and an angular speed obtained by dividing the finite displacement by a finite time. The average angular velocity with respect to America, of Africa in Figure 10.11 is plus half a degree per million years about a pole at 23° W 59° N, or minus half a degree per million years around a pole at 157° E 59° S.

An average angular velocity determined as above may be a close approximation to the individual instantaneous angular velocities during the time interval considered. Geometrical constraints such as long transform segments tend to promote steady plate motion. However, it is possible in some cases that

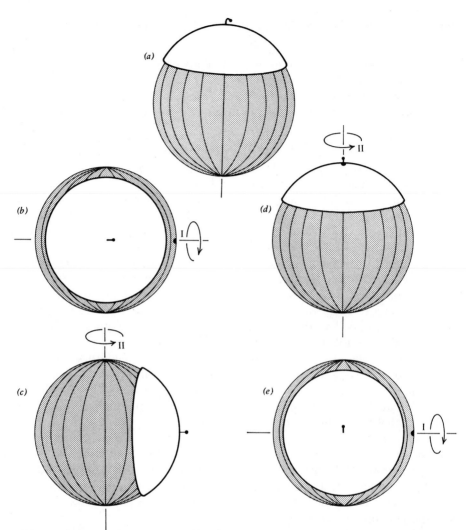

FIGURE 10.12 Successive finite rotations of a spherical cap, demonstrating that finite angular displacements do not add as vectors. The initial position of the cap (a) is altered in (b) by rotation of plus 90° around pole I, followed in (c) by rotation of plus 90° around pole II. In the sequence (a), (d), and (e) the order of the rotations is reversed, with different results.

the instantaneous velocities have been more variable, particularly for small plates and long time intervals. In such cases the so-called average angular speed, for example, may provide a lower limit on the real average speed, and not an average at all. This case is analogous to a one hour journey by car, between cities 60 miles apart as the crow flies, where an "average speed" of 60 miles per hour is the lower limit of possible average speeds.

Angular velocity and infinitesimal angular displacement *are* vector quantities. A proof of this is given by Den Hartog (1948, p. 316). Average angular

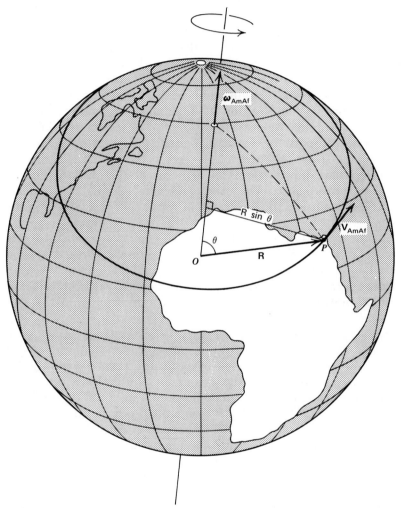

FIGURE 10.13   Relation between angular velocity and particle velocity. Point O is at the center of the earth. Point P is on the African plate, connected to O by the radius vector **R**. The particle velocity vector $\mathbf{V}_{AmAf}$ is perpendicular to the plane defined by ω_{AmAf} and **R**. If ω_{AmAf} is 0.5° per million years (or 0.0087 radians per million years), **R** is 6371 km in length, and θ is 44°, the magnitude of $\mathbf{V}_{AmAf}$ at P is 38.5 km per million years or 3.85 cm per year.

velocity is a vector quantity if we assume steady plate motion, so that the so-called average velocity is really taken to be a constant angular velocity for the time interval considered.

We write angular velocity vectors in the form ω_{AmAf}, meaning the angular velocity with respect to America, of Africa. Notice that ω_{AmAf} is equal to but opposite in sign to ω_{AfAm}.

The arrow representing an angular velocity vector is drawn *parallel to the instantaneous axis of rotation.* Rotations that are clockwise (or positive) when viewed looking away from the center of the earth are represented by arrows pointing outward from the corresponding pole of rotation (Fig. 10.11). This can be remembered using a "right-hand rule": if the fingers of the right hand are held pointing in the direction of rotation, the thumb will point in the direction of the angular velocity arrow. As for all vectors, the length of the arrow is drawn proportional to the magnitude of the vector quantity, in this case the angular speed. Angular velocity vectors are confusing at first because they are represented by an arrow aligned normal to, rather than parallel to, the paths of individual particles. Angular velocity vectors are so-called *axial vectors,* a type familiar in the mechanics of rotating bodies [see Shonle (1965, 1966)].

The vector ω_{AmAf} is drawn at the northern hemisphere pole of rotation in Figure 10.11; it points away from the center of the earth. Drawn at the southern hemisphere pole of rotation, ω_{AmAf} would point toward the center of the earth. On the other hand, ω_{AfAm} would point toward the center of the earth at the northern hemisphere pole and away from it at the southern hemisphere pole.

Relationship of Angular Velocity to Particle Velocity. If we know the instantaneous angular speed of Africa with respect to America, in radians per million years, we can calculate the linear speed of a particle P on the African plate. The distance of P from the axis of rotation is $R\sin\theta$, where R is the radius of the earth and θ is the angle between a radius to P and the axis of rotation (Fig. 10.13). The speed of P then is the product of the angular speed in radians and the distance $R\sin\theta$. The direction in which P is traveling will be tangent to the surface of the earth and tangent to a small circle around the pole of rotation. All this is expressed compactly by the vector equation

$$\mathbf{V}_{AmAf} = (\omega_{AmAf} \times \mathbf{R}) \tag{10.1}$$

where $\mathbf{V}_{AmAf}$ is a vector representing the linear velocity of a particle with respect to America, on Africa; and $\mathbf{R}$ is a radius vector drawn from the center of the earth to P. $(\omega_{AmAf} \times \mathbf{R})$ is a vector cross product (see Spiegel, p. 16) that yields both the magnitude and direction of $\mathbf{V}_{AmAf}$.

The fact that particle velocities such as $\mathbf{V}_{AmAf}$ must be directed tangentially to small circles around the instantaneous pole of rotation gives rise to a method used by Morgan (1968) and LePichon (1968) to locate poles of rotation for plates separated by a ridge. Consider a transform fault along the Mid-Atlantic Ridge; one side moves with Africa, the other is fixed to America. As long as the pole of rotation is fixed, points immediately on the African side of the fault must move on small circles around the pole of rotation. If the fault is to remain a boundary along which no spreading or subduction occurs, then the fault itself must be aligned along a small circle around the pole of rotation. Morgan and LePichon drew great circles normal to segments of transform faults along a ridge, and the pole of rotation was taken as coincident with the point at which the great circles intersect each other. Having established the orientation of an angular velocity vector in this way, the average magnitude was determined from spreading rates based on dated magnetic anomalies.

Interrelationships of Angular Velocities for Several Plates. If the angular velocities of plates A and B are known in some reference frame M, we can determine ω_{AB}. The method involves a vector sum that will be illustrated first by a simple linear velocity problem (Fig. 10.14). An airplane taxis north relative to the map with velocity $\mathbf{V}_{MA}$ of 100 kmph. A boat moves east with velocity $\mathbf{V}_{MB}$ of 20 kmph. We determine $\mathbf{V}_{AB}$, the velocity with respect to the airplane of the boat from the vector equation

$$\mathbf{V}_{AB} = \mathbf{V}_{AM} + \mathbf{V}_{MB}$$

or
$$\mathbf{V}_{AB} = \mathbf{V}_{MB} - \mathbf{V}_{MA} \tag{10.2}$$

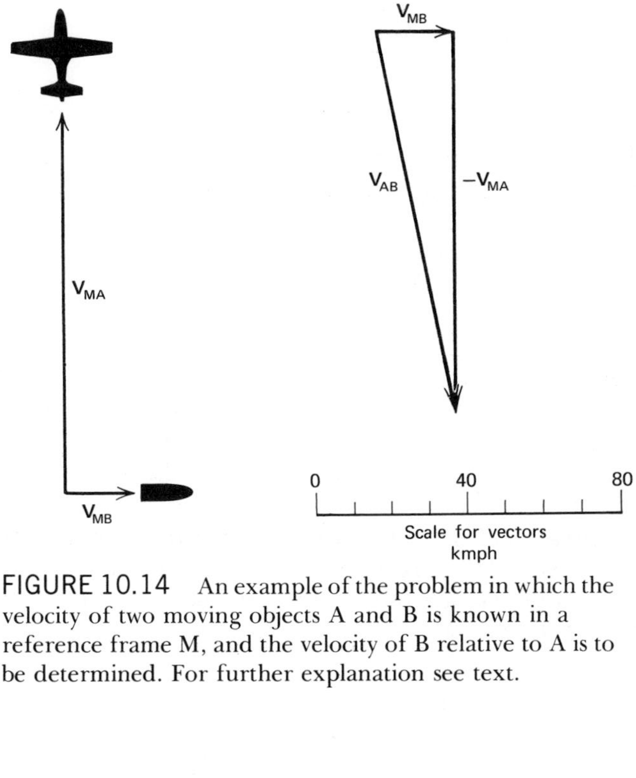

FIGURE 10.14 An example of the problem in which the velocity of two moving objects A and B is known in a reference frame M, and the velocity of B relative to A is to be determined. For further explanation see text.

since $\mathbf{V}_{AM}$ and $\mathbf{V}_{MA}$ are equal but opposite in sign. Equation 10.2 says that the velocity with respect to A of B is the difference between the velocities of B and A in frame M. To solve the equation graphically we draw an east-pointing arrow representing $\mathbf{V}_{MB}$ and then draw a south-pointing arrow representing $-\mathbf{V}_{MA}$, with its tail touching the head of the first arrow, as shown in Figure 10.14. Putting a sequence of arrows in tail-to-head arrangement like this represents addition of the corresponding vector quantities (Spiegel, p. 2), in this case, the vectors $\mathbf{V}_{MB}$ and $-\mathbf{V}_{MA}$. The sum of the vectors $\mathbf{V}_{MB}$ and $-\mathbf{V}_{MA}$ is then $\mathbf{V}_{AB}$, represented by drawing an arrow from the tail of $\mathbf{V}_{MB}$ to the head of $-\mathbf{V}_{MA}$.

We now write equations identical in form to Equation 10.2 for angular velocities of three plates A, B, C in a reference frame M.

$$\omega_{AB} = \omega_{MB} - \omega_{MA}$$
$$\omega_{BC} = \omega_{MC} - \omega_{MB} \qquad (10.3)$$
$$\omega_{CA} = \omega_{MA} - \omega_{MC}$$

The frame M is quite arbitrary. It can be fixed to some part of the mantle below the lithosphere, or to some plate other than A, B, C.

Adding Equations 10.3 together gives the very useful result that

$$\omega_{AB} + \omega_{BC} + \omega_{CA} = \mathbf{0} \qquad (10.4)$$

which is equivalent to

$$\omega_{AB} = -\omega_{BC} - \omega_{CA}$$

or to

$$\omega_{AB} = \omega_{CB} + \omega_{AC} \qquad (10.5)$$

Equation 10.5 says that if we know the instantaneous values of ω_{CB} and ω_{AC} we can add them together vectorially to obtain the instantaneous value of ω_{AB}. This was done by LePichon (1968) to calculate an approximate present-day value of ω_{EurAf} from estimates of ω_{AmAf} and ω_{EurAm} based on sea-floor data. Equation 10.5 for this calculation is

$$\omega_{EurAf} = \omega_{AmAf} + \omega_{EurAm}$$

and the solution is illustrated in Figure 10.15. Notice that *any three angular velocities that sum to zero*, as in Equation 10.4, *must be represented by a closed vector triangle* (because the sum is a vector with zero magnitude, ω_{AA}, ω_{BB} or ω_{CC}). A closed vector triangle is necessarily composed of three coplanar vectors, and therefore, the *three poles of rotation corresponding to ω_{AB}, ω_{BC} and ω_{CA} must lie on a great circle on the earth*. Figure 10.16 shows the necessary general orientation relationships between the six angular velocity vectors in Equations 10.3. There are four coplanar sets of vectors, corresponding to the four equations in (10.3) and (10.4).

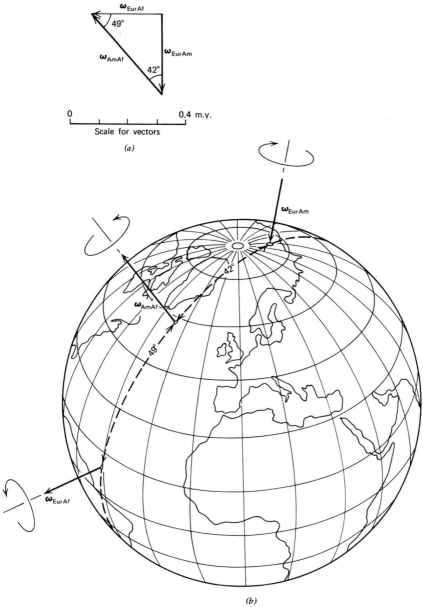

FIGURE 10.15   (*a*) Vector triangle used to determine ω_{EurAf} from knowledge of ω_{AmAf} and ω_{EurAm} as given by LePichon (1968). (*b*) Sketch showing the orientations of these three vectors and the great circle connecting the three poles of rotation.

Interrelationships Between Particle Velocities on Adjacent Plates. Consider a point P on the boundary between plates A and B. If we know ω_{AB}, and therefore ω_{BA}, we can find the tangential velocities $\mathbf{V}_{AB}$ and $\mathbf{V}_{BA}$ from two equations like Equation 10.1

$$\mathbf{V}_{AB} = (\omega_{AB} \times \mathbf{R})$$
$$\mathbf{V}_{BA} = (\omega_{BA} \times \mathbf{R})$$

where $\mathbf{R}$ is the radius vector to point P. It can be seen that $\mathbf{V}_{AB}$ and $\mathbf{V}_{BA}$ will be equal and opposite and that

$$\mathbf{V}_{AB} + \mathbf{V}_{BA} = \mathbf{0}$$

If we now write the velocity sum for a point P at a triple junction, where *three* plates A, B, and C meet, we obtain

$$\mathbf{V}_{AB} + \mathbf{V}_{BC} + \mathbf{V}_{CA} = \mathbf{0} \tag{10.6}$$

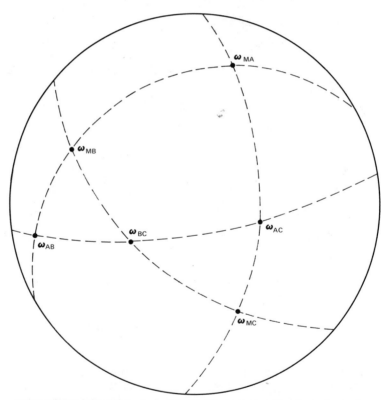

FIGURE 10.16 Stereographic projection showing the orientation relations between a possible set of the six angular velocity vectors in Equations 10.3 and 10.4.

This equation is used by McKenzie and Morgan (1969) to investigate the stability of triple junctions. It can also be used to determine one of the velocities if the other two are known, as for example in Figure 10.17. It is emphasized however that Equation 10.6 is strictly correct only for three *coincident* particles attached to plates A, B, and C. The equation is not exact for three particles at finite distances from one another.

The basic equations [(10.4) and (10.6)] can be written for any number of plates. If there are N plates there should be N velocities in the sum set equal to zero, and each plate symbol should appear twice among the subscripts, once as a reference frame and once as the moving plate.

Tectonic Character of Plate Boundaries. The instantaneous character of the boundary between plates A and B at a particular point will depend on the relative orientations of the boundary and the vector $\mathbf{V}_{AB}$ at that point. Thus, if plate A lies south of plate B and the boundary between them is east-west, the character of the boundary between them will be a pure dextral transform if $\mathbf{V}_{AB}$ is directed from west to east. If $\mathbf{V}_{AB}$ is directed from north to south the boundary will be a pure trench. If $\mathbf{V}_{AB}$ is directed from northwest to southeast, the boundary will have mixed character and could be called a dextral trench. All possible relations can be summarized in a diagram (Fig. 10.18) showing the orientation of $\mathbf{V}_{AB}$ relative to η_{AB}, where η_{AB} is a unit vector (a vector of length 1.0) normal to the plate boundary, tangent to the surface of the earth, and pointing from the reference plate A toward B.

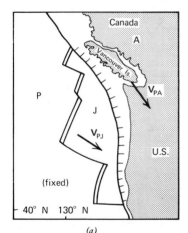

(a)

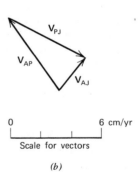

(b)

FIGURE 10.17   (a) Map showing Juan de Fuca plate (J) and parts of the Pacific (P) and American (A) plates. Vectors show possible velocities relative to the Pacific plate according to Atwater (1970). East-west dimension of map is about 650 miles. (b) Vector triangle used to determine $\mathbf{V}_{AJ}$. $\mathbf{V}_{AJ} + \mathbf{V}_{JP} + \mathbf{V}_{PA} = \mathbf{0}$ so that $\mathbf{V}_{AJ} = \mathbf{V}_{AP} + \mathbf{V}_{PJ}$. [From Atwater (1970), with permission of the Geological Society of America.]

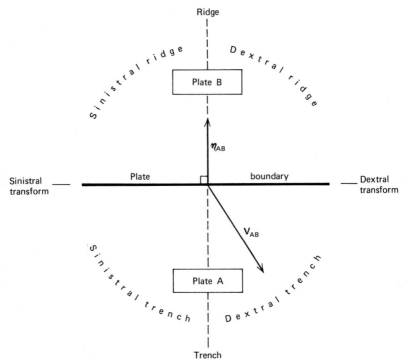

FIGURE 10.18 Diagram representing all possible kinds of plate boundaries. Plane of paper is tangent to the surface of the earth, oriented so that the heavy line is parallel to a plate boundary. η_{AB} is a unit vector normal to the boundary and pointing from plate A toward plate B. The boundary and η_{AB} are fixed in the diagram. $\mathbf{V}_{AB}$, the local particle velocity vector for points on B, may take any orientation depending on the character of the boundary; for the $\mathbf{V}_{AB}$ shown, the boundary is a dextral trench. The tip of the vector $\mathbf{V}_{AB}$ will describe some path in this diagram if the character of the boundary changes in time.

Care should be taken not to assume any particular structure for the boundaries of mixed character shown in Figure 10.18. Thus, for example, if $\mathbf{V}_{AB}$ lies in the upper-right quadrant, the boundary has *overall* dextral ridge character. It may turn out, however, that on a smaller scale such a boundary is made up of alternating segments of pure ridge and pure dextral transform, as appears to be true in the equatorial Atlantic, the Gulf of California, and the Gulf of Aden. Similarly no assumptions should be made about the structure of trenches with a transform component. Several alternative structures for a sinistral trench are drawn in Figure 10.19. The structures shown may not all be mechanically plausible, but they are all geometrically possible.

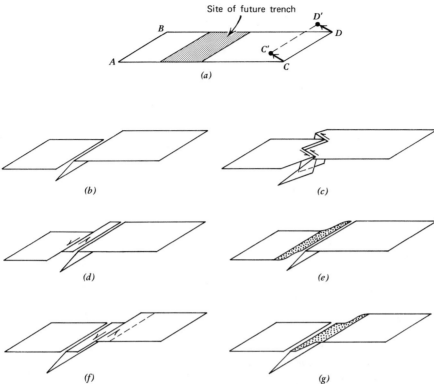

FIGURE 10.19   Several possible structures for a sinistral trench. (*a*) Initial configuration. In all succeeding drawings the edge *C-D* has been displaced to *C′-D′* relative to *A-B*. Structures to accommodate this displacement are drawn in the plate boundary zone (shaded). (*b*) Simple oblique subduction. (*c*) Stepped alternation of pure trench and pure transform segments. (*d*) Down-dip subduction accompanied by transcurrent faulting on the continental side of the Benioff zone [as suggested by Fitch (1972) for certain west Pacific arcs]. (*e*) Same as (*d*) but with transcurrent fault replaced by a broader zone of transcurrent displacements. (*f*) Similar to (*d*) but with transcurrent faulting on the oceanic side of the Benioff zone. New transcurrent fault must form (dashed) if the original one is rendered inactive by subduction. (*g*) Like (*f*) but with a broad zone of transcurrent displacement on the oceanic side.

At a given time, the character of a boundary may be different at different points along its length. This will occur wherever the angle between $\mathbf{V}_{AB}$ and η_{AB} is different at different points or, in other words, where a boundary makes different angles at different points with the family of small circles around the instantaneous pole of rotation. An example is the boundary south and west of the Aleutian islands, which is said to be more or less pure trench at

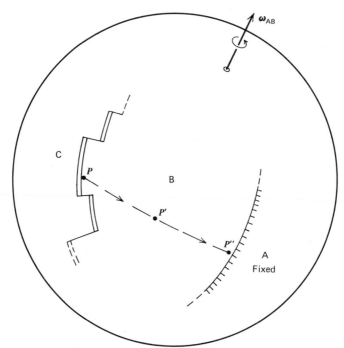

FIGURE 10.20 Upper hemisphere stereographic projection illustrating change in tectonic setting of P as it moves from ridge to trench (example 1 in text). Ridge C-B shown for time represented by P only; transform faults between C and B follow small circles around ω_{CB} (not shown). Angular velocity vectors in this and succeeding figures are drawn in on the projection as they would appear in a perspective view.

its eastern end and more or less pure transform at its western end. In between it has mixed trench-transform character [Stauder (1968); McKenzie and Parker (1967)]. A hypothetical example of this type is represented by the boundary D-E in Figure 10.8.

Changes in Tectonic Setting with Time. The geology of mountain belts typically indicates that rocks have experienced a number of different tectonic settings, and so we ask how the plate model might explain such changes. Are there situations, for example, under which rocks could spend their early history in the vicinity of a trench and their later history in the different regime near a transform fault? Could rocks adjacent to a rapidly subducting, dextral trench experience an abrupt change of setting to that of a slow, sinistral trench? Plate theory suggests answers to such questions through analysis of the necessary geometric interactions between moving spherical caps.

The most obvious causes for change of setting at a point are fundamental changes in plate motions or in boundary configurations, that is, establishment of new angular velocity vectors or new boundary positions. Such fundamental changes may be responsible for some of the grosser features of the earth's tectonic history, but they cannot be discussed very usefully until more is known of the driving mechanisms for plate motion. Meanwhile there are a surprising variety of ways in which plate theory predicts temporal changes in tectonic setting that are *independent* of fundamental changes in angular velocities or plate boundary positions. Four examples are discussed below.

Example 1. The simplest example is of the type shown in Figure 10.20. A rock body originating at a point (*P*) on a ridge, moves away from the ridge (*P'*) and ultimately encounters a trench (*P"*). The body, which can be taken for example, as a piece of peridotite beneath the ridge, or as the whole sequence of rocks above the piece of peridotite (including any components added during its journey to the trench), experiences at least three different tectonic settings and may ultimately display structural, petrological, or stratigraphic features recording all parts of this history. This example is taken directly from Hess (1962) who first suggested that rocks of the ophiolite suite, now found in fold belts, might have been carried from ridges and across ocean basins to the sites of mountain building at ocean margins.

Example 2. Next consider a body of rock that occupies a position near a plate boundary for a considerable period of time. Such a body might be one situated on the hanging wall side of a trench. Figure 10.21 shows a body in such a position (*P*) on a major north-south trench. Intersecting this trench is another trench, oriented east-west. At the beginning of the history investigated here, the triple junction (where the plates A, B, and C intersect) is located south of *P*. We assume that plate A is fixed in our projection frame of reference and that the angular velocity vectors ω_{AB} and ω_{AC} are also fixed in orientation and magnitude for the time interval discussed. By Equation 10.4, ω_{CB} must similarly be fixed with respect to the projection frame for this interval. Plates B and C will be moving in a uniform way over the projection, as indicated by ω_{AB} and ω_{AC}, respectively. Paths for several particles on each plate are indicated by arrows parallel to small circles about the appropriate poles of rotation. It can be seen that the east-west trench must move gradually north, along with the rest of plate C to which it is attached. The triple junction will migrate up the North-South trench and eventually pass point *P*. As this happens the character of the trench at *P* will change rather rapidly from a trench with dextral strike-slip component (Fig. 10.21*a*) to a trench with sinistral strike-slip component (Fig. 10.21*b*). Furthermore, there will be a rapid increase in the rate of displacement of the plate moving beneath A (first B, then C). Any rock body at, or evolving at, *P* during this period may record evidence of three different stages in the local tectonic history: a protracted early stage in the setting of a slow, dextral trench; a relatively short middle stage when the east-west trench complex is passing P (quite possibly causing a short interval of distinctive vulcanicity, sedimentation, and vertical movements at P); and a protracted late stage in the setting of a relatively fast, sinistral trench. Notice that the distinctive events occurring as the triple junction passes P will occur at progressively later times along points farther and farther north along the north-south trench. Any geological events that migrate in some systematic way like this are called *diachronous*. The rock products of such events have the same or similar character at all points but *different ages at different points*. Most kinds of geological events are likely to be diachronous to some degree.

This example is based on one of many cases involving triple junctions discussed and analyzed by McKenzie and Morgan (1969). In addition, these authors make an important general observation that is illustrated in Figure 10.21. Note that whereas ω_{AB} and ω_{AC} are fixed for a finite interval of time with respect to the plates A, B and A, C, respectively, ω_{BC} is not fixed to either plate B or C for any finite interval. Thus plates B and C are rotating with respect to one another about a pole *that is continually changing position relative to these two plates*. As McKenzie and Morgan note, *finite rotations cannot take place about the instantaneous material positions of all three poles of rotation*. One or more poles of rotation must migrate with respect to its plates, and in the general situation all three poles should be considered free to move with respect to their plates. The special situations discussed in examples 1 and 2, in which as many poles are fixed as is permissible, make it easier to demonstrate temporal changes in tectonic settings at a point and may be common in nature. Long segments of transform fault across boundaries, for example, may put limits on the ability of poles of rotation to migrate [McKenzie (1972); Menard and Atwater (1968); Lachenbruch and Thompson (1973)].

Example 3.   Here we consider a situation in which a gradual change in the rate of subduction at a trench may occur, independent of triple junction behavior and again, of course, independent of fundamental changes in plate motion. This case is illustrated by the history of point Q in Figure 10.21. In Figure 10.21*a* Q is on the hanging wall side of a dextral trench. The rate of subduction beneath Q is moderate because Q is about 50° from the pole of rotation for plates C and B. With the passage of time, however, Q moves farther from its pole of rotation, reaching 70°, for example, in Figure 10.21*b* and since ω_{CB} is fixed, V_{CB} increases. There is accordingly a gradual increase in the rate of subduction beneath Q. The attainment of any given subduction rate will be a diachronous event progressing from west to east along the trench.

Example 4.   The final example illustrates how a boundary may gradually change character altogether, in this case from a trench to a transform fault. As in example 3 the boundary evolves because there is relative movement between a pair of plates and their pole of rotation. In Figure 10.22 plate A is again fixed to the projection frame and ω_{AB} and ω_{AC} are fixed, with ω_{BC} in

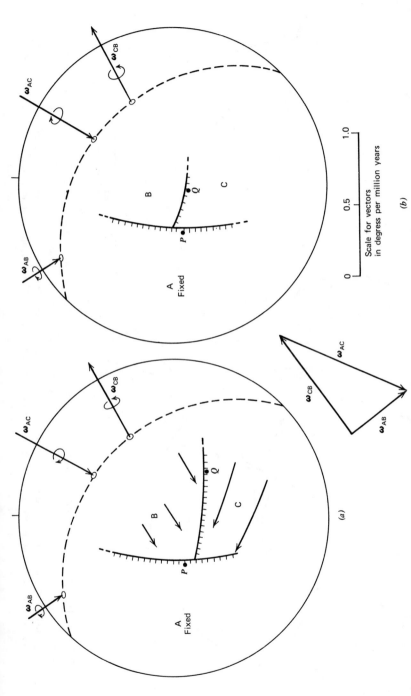

FIGURE 10.21 Stereographic projections illustrating two types of plate boundary evolution (examples 2 and 3 in text). (*a*) Intersecting trenches. Arrows on plates B and C follow particle tracks on these plates relative to plate A, and are porportional in length to displacement rates. *P* is adjacent to a dextral trench. *Q* is adjacent to a trench with subduction at a moderate rate. (*b*) The same boundaries after an interval of time. *P* is now adjacent to a sinistral trench, and *Q* is influenced by subduction at a more rapid rate. Vector triangle shows the angular velocities prevailing throughout the interval between (*a*) and (*b*).

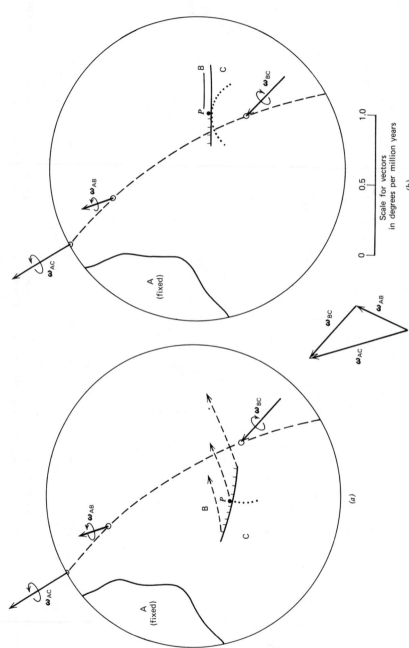

FIGURE 10.22 Stereographic projections illustrating plate boundary evolution from trench through transform fault to ridge (example 4 in text). (*a*) Initial configuration. Point *P* is adjacent to a pure trench. Light dashed lines show particle tracks for points on the hanging wall side of the trench, following small circles about ω_{AB}. (*b*) The same boundary as in (*a*) after an interval of time. *P* is now adjacent to a pure transform fault while points to the west and east have trench and ridge character, respectively. Vector triangle shows angular velocities prevailing throughout the interval between (*a*) and (*b*).

consequence fixed in the projection frame, but not fixed with respect to plates B and C for any finite period of time. Consider the history of point P on the hanging-wall side of the trench between plates B and C. Initially the character of the boundary is pure trench. The trench then migrates eastward along with plate B. The character of the boundary at P becomes increasingly transform, until the configuration in Figure 10.22b is reached. At this stage the boundary at P is parallel to a small circle about ω_{BC} so that the boundary is a pure sinistral transform. Note that points on the boundary west of P have sinistral trench character, whereas those east of P have sinistral ridge character. The attainment of pure transform character, as at P in Figure 10.22b, is a diachronous event that migrates from east to west along the boundary. Plate boundary evolution of the types discussed in examples 3 and 4 was recognized by McKenzie and Morgan (1969) but has not yet been systematically analyzed.

Relative Motion of Lithosphere and Mantle. Angular velocity vectors, used above to describe the motion of spherical *caps*, are also appropriate to describe the motion of concentric spherical *shells*. Thus, for example, there will be an

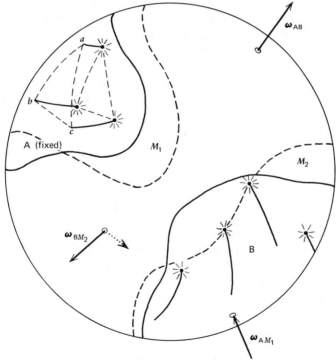

FIGURE 10.23 Stereographic projection showing two lithosphere plates (A and B) and "hot spots" (stars) on two mantle segments (M_1 and M_2). For further explanation see text.

angular velocity vector that describes the instantaneous motion of a lithospheric plate with respect to an underlying spherical shell segment in the mantle. The only requirement is that the mantle segment, like the lithospheric plate, be rigid. In fact, we can relax this requirement somewhat and still have a useful angular velocity vector if there are at least three, noncolinear points on the mantle shell that maintain fixed distances and angles with respect to one another. In this case the angular velocity vector relates motion of the lithospheric plate to the spherical triangle defined by the three points, and not necessarily to other parts of the spherical shell segment. Figure 10.23 shows a hypothetical case illustrating this further application of vector methods in plate tectonics. Plate A is fixed in the projection. The three stars on plate A represent the present positions beneath plate A of "hot spots" or "plumes" (see Section 10.3.4) on mantle segment M_1. M_1 (or at least the triangle of hot spots) is moving with respect to plate A in the manner described by ω_{AM_1}. The curves drawn behind each star on A are "plume tracks", which connect sites on A previously over the hot spots on M_1. The tracks follow small circles about ω_{AM_1} if this vector has been fixed relative to A and M_1 for the time represented by the plume track. This has been assumed in constructing Figure 10.23. The rigidity of the hot spot triangle and of plate A is reflected by the fixed geometry of the spherical triangle a-b-c along the plume track. Another plate, B, is also shown, together with its underlying mantle segment M_2 and the angular velocity vector relating them, ω_{BM_2}. Note that this vector, which is assumed fixed to B for the time interval represented, is moving in the projection frame in the direction shown by the dotted arrow. Plume tracks on B follow small circles around ω_{BM_2}. The motion of the two lithospheric plates is described by ω_{AB}.

Extending Equation 10.4 to four vectors we may write

$$\omega_{AB} + \omega_{BM_2} + \omega_{M_2M_1} + \omega_{M_1A} = \mathbf{0}$$

or,

$$\omega_{M_1M_2} = \omega_{AB} + \omega_{BM_2} + \omega_{M_1A}$$

By vector addition, therefore, we can determine the instantaneous relative motion of one mantle segment relative to another if we know the other three vectors from spreading rates and suitable dated plume tracks. Study of mantle motions has just begun, and there is still disagreement as to whether or not groups of hot spots have moved with respect to one another [see Morgan (1972b); Burke, et al. (1973)].

10.2.6   Plate Tectonics and Mountain Belts

In this section we discuss several ways in which plate tectonics suggests revision or refinement of previous ideas about geosynclines and orogeny. A *geosyncline* in the broadest sense is a depositional region within which sediments and volcanic rocks accumulate, commonly to thicknesses measured in thousands of meters. *Orogeny,* in an equally broad sense, is a complex of deformational processes by which rocks become folded, faulted, and incorporated in a mountain belt. A *mountain belt* is an extensive linear or arcuate region characterized most essentially by zones of strongly deformed rocks, and, commonly, by extensive metamorphism. Mountain belts may or may not be accompanied by physiographic mountains. [There have been numerous more restricted definitions of the terms geosyncline and orogeny over the past hundred years, each reflecting somewhat different ideas about the origin of mountain belts. Reference should be made to Aubouin (1965); Dennis (1967); Hsu (1972); Cebull (1973); Gilluly, (1973); and Dott, (1974.)]

Orogenic Settings.   Plate theory suggests that at present and for some undetermined time in the past, orogeny has been occurring at or near plate boundaries across which there is a shortening component (i.e., trenches and seismically active young mountain belts). At least three types of orogenic settings are recognized [Mitchell and Reading (1969); Dewey and Bird (1970a)], which we can refer to as the *island-arc type,* the *Andean type,* and the *Himalayan type* (Fig. 10.24). Island-arc orogeny involves oceanic crust on both sides of the orogen. Andean orogeny involves oceanic crust on one side and continental crust on the other. In the Himalayan type an oceanic plate has been completely consumed so that one continent collides with another. The three

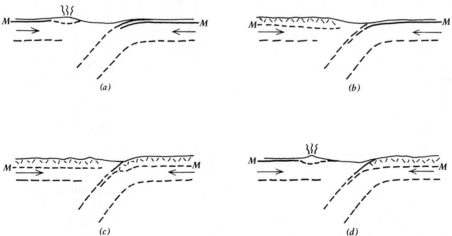

FIGURE 10.24   Orogenic settings involving convergent plate boundaries according to Mitchell and Reading (1969) and Dewey and Bird (1970a). (*a*) Island arc type. (*b*) Andean type. (*c*) Himalayan type. (*d*) Type involving continent-island arc collision. The terms for (*a*), (*b*), and (*c*) are those of Mitchell and Reading. Arrows represent component of relative movement normal to the plate boundary.

names refer to present-day examples of such orogens. Dewey and Bird (1970a) suggest a fourth type, in which a continent collides with an island arc (Fig. 10.24*d*), and cite the mountains of New Guinea as a recent example.

Figure 10.24*c* suggests that orogeny of Himalayan type should be preceded by orogeny of Andean type. An example of this is seen in the history of the northern Appalachians according to Bird and Dewey (1970). The interpretation is shown schematically in Figure 10.25. A proto-Atlantic ocean was initiated in late pre-Cambrian or early Cambrian time by rifting (Fig. 10.25*a*) of an originally continuous plate [Wilson (1966)]. Thick sedimentation then occurred on the continental margins [Dietz (1963)] either side of this basin (Fig. 10.23*b*). Orogeny of Andean type and possibly island-arc type began on the western margin of the basin when a trench (Fig. 10.23*c*) or series of trenches developed there. This prolonged Andean phase (the "Taconic" orogeny) was followed in middle Paleozoic time by a relatively brief Himalayan phase (the "Acadian" orogeny) when continental crust on the eastern margin of the closing ocean collided with continental crust on the western margin (Fig. 10.25*d*).

Geosynclines.   It can be seen from Figure 10.25 that, in principle, sediments deposited anywhere on the sea floor can ultimately be carried to an orogenic zone and incorporated in a mountain belt. Rock sequences now adjacent to one another in mountain belts may accordingly have originated in widely separated depositional sites that are subsequently "sutured" together by convergent plate motion and subduction of intervening lithosphere (Figure 10.25 *c-d*). This conception of mountain belts as accumulations of various combinations of island-arc rocks, continental margin rocks, deep-sea sediments, and slices of oceanic crust and mantle, all swept together into narrow zones by closing oceans [Wilson (1966, 1968b)] contrasts with earlier conceptions of geosynclines and mountain building. The contrast is strongest with earlier ideas developed in North America [see Kay (1951)], in which rocks of a mountain belt were supposed to have originated together, in a belt no more than a few times as wide as the eventual mountain belt. The contrast with earlier ideas is less marked in the European literature however, where broad complex geosynclines and even subduction on a moderate scale [Ampferer and Hammer (1910); see White et al. (1970)] were conceived of many years ago.

Another modification of geosynclinal theory concerns the genetic relation between thick sedimentation and deformation. The plate model indicates that there need be no simple, routine connection between sedimentation and deformation. Whether a sequence of strata becomes deformed or not, and how soon after sedimentation this occurs, depends on the type of depositional site and accidental factors that determine whether and when the rocks are carried into an orogenic zone. Thus, sediments fringing a continental margin like the present east coast of North America (remote from a trench) can accumulate

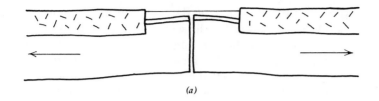

(a)

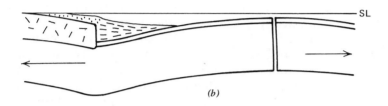

(b)

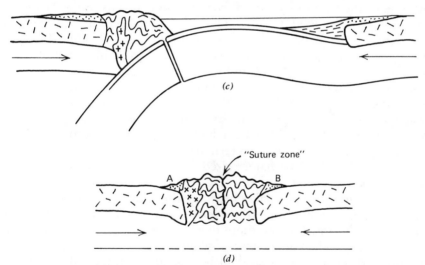

(c)

(d)

FIGURE 10.25   *Schematic* plate tectonic interpretation of the northern
Appalachians, according to Bird and Dewey (1970) and Dietz (1972). (*a*) Rifting
of continental crust in late pre-Cambrian time. (*b*) Opening of the
proto-Atlantic ocean and continental margin sedimentation. (*c*) Closing of
proto-Atlantic ocean with orogeny of Andean type during early Paleozoic.
(*d*) Disappearance of proto-Atlantic ocean with orogeny of Himalayan
type as opposed continents collide. [From R.S. Dietz (1972), "Geosynclines,
Mountains, and Continent-building." Copyright © 1972 by Scientific
American, Inc. All rights reserved.]

for a long period of time and escape deformation indefinitely. On the other hand, sediments accumulating off the west coast of South America (adjacent to a trench) are destined by their setting to undergo early deformation.

The reader is referred to Drake, Ewing, and Sutton (1959), Dietz (1963), Dietz and Holden (1966), Mitchell and Reading (1969), Coney (1970), Dewey and Bird (1970a, b), Dickinson (1971, 1972) Dietz and Holden (1974), and Dott (1974) for further discussion of geosynclines in the context of plate tectonics.

Displacement History. Plate theory is distinguished from earlier orogenic theories in that it more clearly specifies what the relative movements of opposite sides of an orogen may be. This is particularly true for present-day or recently active belts, across which the displacement history can be pieced together quantitatively from magnetic anomaly patterns. Thus, for example, the anomaly patterns in the North Atlantic permit detailed analysis of the movement of Europe and Africa away from North America during Cretaceous and Tertiary time [Pitman and Talwani (1972)]. These results have been used to establish a chronology for displacements between Europe and Africa over the past 180 million years, that is a history of relative plate motions either side of the Cretaceous and Tertiary mountain belts of the Mediterranean region [Dewey et al. (1973)]. The degree to which this approach can throw light on the detailed displacement history of individual European mountain chains is however limited because there are many small plates between the major plates of Africa and Europe.

A second example, on a scale where more detailed quantitative interpretation is possible, is provided by the work of Atwater (1970), which relates to the Cenozoic history of the plate boundary along the western margin of North America. We simplify her argument below in order to illustrate its main features. The magnetic anomaly pattern in the northeast Pacific ocean [Atwater and Menard (1970)] is characterized by anomalies that become *younger* toward the coast of North America (Fig. 10.26a). This is the inverse of the usual situation, in which anomalies become *older* toward continental margins. It is readily explained in plate tectonic terms by supposing that the anomalies young toward the east because there once existed a ridge in this direction. Since spreading at ridges generates two symmetrically equivalent plates, there must also have existed an oceanic plate east of the ridge [McKenzie and Morgan (1969)]. This easterly plate and most of the ridge have since vanished. We refer here to the vanished plate as F, to the Pacific plate as P, and to the American plate as A. The simplest explanation for the disappearance of F is that it was consumed by a trench at the boundary between A and P and that this trench was roughly coincident with the present continental margin (Fig. 10.26b). It is evident then that we can work out the displacement history across the boundary F-A if we know enough about the motions of F relative to P and of A relative to P. The motion of F relative to P is determined quite exactly

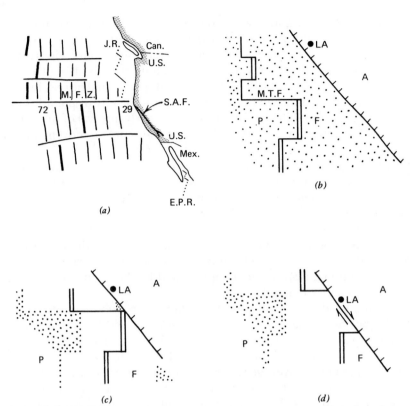

FIGURE 10.26   Possible evolution of the San Andreas fault, based on McKenzie and Morgan (1969) and Atwater (1970). (*a*) Present magnetic anomaly pattern. M.F.Z. Mendocino fracture zone; S.A.F. San Andreas fault; E.P.R. East Pacific Rise; J. R. Juan de Fuca Ridge. Numbers give anomaly ages in millions of years just south of M.F.Z. Heavy line is anomaly 53 m.y. old. (*b*) Inferred relations 53 m.y. ago. M.T.F. Mendocino transform fault. (*c*) About 30 m.y. ago, when ridge PF first intersects trench FA. (*d*) About 10 m.y. ago, when a segment of the trench has been converted into a transform fault. LA marks a site on P at about the present latitude of Los Angeles. Sketches based on figures in Atwater (1970), with simplifications. Stippled areas in (*b*), (*c*), and (*d*) represent oceanic crust greater than 53 m.y. old. For further explanation see text. [After Atwater (1970), with permission of the Geological Society of America.]

from the directions of the transform fault extensions and the spacing of dated magnetic anomalies on plate P. The motion of A relative to P is much less certain. In Atwater's analysis, therefore, two possible models are studied, each assuming a different displacement history for A relative to P. The simpler model (the one presented here) assumes that A has moved relative to P at a constant rate of 6 cm/yr and parallel to the present San Andreas fault. Figure 10.26*c* and *d* are results based on this assumption. Figure 10.26*c* shows relations when the P-F boundary first intersects the trench, and Figure 10.26*d* shows the pattern about 20 m.y. later when a long stretch of the trench has been converted into a transform fault, the developing San Andreas fault. The displacement history shown in these figures for the vicinity of point LA (at the present latitude of Los Angeles) is as follows: an early interval of underthrusting by plate F followed by an abrupt change to transcurrent movements about 20 m.y. ago.

Plate theory gives new estimates of the displacements across mountain belts, but at the same time it implies that *some of the displacements can never be worked out from the structures alone.* Thus, for example, in Figure 10.25*d*, we might be able to determine that the folded sediments on either side of the suture zone were shortened, for example, 50 percent, and that point B had therefore been displaced with respect to point A by at least the present width of the fold belt. But we would have no way of determining from structural evidence alone the very large additional displacement represented by the suture zone itself, the width of an ocean that closed.

Plate theory yields new estimates of the average strain rate in orogens if we divide displacement rates across trenches by *assumed* widths of strata through which the displacement is absorbed as strain. Thus, if a plate is traveling into a trench at 5 cm/yr and strata in the vicinity of the trench are shortening at 5 cm/yr, we can obtain an average strain rate for any affected width of strata. If the width is 100 km, the strain rate is $1.6 \times 10^{-14} \ sec^{-1}$. If the width is 10 km, the strain rate is $1.6 \times 10^{-13} \ sec^{-1}$. Similar numbers are obtained if we observe that horizontal displacements of many types occur at a rate of around 1 cm/yr [Sutton (1969)] and if we again *assume* a width of rock through which the displacement is absorbed as strain. Distributed through 100 km, 1 cm/yr gives a strain rate of $3.2 \times 10^{-15} \ sec^{-1}$. These values are all very uncertain because there is not yet any good basis for estimating the affected width of strata. At best they approximate average values for long times and large distances. They probably have little relevance to the rates at which deformation actually occurs at individual localities within orogens. Heard (1968, p. 443) suggests that rocks flow in nature at widely variable rates, commonly from $10^{-12} \ sec^{-1}$ to $10^{-18} \ sec^{-1}$. This range of strain rates brackets a figure of $10^{-14} \ sec^{-1}$ that is widely but unjustifiably quoted as *the* typical natural strain rate, based largely on geodetic measurements of present-day strain near the San Andreas fault in California [Whitten (1956); Burford (1968); Scholz and Fitch (1969)].

A feature of orogenic displacements deserving separate attention is the role played by transcurrent movements during the main phases of deformation. Do the boundaries of orogens commonly move *parallel* to each other as well as perpendicular to each other during intervals of folding? Plate tectonics suggests more clearly than previous theories that a transcurrent component *is* a normal part of the total displacement across an orogen. It suggests further that the rates of the transcurrent and perpendicular components may change very markedly during the lifespan of an orogen, as suggested by some of the examples in Section 10.2.5. This is interesting because petrological and stratigraphic features of some fold belts are increasingly being interpreted as possible indicators of transcurrent displacements (e.g., see Ernst et al. (1970, p. 232); Landis and Bishop (1972)]. What remains unclear is the nature of structures that may be expected where transcurrent displacements are important. *En échelon* fold and fault-block patterns are one possible expression of transcurrent displacements. Transcurrent faults parallel to an orogen are another. *En échelon* fault blocks having this origin may be represented by the *en échelon* extensional troughs in the southern part of the New Hebrides island arc system [Karig and Mammerickx (1972)].

A final point, and an important simplifying feature of plate tectonics, is that the displacements occurring across a given orogen at a given time must be related to the whole integrated system of displacements across all plate boundaries at that time. Thus, the events at one orogen may well be related in some simple fashion to events at other active boundaries thousands of kilometers away. This "global" aspect of plate theory renders it powerful and testable as a predictive tool.

Multiple Deformation.   Modern structural mapping has shown that the rocks of orogens typically display more than one generation of structures (see Chapters 8, 9). This can be accounted for, in part, by the complex response of rocks to even a single simple deformation (as for example when the development of a schistosity due to flattening is followed by development of conjugate crenulation cleavages in response to continued flattening). It is also probable, however, that several generations of structures in many cases reflect a sequence of different tectonic settings. The plate model offers at least three explanations for multiple deformation of this type.

1. Change in *type of orogeny,* without change in positions of plate boundaries or displacement directions. An example may be the occurrence of Andean orogeny followed by Himalayan orogeny proposed for the northern Appalachians by Bird and Dewey (Fig. 10.25).
2. Change in *direction of displacement* across orogenic belts, with or without fundamental changes in relative plate motions. This is illustrated by Figure 10.26 if we follow Atwater in speculating that the transition from underthrusting to transcurrent displacements at the western margin of North America occurred not only in the vicinity of the San Andreas fault but also over a wide soft zone extending inland for 100 km or more.
3. Change in *rate of displacement* across boundaries caused either by fundamental changes in plate motions or by effects of the type discussed in Section 10.2.5. If the rate of subduction changes gradually at a trench,

for example, there should be corresponding changes in the thermal regime of the trench and perhaps also in the strain rates of any deforming rocks. If changes in temperature or strain rate are large enough, then changes in the structural response should occur, possibly initiating a new generation of structures.

12.2.7 Further Problems in Plate Tectonics

We list below some of the numerous interesting problems that remain for further development of plate theory. All are the subject of current research.

1. *Elucidation of the driving mechanism.* No one knows what makes plates move, though every possibility has probably been suggested. These include the ideas that plates are pushed from ridges, pulled from trenches, dragged from beneath by the mantle or from above by the moon, or that they slide downhill from ridges under the influence of gravity. The driving mechanism is discussed by McKenzie (1969b, 1972) and Richter (1973).

2. *Determination of plate motions relative to the deeper mantle.* There is evidence in places that plates move relative to the underlying mantle [e.g., from lines of intraplate volcanoes possibly caused by motion of plates over mantle "hot spots," Morgan (1972a)]. But there is also evidence that some regions have been over the same part of the mantle for long periods [e.g., large depositional basins that have been subsiding for hundreds of millions of years, Beloussov (1968)]. A more complete picture of plate motions relative to the mantle is obviously desirable and could lead to better understanding of mantle motions and the coupling, if any, between lithosphere and deeper mantle.

3. *Geometrical consequences of plate motion.* The purely geometrical aspects of plate motion and interactions of the type discussed in Section 10.2.5 have not been fully worked out. This needs to be done with and without the assumption of ideal rigidity of plates.

4. *Plate configuration in three dimensions.* The base of plates is commonly taken as the upper part of the seismic low-velocity channel [Isacks et al. (1968)]. However, this horizon is by no means well defined or well understood. In some places it has not even been detected.

5. *Vertical movements.* Additional development of the plate model is required in order to see how and where its fundamental horizontal movements are likely to be related to vertical movements [see Menard (1973)].

6. *Tectonics of present-day trench complexes.* Knowledge of the displacements and structural phenomena in and around trenches and marginal basins is still very limited although some progress has been made [see Plafker (1972)]. Understanding these regions better is obviously essential for relating plate tectonics to mountain building.

7. *Initiation of plate boundaries.* Little is known about the factors controlling the orientation and nature of newly developed plate boundaries. To what extent are new boundaries controlled by conditions in the lithosphere itself rather than by conditions in the underlying mantle?

9. *Age of the first plates.* How long ago did plate motions and interactions begin to play a significant role in tectonics?

Skeptics say the answer to the last question is about 10 years. There are, in fact, many thoughtful geologists who que~tion the importance of plate tectonics as a guide to the past behavior of the earth. It is pointed out that while some kinds of past tectonic activity can be explained by the plate model, there are many others that are not so well explained. Examples are vertical movements of many kinds and structural and volcanic events that occur remote from probable plate boundaries. Some claim that plate tectonics can explain recent features of the sea floor, but that it has little or no connection with past orogeny and geosynclines. The theory is said by some to be ill-founded even for oceanic regions because magnetic anomaly patterns have been oversimplified and misinterpreted, or because we still do not know for sure that the ocean basins are not underlain by Paleozoic rocks. These and other critical arguments can be found in papers by Beloussov (1970), Meyerhoff (1970), Wesson (1971), Gilluly (1972), Cady (1972), and Meyerhoff and Meyerhoff (1972). Perhaps it will be many years before plate tectonics can become more than a working hypothesis when applied to older deformed terrains.

10.3 OTHER EXPLANATIONS FOR CRUSTAL DEFORMATION

Several other theories have been advanced for the tectonic behavior of the earth. Each of these has had something to say about the origin of deformed rocks. The main proposals involve global volume change, continental drift, vertical block movements, and mantle convection. In what follows we briefly explain the possible connections between these processes and deformation. Questions regarding the detailed nature and plausibility of the processes are beyond the scope of the book. It is important to stress here that some of these processes are partly or wholly compatible with the existence and movement of lithospheric plates. These processes may, therefore, play significant or even dominant roles locally even if the mechanisms proposed by plate theory are also operative.

Global Contraction or Expansion.   The first major scientific proposal to explain crustal deformation was the proposal by Dana (1847) and de Beaumont (1852) that the earth's deep interior was cooling and shrinking and that its outer layers became puckered and folded like the skin of a dried-out apple. The discovery of radioactive heating, however, led to the conclusion that the earth was probably not cooling at all and that general contraction, if it occurs, must have another origin.

The idea of general global contraction received a further setback when it was recognized that there were large areas on earth (the oceanic ridge system) that suggested local horizontal extension in some directions rather than shortening. In fact this system was first interpreted as an indication of global expansion [Heezen, (1960)]. Other arguments for general expansion have been given by Egyed (1957), Carey (1958), Dicke (1962), and Meservey (1969). After reviewing the evidence, Bott (1972, p. 273) concludes that very slow expansion may be occurring but that its tectonic effects would be negligible compared with effects due for example to subduction.

Continental Drift. Continental drift, or relative movements of continents on a global scale, deserves separate mention here despite the fact that relative movement of continents is an integral part of plate tectonics. The pre-1960 versions of this theory played a major role in preparing many geologists for tentative or even enthusiastic acceptance of plate tectonics. Virtually all the evidence once cited for continental drift is now used as supporting or even

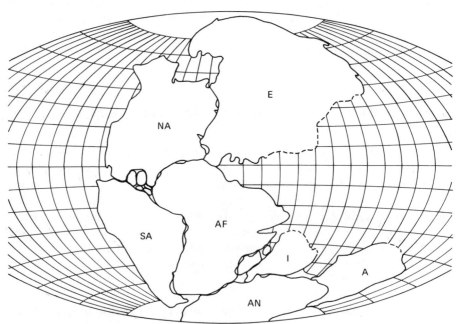

FIGURE 10.27   One possible reconstruction of the supercontinent Pangaea, before the present cycle of continental drift. E, Eurasia; NA, North America; SA, South America; AF, Africa; I, India; A., Australia; AN, Antarctica. [From Dietz and Holden (1970), with permission of the American Geophysical Union.]

vital evidence for plate tectonics, for example, the fit of both continental margins and geology across the Atlantic.

The difference between earlier theories of continental drift and the plate tectonics version is concerned with the coupling or lack of coupling between continents and the underlying basic rocks. Pre-1960 continental drift supposed that the *si*lica- and *al*umina-rich *(sialic)* continental rafts moved freely over the underlying *si*lica- and *ma*gnesia-rich *(simatic)* rocks of the lower continental crust and ocean basins. Plate tectonics holds that the whole of the continental crust is coupled firmly to the upper hundred kilometers of the mantle and that continents are carried passively with the lithospheric plates. Furthermore, in pre-1960 drift theories, continents were separated by *wide zones* (the oceans) so that there were minimal geometric constraints on the freedom of each continent to move independently of the others. In plate tectonics, on the other hand, the continents are effectively separated by *narrow zones* (the plate boundaries) and are thus part of an interlocking mosaic of plates. In such a system the continents are less free to drift independently of one another.

The continental drift theory, as developed for example by Holmes (1931), Carey (1958), and Wilson (1963b) anticipated much of what plate tectonics has

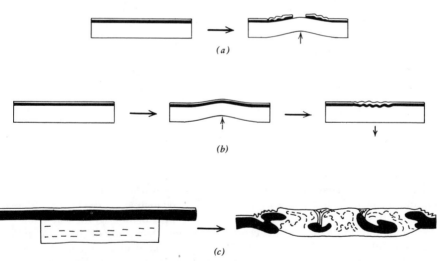

FIGURE 10.28   Models for regional deformation related to fundamental vertical movements. (*a*) Gravity sliding of superficial layers down flanks of a basement arch. (*b*) Basement arch, stretching superficial layers, followed by collapse of arch and crumpling of stretched layers. (*c*) Deformation associated with buoyant rise of a low density mass (white) through a layer of higher density (black). (*a*) Based on Haarman (1930) [see Hills (1963, p. 338)]: (*b*) Based on Beloussov (1962, p. 527): (*c*) Simplified from centrifuge experiment S114 by Ramberg (1967, p. 150) with permission of the Academic Press, Ltd.

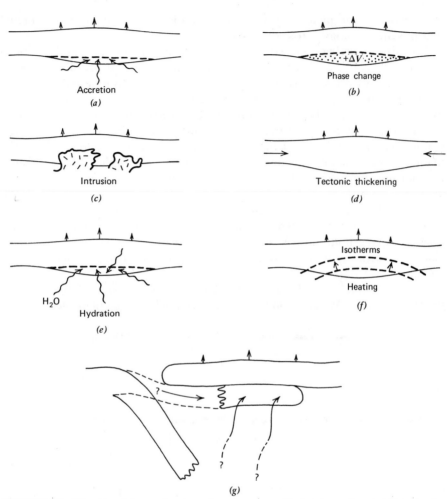

FIGURE 10.29 Possible mechanisms for plateau uplift. (*a*) Accretion (addition) of low density material to the base of the crust from the mantle. (*b*) Isochemical phase change with volume increase in the lower crust or upper mantle. (*c*) Intrusion of low density plutons in the base of the crust. (*d*) Tectonic thickening of the crust and/or upper mantle, without subduction. (*e*) Hydration or other metasomatic alteration involving volume increase in the lower crust or upper mantle. (*f*) Heating of crust and upper mantle resulting in volume increase (*g*) Uplift due to thickening of lithosphere by underthrusting or by float-up of a previously subducted block.

to say about the location of orogenic belts. Such belts were seen to be associated with island arcs and continental margins and were ascribed to deformation of and near the leading edges of horizontally moving parts of the crust.

One aspect of many pre-1960 drift theories that may be revived as plate theory matures is the acceptance of greater mobility in the oceanic and continental crust than is permissible in the theory of highly rigid plates. This is seen to some extent already in the recognition that areas of Cenozoic continental deformation may be very wide soft zones within or between plates.

Figure 10.27 shows a possible reconstruction of the supercontinent Pangaea 225 m.y. ago, according to Dietz and Holden (1970). Other reconstructions are given by Wilson (1963b), Creer (1964), Hurley and Rand (1969), and Smith and Hallam (1970).

Vertical Movements and Gravity Tectonics.   There has been a long argument in tectonics about whether regional deformation is a consequence of large-scale horizontal or vertical movements [e.g., see the discussion between Wilson (1968a) and Beloussov (1968) in *Geotimes*]. Plate tectonics and continental drift are obviously theories emphasizing the importance of horizontal movements. But these theories in their present form are not very successful in accounting for long-continued vertical movements. Such movements are known, for example, subsidence of the Moscow and Michigan basins over periods of hundreds of millions of years. Such motions are ascribed to fundamental vertical movements in the underlying mantle. Some geologists have proposed that similar deep movements are also responsible for regional deformation in mountain belts [Haarmann (1930); vanBemellen (1960); Beloussov (1961); Ramberg (1967); Maxwell (1968)].

The process usually called on to produce deformed rocks in association with fundamental vertical movements is some kind of adjustment of rock bodies toward minimum potential energy in the gravity field. This may occur by downslope sliding of materials in the shallower parts of orogens, *gravity tectonics,* or by movements driven by bouyant forces at deeper levels [see DeJong and Scholten (1973)]. Both processes can occur and have important local structural effects even if the fundamental movements are horizontal, so long as the necessary slopes or density inversions exist.

Three models for folding as a consequence of dominantly vertical movements and gravitational adjustments are shown diagrammatically in Figure 10.28.

Vertical movements without accompanying large strains, as in plateau uplift or basin subsidence, have been explained in many ways, all of them very speculative. Possible explanations for plateau uplift are shown in Figure 10.29. The inverse of many of these processes might account for subsidence.

Mantle Convection.   Mantle movements, in particular, thermally driven convective movements, were considered important for regional deformation in the crust well before the advent of sea-floor spreading and plate tectonics [Holmes (1931); Griggs (1939)]. Figure 10.30 shows two models expressing these ideas. Orogenic deformation was believed to occur in the crust in response to compressive stress generated there by convergent horizontal flow in

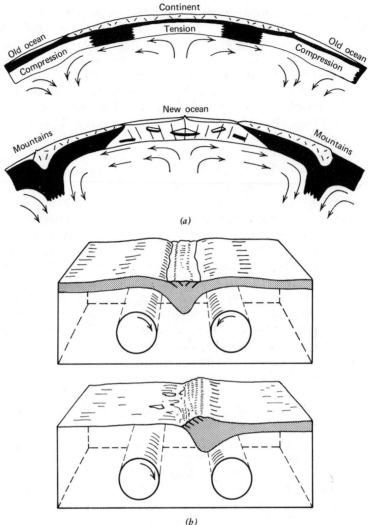

FIGURE 10.30 Two early models for crustal deformation due to convection. (*a*) After Arthur Holmes (1945), *Principles of Physical Geology.* Copyright © 1945, The Ronald Press Co., New York. Mantle convection (arrows) rifts continents and creates new oceans. Mountain belts form above descending currents. (*b*) After Griggs (1939, *Am. Jour. Science,* Vol. 237 (9), with permission of the *American Journal of Science.* This is a working model, in which drums are rotated as shown, in a box full of water glass. A floating layer of sand and oil is pulled down and deformed symmetrically if both drums are rotated. If only one drum is rotated, the originally continuous floating layer is swept from one side only and piled up near the center of the box.

the underlying mantle. At the present time, convection in the mantle is widely regarded as playing an essential role in driving plate motion [see McKenzie (1969b, 1972)], but the details of the form and position of the currents are unknown. It is not considered likely that there is a simple correspondence between the positions of ridges and the positions of upward-diverging currents as shown in Figure 10.30a, and the convective pattern around trenches is also obscure. Even the direction of movement of the asthenosphere immediately beneath plate interiors is uncertain; it may be moving in the same direction as the plates and helping to drive them, or it may be moving slowly in the opposite direction, while the plates slide gravitationally or are pushed or pulled over it [see Bott (1971, p. 279)].

A recent development in regard to mass transfer in the mantle is the proposal by Morgan (1972a) that local "hot spots" on the mantle beneath plates [cf. Wilson (1963a)] are underlain by rising "plumes" of material roughly 150 km in diameter. The motion of plates over such mantle plumes is supposed to account for aseismic oceanic ridges and for chains of intraplate volcanoes (like the Hawaii-Emperor chain in the Pacific). A related proposal has been made by Ramberg (1972), who provides theoretical and experimental arguments for ascent and melting of low density diapirs in the mantle. Kidd et al. (1973) estimate that the present world population of mantle plumes is at least 150. Burke and Dewey (1973) believe that rifting at ridges is commonly initiated above mantle plumes and that uplift and three-armed graben structures are the usual initial expressions of plume-generated deformation (see also Section 9.3).

APPENDIX

In many instances during the analysis of complicated geological structures it is convenient to be able to represent data in a simple two-dimensional manner. The stereographic and equal area projections are powerful graphical tools of use in portraying three-dimensional orientation data in two dimensions or in graphically solving complicated three-dimensional problems. The problems capable of being solved using these projections are those involving the *angles between lines and planes* rather than those concerned entirely with the *relative positions* of lines and planes in space.

THE STEREOGRAPHIC OR EQUIANGULAR PROJECTION

The essential features of the stereographic projection are illustrated in Figure A1. The orientation of a plane such as a bedding, cleavage, schistosity or fault surface may be represented by imagining that plane to pass through the center, O, of a sphere of radius R. This sphere is known as the *projection sphere*. The dipping plane of interest intersects the projection sphere in a circle that has the same radius as the sphere itself and is accordingly called a *great circle*. This great circle is the *spherical projection* of the dipping plane (Fig. A1a). For each orientation of a plane there is a unique great circle.

To obtain a two-dimensional projection of the dipping plane a convenient *projection plane* is selected through the center O of the sphere. In structural geology this is commonly the horizontal plane although in some applications it may be more convenient to select a plane with some other orientation as the

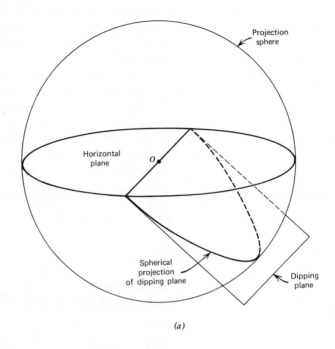

(a)

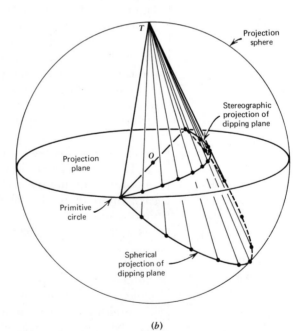

(b)

FIGURE A1   (a), (b) Principle of the stereographic projection.

projection plane. In what follows it is assumed that the projection plane is horizontal. The intersection of the projection plane with the projection sphere is known as the *primitive circle*. Now if each point on the lower hemisphere part of the spherical projection is joined by a straight line to the topmost point T of the projection sphere (Fig. A1b) then part of a circular cone is generated*. This cone intersects the horizontal plane through the centre of the sphere in a circle known as the *stereographic projection* of the dipping plane.

The circle that always results as the stereographic projection of a plane is known as the *cyclographic trace* of that plane. This circle is also loosely referred to as a great circle because it corresponds to a great circle on the projection sphere.

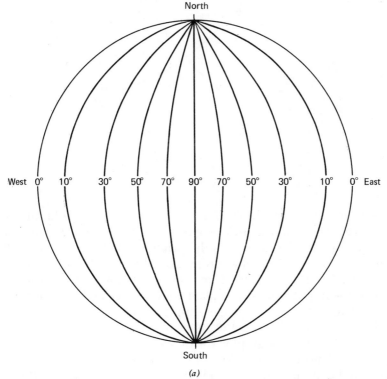

(a)

FIGURE A2   (a) Stereographic projections of planes dipping due east or due west at the various indicated angles. (b) Stereographic projections of variously oriented planes.

*Note that points on the upper hemisphere part of the spherical projection could equally well be joined to the lowermost point on the projection sphere. This is the convention adopted by crystallographers. It is important to state which convention is being used.

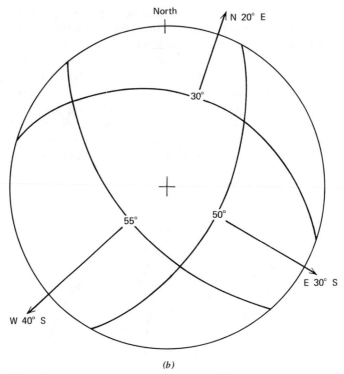

(b)

FIGURE A2 (Continued)

The stereographic projection of a vertical plane is a straight line through the center of the primitive circle whereas the stereographic projection of a horizontal plane is coincident with the primitive circle. Planes with all other orientations have stereographic projections that lie between these two extremes (Fig. A2a). If geographical coordinates are now added to the primitive circle then the orientation of any plane may be specified by a circular trace. Thus, in Figure A2a, various planes dipping either due east or due west are projected. Similarly, in Figure A2b, other planes with diverse orientations are represented by their cyclographic traces.

Figure A1b shows that the cyclographic trace of a plane intersects the primitive circle at the strike of the plane. Figure A3a is a vertical section through the center, O, of the projection sphere. A plane dipping at an angle ϕ, with the true dip in the plane of the section, intersects the section plane in the line OB. The shortest distance from the center of the sphere to the cyclographic trace is OX and is equal to $R \tan (\pi/4 - \phi/2)$. Thus, the cyclographic trace of a plane may be constructed as in Figure A3b; it is a circle passing through the two points on the primitive circle that represent the strike of that plane. It also

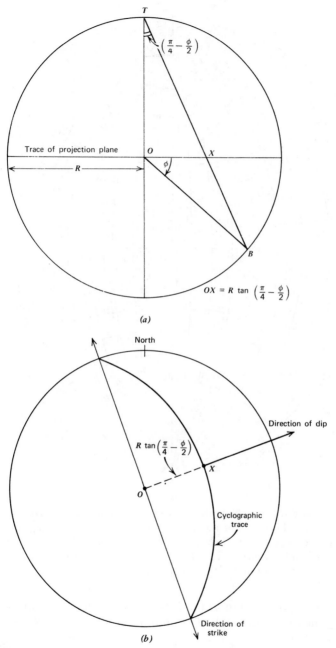

FIGURE A3   (a) Cross section through the projection sphere showing the trace of a plane, OB, dipping at ϕ. (b) Stereographic projection of a plane dipping at an angle ϕ.

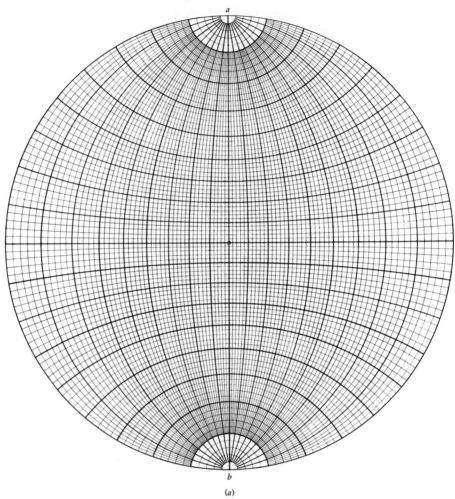

a

b

(a)

FIGURE A4   (*a*) Wulff net. Note that there is no North direction reference mark on the net. *North* is fixed on the tracing paper used over the net. (*b*) Diagram illustrating that a family of concentric small circles on the projection sphere is the spherical projection of a family of coaxial circular cones. The family of great circles is the spherical projection of a family of cozonal planes.

passes through a point distant $R \tan (\pi/4 - \phi/2)$ from the center of the primitive circle in the direction of dip. The amount of dip is ϕ and R is the radius of the primitive circle.

Using these relationships, a protractor such as the one illustrated in Figure A2*a* may be constructed. Such a protractor is known as a *stereographic net* once some way of graduating the angular distance along these cyclographic traces is

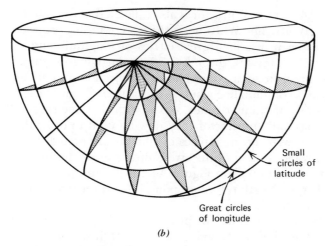

(b)

Small
circles of
latitude

Great circles
of longitude

FIGURE A4 (Continued)

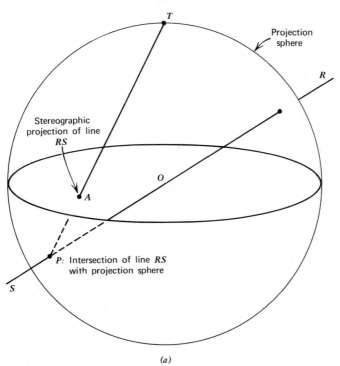

(a)

FIGURE A5   (*a*) Stereographic projection of the line, *RS*.
(*b*) Stereographic projections of variously oriented lines.

added (Fig. A4a). This protractor is also called a *Wulff net* after the crystallographer G. V. Wulff. The Wulff net is the stereographic projection of a sphere on which has been inscribed great circles of longitude and small circles of latitude all at 2° intervals, the projection point lying on the equator of the sphere. The great circles are the spherical projections of a family of planes all intersecting in a common line whereas the *small circles* are the spherical projections of a family of circular cones coaxial with that line (Fig. A4b).

This stereographic protractor may now be used to draw the cyclographic trace of any plane by laying a piece of suitably orientated tracing paper over the net and rotating it until the strike direction of the plane in question is parallel to the direction ab in Figure A4a. It is important to remember when doing this that the North reference point is fixed relative to the tracing paper and not relative to the net.

In a similar manner to that of planes, the orientation of any line such as a fold hinge, a lineation, or the axis of a diamond drill hole may be represented on the plane of projection. Thus, in Figure A5a, the line under consideration,

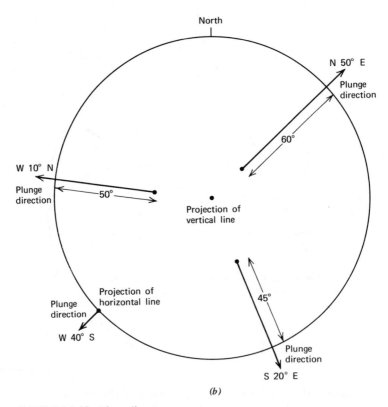

(b)

FIGURE A5 (Continued)

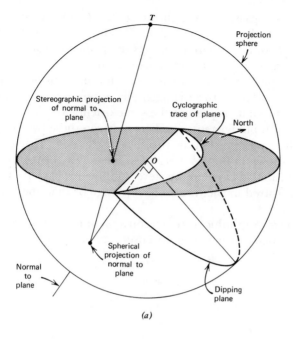

(a)

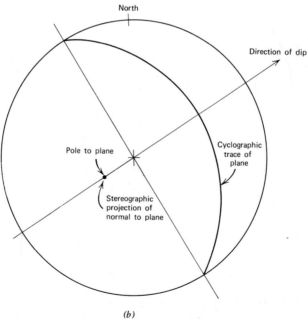

(b)

FIGURE A6   (a), (b) The stereographic projection of a plane and of
the normal to that plane.

RS, is imagined as passing through the center, *O*, of the projection sphere. This line intersects the sphere in a point *P*. The straight line joining *P* to the topmost point, *T*, of the sphere intersects the plane of projection in a point *A* which is the stereographic projection of the line *RS*.

A vertical line projects as a point at the center of the projection whereas any horizontal line projects as two diametrically opposite points on the primitive circle. Since both points represent the same orientation only one is generally plotted. Lines with plunges intermediate between these two project as points with intermediate positions on the projection (Fig. A5*b*). Using a construction similar to that shown in Figure A3*a*, it can be seen that the distance from the center of the primitive circle to the stereographic projection of a line is $R \tan (\pi/4 - \phi/2)$ measured in the direction of plunge of that line. The amount of plunge is ϕ.

Although the cyclographic trace is a convenient method of representing the orientation of a plane, it may be more useful at times to project the *normal* to

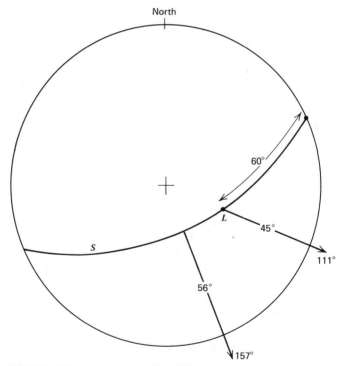

FIGURE A7   Stereographic projection of a plane dipping at 56° towards 157°. A lineation, *L*, pitches 60° NE in this plane. The plunge of the lineation may be read from the projection as 45° toward 111°.

that plane. This is depicted in Figure A6 where the two different ways of representing the orientation of a plane are illustrated. On one hand, the *circle,* which is the cyclographic trace of that plane, may be drawn or, on the other hand, the *point,* which is the stereographic projection of the normal to that plane, may be plotted. This point is commonly referred to as the *pole* to the plane to be represented, and diagrams representing poles to a given surface, for example, S, are referred to as poles-to-S diagrams or more commonly⊥S or πS diagrams. The distance from the center of the primitive circle to a pole is $R \tan(\phi/2)$ where ϕ is the dip of the plane. This distance is measured in a direction *opposite* to that of the dip.

Thus the stereographic net (Fig. A4*a*) may be used to construct the orientation of any linear feature or of the normal of any plane.

As an example, in Figure A7, a schistosity plane, S, is shown, dipping at 56° toward 157°. Also shown is a lineation, L, lying within the schistosity plane and pitching 60° NE. This lineation may be plotted by counting 60° along the

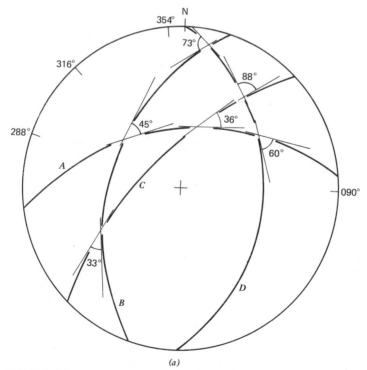

(*a*)

FIGURE A8   (*a*) The equiangular property of the stereographic projection. (*b*) The distortion of area by the stereographic projection. (*c*) The distortion of angles by the equal area projection. (*d*) The equiarea property of the equal area projection.

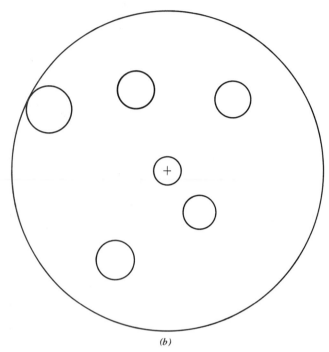

(b)

FIGURE A8 (Continued)

cyclographic trace of S using the small circle graduations of the Wulff net starting at the NE end of the trace. The plunge of this lineation may then be read directly from the net as 45° toward 111°.

Two important geometrical features of the stereographic projection may now be established. Figure A8a shows the cyclographic traces of four planes A, B, C and D. The orientation of each plane and the angles between these planes taken two at a time are given in Table A1. In Figure A8a the angles between the tangents to the cyclographic traces at each point of intersection have also been indicated. It can be seen that these angles as measured in the plane of projection are the same as those recorded in Table A1. Thus, the first geometrical property of the stereographic projection is that *the procedure preserves the angles between curves on the projection sphere*. In this particular example, the curves referred to are the spherical projections of the various planes A, B, C, and D; the angle between two of these planes is equal to the angle between their great circle traces on the projection sphere. The stereographic projection is therefore referred to as *equiangular*.

The second geometrical feature of the stereographic projection is illustrated in Figure A8b where the various circles, representing the projections of identical circles inscribed on the projection sphere, differ in area by up to a factor of about two. *Thus, the stereographic projection fails to conserve area.*

TABLE A1

Plane	Orientation of Plane	
		Angle between A and B : 45°
		Angle between C and A : 36°
A	50° - 354°	Angle between A and D : 60°
B	41° - 288°	Angle between B and C : 33°
C	65° - 316°	Angle between B and D : 73°
D	34° - 090°	Angle between C and D : 88°

THE EQUAL AREA PROJECTION

For some applications in structural geology it is convenient to use a projection that does not distort area. Thus, if a large number of poles have been plotted on a projection, representing the orientation of lineations, or of poles to schistosity, or of c axes of quartz, the interest is commonly in where the areas of maximum concentration are in the distributions appearing in projection.

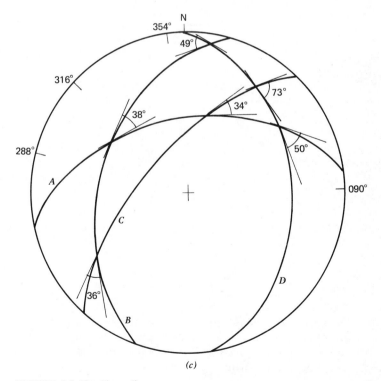

(c)

FIGURE A8 (Continued)

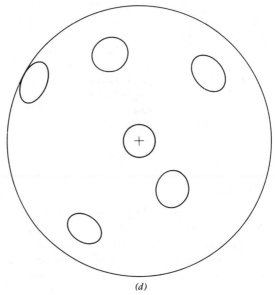

(d)

FIGURE A8 (Continued)

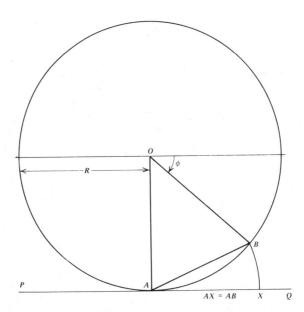

FIGURE A9 The principle of the equal area projection. O is the center of the projection sphere, radius R. OB is the trace of a plane dipping at angle ϕ. X is a point on the equal area projection of this plane.

Since the stereographic projection procedure distorts area, equal concentrations of points on a spherical projection will appear as unequal concentrations on the stereographic projection. For this reason the equal area projection is commonly used in structural geology. A property of this method of projection is that equal areas on the projection sphere plot as equal areas on the projection plane although the shapes of these areas are greatly distorted (Fig. A8*d*). The equiangular property of the stereographic projection is no longer maintained for the equal area projection (Fig. A8*c*).

The method of constructing an equal area projection is illustrated in Figure A9 where a section through the projection sphere is shown. *OB* is the trace of a plane dipping at an angle ϕ. This plane passes through the center, *O*, of the

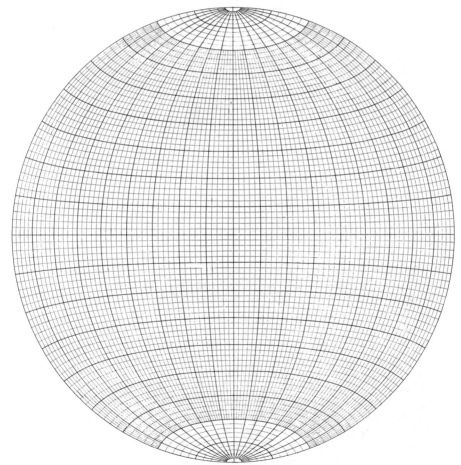

FIGURE A10   Schmidt net.

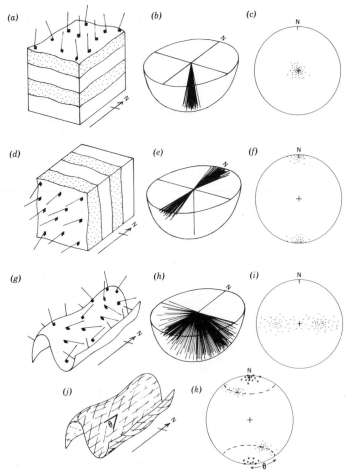

FIGURE A11 Diagrammatic representation of equal area projection of various fabric elements. (*a*) Horizontal bedding and poles have been drawn perpendicular to this surface. Since the bedding is not truly planar the poles vary in orientation. (*b*) The poles within the projection hemisphere and (*c*) the projection of the poles on an equal area diagram are shown. (*d*), (*e*), and (*f*) In the same manner a series of vertically dipping beds is represented and (*g*), (*h*), and (*i*) represent folded bedding. (*c*), (*f*) and (*i*) are referred to as poles-to-bedding figures or, if S is used to denote bedding, they are commonly referred to as $\perp S$ of πS figures. In (*j*) a prefolding lineation lies in the folded plane and there is a second lineation parallel to the fold hinge line. The angle between the two lineations (θ) is approximately constant. (*k*) represents both lineations in equal area projection. The early lineation (dots) defines a small circle girdle indicating that, within the projection hemisphere, it lies on the surface of a cone of 2θ apical angle. The later lineation (crosses) lies parallel to the cone axis.

projection sphere and intersects this sphere at B. The plane PQ, tangent to the sphere at A is the plane on which an equal area projection is to be constructed. The equal area projection of the point B is marked off at X on PQ such that $AX = AB$. In terms of the radius R, of the projection sphere and of the dip, ϕ, of the plane to be projected

$$AX = 2R \, \sin \left(\frac{\pi}{4} - \frac{\phi}{2} \right)$$

When $\phi = 0$ in this equation, $AX = R\sqrt{2}$. Thus, if a horizontal plane is to project as a circle with the same radius as the projection sphere then the distance AX should be scaled such that

$$AX = \sqrt{2}\,R \, \sin \left(\frac{\pi}{4} - \frac{\phi}{2} \right)$$

The similarities and differences between the stereographic and equal area projections are summarized in Table A2.

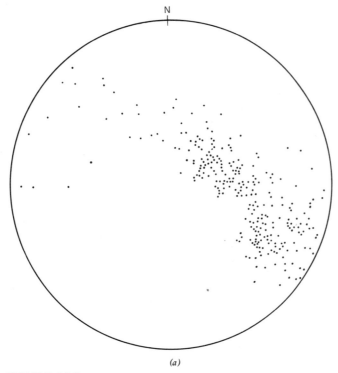

(a)

FIGURE A12   (a) "Scatter" diagram of poles to foliation planes. (b) The data of diagram A12a is contoured according to density per unit area of the projection. Contours are 1, 5, 10, and 12 percent per 1 percent area.

An equal area net may now be constructed as illustrated in Figure A10. The great circles of the projection sphere become curves that are fourth-order quadrics in projection as also do the small circles. This net is commonly called a *Schmidt net* after the German geologist who first introduced it to structural geology.

The Schmidt net may be used to construct the projections of lines and planes in exactly the same manner as the Wulff net is used. This net is more commonly used to represent patterns of distributions of data where a large number of readings of poles to planes, lineations, or crystallographic directions have been assembled. Thus, for example, in a given area bedding may be approximately planar or it may be folded; in either case if measurements of dip are made throughout the area and if they are plotted on a Schmidt projection (Fig. A11) they will define a pattern diagnostic of the structure (see Section 8.5.2). The poles are then contoured according to density per unit area of the projection as shown in Figure A12. Details of these contouring procedures may be found in Phillips (1971, pp. 64–65), Turner and Weiss

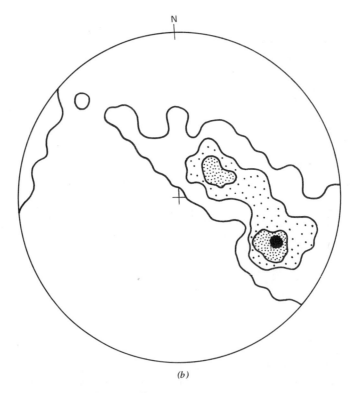

(b)

FIGURE A12 (Continued)

(1963, pp. 58–64) and Ramsay (1967, pp. 8–9). For further discussion of the stereographic and equal area projections reference should be made to the above books and to Vistelius (1966).

TABLE A2   Summary of Geometrical Properties of Stereographic and Equal Area Projections. R is radius of primitive circle. ϕ is amount of dip of a plane of plunge of a line.

Property	Stereographic Projection	Equal Area Projection
Projection fails to preserve	Areas	Angles
Projection preserves	Angles	Areas
Planes project as	Circles	Fourth-order quadrics
Lines project as	Points	Points
All great circles project as	Circles	Fourth-order quadrics
All small circles project as	Circles	Fourth-order quadrics
Distance from center of primitive circle to cyclographic trace measured in direction of dip.	$R \tan (\pi/4-\phi/2)$	$R\sqrt{2} \sin (\pi/4-\phi/2)$
Distance from center of primitive circle to pole of the normal to a plane *measured in the direction opposite to that of the dip*.	$R \tan \phi/2$	$R\sqrt{2} \sin \phi/2$
Distance from center of primitive circle to a lineation or other linear feature — measured in direction of plunge	$R \tan (\pi/4-\phi/2)$	$R\sqrt{2} \sin (\pi/4-\phi/2)$

ANSWERS TO PROBLEMS IN CHAPTER 1

1. Force (mass × acceleration) has the dimensions $[MLT^{-2}]$. Area has the dimension $[L^2]$. Therefore, stress (force per unit area) has the dimensions

$$[MLT^{-2}L^{-2}] = [ML^{-1}T^{-2}].$$

Commonly used units for stress are given in Table 1.1.

Since strain is the ratio of a change of length to an initial length it has no dimension. Hence there are no units for strain. A strain is generally expressed as a percentage change or as a fractional change. Thus, if a line initially 10 cm long is shortened until it becomes 8 cm the change in length is 2 cm and the strain (shortening) is 0.2 (no units) or 20 percent (no units).

The dimensions of strain-rate are $[T^{-1}]$ so that a strain rate might be written as 10^{-5} sec^{-1}, for example. If the above shortening of 20 percent takes place in one year (3.1536×10^7 sec) then the strain rate is

$$0.2/(3.1536 \times 10^7) = 6.3 \times 10^{-9} \text{sec}^{-1}$$

2. A cross section through the plane parallel to the applied force is shown in Figure B1.
   (i) The component of force resolved normal to the plane is $10^{12} \cos 72° = 0.31 \times 12^{12}$ dynes.

The normal stress is the normal force divided by the area. This is $(0.31 \times 10^{12})/10$ dynes cm^{-2}
or

$$0.31 \times 10^{11} \text{ dynes cm}^{-2}$$

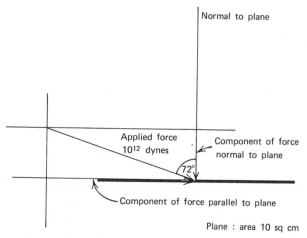

Normal to plane

Applied force
10^{12} dynes

Component of force
normal to plane

72°

Component of force parallel to plane

Plane : area 10 sq cm

FIGURE B1

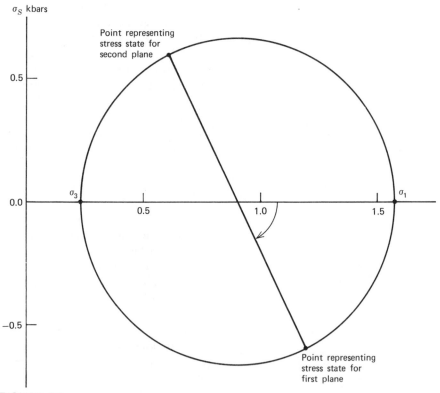

σ_S kbars

Point representing
stress state for
second plane

0.5

σ_3

0.0

0.5   1.0   1.5

σ_1

-0.5

Point representing
stress state for
first plane

FIGURE B2

(ii) The component of force resolved parallel to the plane is $10^{12} \sin 72° = 0.95 \times 10^{12}$ dynes.

The shear stress is the shear force divided by the area. This is $(0.95 \times 10^{12})/10$ dynes cm^{-2}

or 0.95×10^{11} dynes cm^{-2}

From Table 1.1,

$$0.31 \times 10^{11} \text{ dynes cm}^{-2} = 31.0 \text{ kbar}$$
$$= 4.5 \times 10^5 \text{ pounds ins}^{-2}$$
$$= 3.1 \text{ GPa}$$

3. Examination of Fig. 1.3a and comparison with Equations 1.7 shows that a stress state in which the normal stress is σ_N and the shear stress is σ_S may be represented as a point with coordinates $(\sigma_N, -\sigma_S)$ on a Mohr diagram. The important point to notice is the minus sign. The stress states for each of the planes in the question may therefore be plotted as in Figure B2. Since σ_2 lies parallel to the intersection of the planes, if the

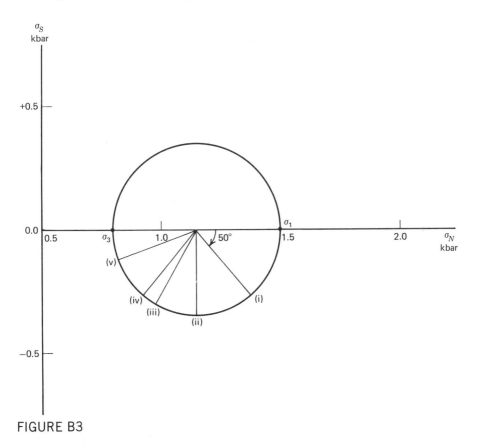

FIGURE B3

normal to one plane is inclined at θ to σ_1, the normal to the other plane is inclined at $\left[\theta + (\pi/2)\right]$ to σ_1. Twice these angles are 2θ and $(2\theta + \pi)$ respectively. In addition, the shear stresses on the two planes are equal in magnitude but opposite in sign. Hence, the two plotted points lie on opposite ends of a diameter of the Mohr circle. The line joining these two points therefore intersects the σ_N axis in a point that is the center of the Mohr circle. The Mohr circle may now be constructed as shown and the values of σ_1 and σ_3 read off as 1.57 kbar and 0.23 kbar, respectively. Twice the angle between σ_1 and the normal to the first plane is also measured as 63.4° giving the angle as 31.7°.

The most important assumption made in the practical determination of such a stress state is that the stress is homogeneous throughout the region in which the measurements were made.

4.   A Mohr circle is constructed as in Figure B3. This is a circle with the σ_N axis as diameter and passing through the points σ_1: (1.5,0) and σ_3: (0.8,0) on this axis. Lines through the center of the circle and inclined to the σ_N axis at *twice* the angles between σ_1 and the normals to the planes are drawn as in Figure B3. These angles are measured clockwise from the σ_N axis.

The required stress states may then be read off as the coordinates of the intersections of these lines with the circle (see Fig. 1.3a for comparison). These stress states are:

Plane	σ_N kbar	σ_S kbar
(i)	1.37	−0.27
(ii)	1.15	−0.35
(iii)	0.98	−0.30
(iv)	0.93	−0.27
(v)	0.82	−0.12

5.   The mean stress is $(\sigma_1 + \sigma_2)/2$.

The deviatoric normal stress is obtained by subtracting the mean stress from the value of the normal stress (see Equation 1.10). The deviatoric shear stress has the same value as the shear stress and is independent of the mean stress.

For question 3, the mean stress is (1.57 + 0.23)/2, or 0.9 kbar.

For question 4, the mean stress is (1.5 + 0.8)/2, or 1.15 kbar.

The deviatoric stresses for the various planes are:

	Deviatoric Normal Stress, kbars	Deviatoric Shear Stress, kbars
Question 3, plane 1	0.3	−0.6
Question 3, plane 2	−0.3	0.6
Question 4, plane (i)	0.22	−0.27
Question 4, plane (ii)	0.00	−0.35
Question 4, plane (iii)	−0.17	−0.30
Question 4, plane (iv)	−0.22	−0.27
Question 4, plane (v)	−0.33	−0.12

The deviatoric maximum principal stresses for questions 3 and 4 are 0.67 and 0.35 kbar, respectively. The deviatoric minimum principal stresses for questions 3 and 4 are −0.67 and −0.35 kbar, respectively (cf. Fig. 1.3d and e).

6. Shortening, ϵ, is given by $\epsilon = \dfrac{\text{initial length} - \text{final length.}}{\text{initial length.}}$

The natural strain, $\bar{\epsilon}$, is given by $\bar{\epsilon} = \log_e (1 + \epsilon)$. In this case ϵ is expressed as a fraction rather than as a percentage and is taken as negative if the strain is a shortening.

The answers, therefore, are as tabulated.

The average strain rate is obtained by dividing the total shortening (expressed as a fraction) by the time taken to accomplish that shortening. These are also tabulated.

	Shortening	Shortening (percent)	Natural Strain	Average Strain Rate
(i)	0.5	50.0	−0.69	5×10^{-3} sec^{-1}
(ii)	0.6	60.0	−0.92	6×10^{-4} sec^{-1}
(iii)	0.7	70.0	−1.20	7×10^{-5} sec^{-1}
(iv)	0.8	80.0	−1.61	8×10^{-6} sec^{-1}
(v)	0.9	90.0	−2.30	9×10^{-7} sec^{-1}

7. The shear strain, γ is given by $\gamma = \tan \psi$ where ψ is the *change* in the angle between two lines initially at right angles. The answers, therefore, are as tabulated.

	Shear Strain	Shear Strain (Percent)	Average Strain Rate
(i)	0.09	8.75	8.75×10^{-4} sec^{-1}
(ii)	0.36	36.40	3.64×10^{-4} sec^{-1}
(iii)	1.00	100.00	1.00×10^{-4} sec^{-1}
(iv)	2.75	274.75	2.75×10^{-5} sec^{-1}
(v)	11.43	1143.01	1.14×10^{-5} sec^{-1}

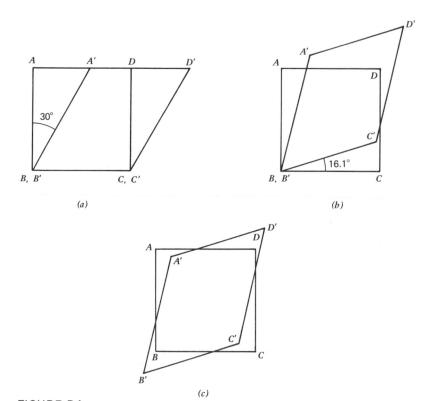

(a)

(b)

(c)

FIGURE B4

8. A shear strain of 0.5774 corresponds to a change in an initial right angle
    of $\tan^{-1}(0.5774)$ or 30°. Therefore an initial right angle becomes 60°. In
    Figure B4a an initial square $ABCD$ is deformed by a simple shear to
    produce the parallelogram $A'B'C'D'$, the angle of shear being 30°. In
    Figure B4b, the parallelogram $A'B'C'D'$ is rotated bodily through 16.1°
    in an anticlockwise sense. In Figure B4c, a small translation has been
    added to produce the required pure shear configuration.

    In this example, the angle of shear for the simple shear is 30°. Hence

    $$\tan^{-1}(\tfrac{1}{2}\tan\psi) = \tan^{-1}(\tfrac{1}{2}\tan 30°)$$
    $$= 16.1°$$

    Hence, the graphical result is compatible with the expression for the
    rotation.

9. Consider a simple shear in which the angle of shear, ψ, is 35° (Fig. B5a).
    Then, $S = \tfrac{1}{2}\tan 35° = 0.35$
    and $\sqrt{S^2 + 1} \pm S$ has the two values 1.41 and -0.71.

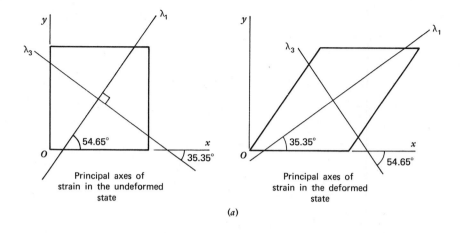

Principal axes of
strain in the undeformed
state

Principal axes of
strain in the deformed
state

(a)

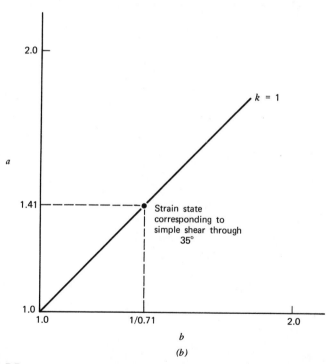

(b)

FIGURE B5   (a), (b) The principal axes of strain in the undeformed state are lines of material particles, mutually orthogonal, that remain orthogonal in the strained state and correspond to the same lines of material particles that define the principal axes of strain in the deformed state.

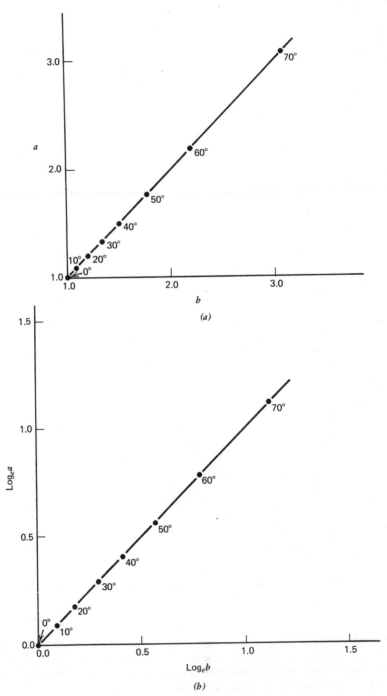

FIGURE B6

(a) The principal axes of strain in the undeformed state make angles (9.65° ± 45°) with Ox. That is, 54.65° and −35.35°, respectively.

(b) The principal axes of strain in the deformed state make angles (±45° − 9.65°) with Ox. That is, 35.35° and −54.65°, respectively.

These axes are drawn in Figure B5a.

(c) The major principal axis of strain has magnitude 1.41 and the least principal axis of strain has magnitude 0.71. The intermediate axis of strain has magnitude 1.0.

(d) This strain state is plotted in Figure B5b.

10. Using the relationships quoted in questions 8 and 9, the angle between $B'C'$ and the major principal axis of strain both in the undeformed and deformed states is summarized below along with the values of the principal strains, $\sqrt{\lambda_1}$ and $\sqrt{\lambda_3}$, and the rigid body rotation.

Angle of Shear, ψ	Angle Between Major Principal Axis of Strain and BC		$\sqrt{\lambda_1}$	$\sqrt{\lambda_3}$	Rigid Body Rotation
	Undeformed State	Deformed State			
0°	—	—	1.00	1.00	0°
10°	47°31′	42°48′	1.09	0.92	5° 2′
20°	50° 9′	39°51′	1.20	0.83	10°19′
30°	53° 3′	36°57′	1.33	0.75	16° 6′
40°	56°23′	33°37′	1.50	0.66	22°46′
50°	60°23′	29°37′	1.76	0.57	30°47′
60°	65°27′	24°33′	2.19	0.46	40°53′
70°	71°58′	18° 2′	3.07	0.33	53°57′

Some of these relationships are illustrated in the progressive simple shear shown in Figure 1.14b. The strains are plotted on Flinn diagrams in Figure B6a and b. For comparative purposes these include both a plotted against b and $\log_e a$ plotted against $\log_e b$ (cf. Fig. 1.16).

The angle θ between the major principal axis of strain and the direction BC is plotted against $\log_e a$ in Figure B7 for both the undeformed and deformed states.

The line AB is progressively extended throughout the deformation history.

The line BC remains of constant length throughout the deformation history.

The line AC is progressively shortened until it becomes normal to $B'C'$. At that stage it is progressively extended. It reaches its initial length again when, in the deformed state it reaches the orientation symmetrical with respect to its starting orientation. It continues to be extended throughout the remainder of the deformation history.

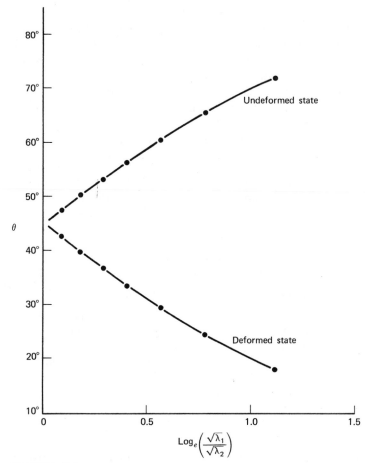

FIGURE B7

The line *BD* is progressively extended throughout the deformation history.

The general conclusion is that all lines initially inclined at greater than 90° to the direction of shear are shortened until they reach that orientation and then are progressively extended. All lines initially oriented at less than 90° to the shear direction are progressively extended throughout the deformation history. The exception is the line *BC*, parallel to the shear direction, which remains unstrained throughout.

The general situation for pure shear is discussed by Jaeger (1969, p. 236) and by Paterson and Weiss (1968).

ANSWERS TO PROBLEMS IN CHAPTER 7

1. The angle between the fault plane and the σ_1 direction is shown in Figure 7.21*b*. It is about 15°. The normal to the fault is therefore inclined at 75° to the σ_1 direction. A point F representing the fault plane is then plotted on Mohr circle III in Figure 7.21*d*, by drawing a radius inclined at 2 × 75 or 150° from the positive end of the σ_N axis. (Notice that point F can be located equally well by drawing a radius at 2 × 15 or 30° to the negative end of the σ_N axis, where 15° is the angle between the fault plane itself and the σ_1 direction.) The absolute value of the σ_S coordinate of point F is the shearing resistance of the specimen (Fig. B.8). The value obtained is about 1.4 kbar. Alternatively, the shearing resistance can be calculated using Equation 1.7. σ_1 is 6.25 kbar (read off Mohr circle III), σ_2 is 0.55 kbar, and 2θ is 150°. Thus the shearing resistance is

$$|\sigma_S| = |(0.55 - 6.25) \sin 150°|$$
$$= 1.42 \text{ kbar}$$

2. The stress-strain diagram (Fig. B.9*a*) is constructed by plotting differential stress against the strain and drawing a smooth curve through the origin and the two given points. Extending this curve slightly beyond the upper point brings it to a differential stress of 15400 bars at which the specimen faulted. The Mohr diagram (Fig. B.9*b*) is constructed by drawing circles centered on the σ_N axis, passing through the point $\sigma_N = 5050$

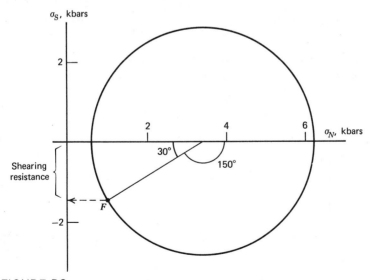

FIGURE B8

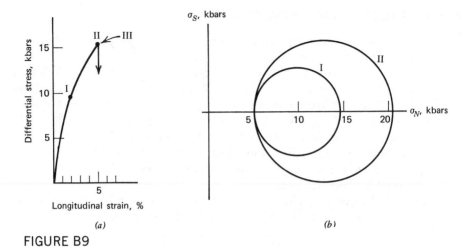

FIGURE B9

bars. The circles corresponding to 2 percent strain (circle I), 5 percent strain (circle II), and the moment of faulting (circle III) have diameters of 9650 bars, 15300 bars, and 15400 bars, respectively. (Circle III is not drawn in Fig. B.9*b* because it is so nearly the same size as circle II.)

3. The angles are found by locating the center of each Mohr circle and drawing a line that connects each center with the point representing the fault plane (Fig. B.10). The angles between these lines and the σ_N axis are *twice* the angles being sought. The angles are, from the lowest to the highest confining pressures, 25°, 25.5°, 22.5°, 29.5°, 33°, and 33°.

4. A diagram is drawn, similar to Figure 7.23*a*, showing a straight-line Mohr envelope with a slope of 1.1 and an intercept on the σ_S axis of 170

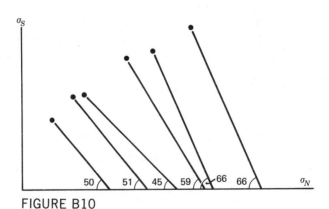

FIGURE B10

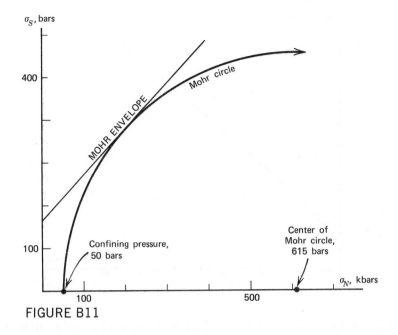

FIGURE B11

bars. The ultimate strength at confining pressure of 50 bars is found by constructing the Mohr circle that passes through 50 bars on the σ_N axis and just touches the Mohr envelope (Fig. B.11). The diameter of this circle is the ultimate strength: 1130 bars.

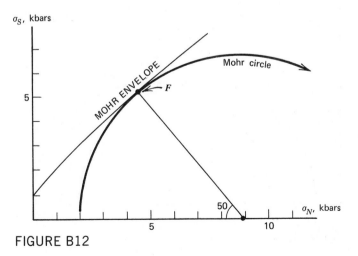

FIGURE B12

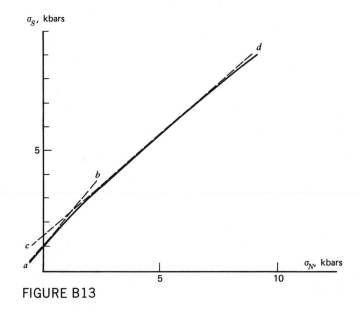

FIGURE B13

5. The Mohr envelope of Figure 7.24 is traced and a Mohr circle is con-
    structed that passes through 2 kbars on the σ_N axis and just touches the
    Mohr envelope at point F. The diameter of this circle is the ultimate
    strength, 13.8 kbar. The line connecting point F to the center of the
    Mohr circle intersects the σ_N axis at 50° (Fig. B.12). The predicted angle
    between the fault and the σ_1 direction is, therefore, 25°.

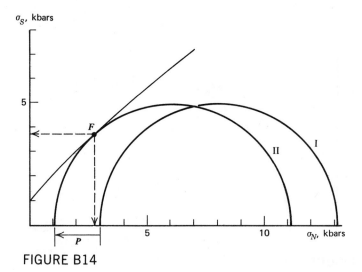

FIGURE B14

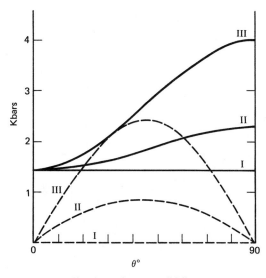

Shearing resistance : solid lines
Shear stress : dashed lines

FIGURE B15

6. The Mohr envelope to the right of the σ_S axis is not a straight line, so values of the cohesion and of the angle of internal friction need to be given separately for different parts of the envelope (Fig. B.13). Thus at normal stresses on the fault plane of less than about 1 kbar, the envelope is approximated by straight line a-b, with a σ_S intercept of 1.0 kbar and slope of 1.2. At normal stresses less than about 1 kbar on the fault plane, the cohesion is therefore 1.0 kbar and the coefficient if internal friction is 1.2. At normal stresses between about 3 kbar and 7 kbar, the envelope is approximated by straight line c-d with σ_S intercept 1.4 kbar and slope about 0.9. At these higher levels of normal stress the cohesion is therefore about 1.4 kbar and the coefficient of internal friction is 0.9. For normal stresses between 1 kbar and 3 kbar the cohesion and coefficient of internal friction have values lying in the ranges 1-1.4 kbar and 1.2-0.9 kbar, respectively.

7. The Mohr envelope of Figure 7.24 is traced and a Mohr circle (I) is constructed with $\sigma_3 = 3$ kbar and $\sigma_1 = 13$ kbar. This is the stable state with zero pore pressure. The unstable condition is found by shifting this circle to the left along the σ_N axis (cf. Fig. 7.27) until it just touches the Mohr envelope at point F. The pore pressure P necessary to bring about this condition is indicated in Figure B.14 and is equal to about 1.9 kbar. The effective normal stress and shear stress on the fault plane are found,

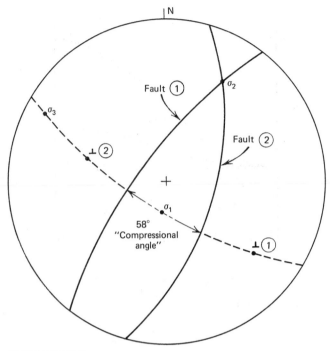

FIGURE B16

by reading off the σ_N and σ_S coordinates of F, to be 2.8 and 3.7 kbar respectively.

8. At the start of the loading, the normal stress on all planes is 0.75 kbar, so the shearing resistance on all planes, given by equation 7.1, is 1.05 kbar + (0.53)(0.75) = 1.45 kbar. The shear stress is zero on all planes at the start of the deformation. This stage is indicated by lines labeled I in Figure B.15.

About halfway through the loading, when $\sigma_1 = 2.4$ kbar, the shearing resistance for planes perpendicular to σ_1 is

$$1.05 \text{ kbar} + (0.53)(2.4) = 2.32 \text{ kbar}$$

and values for planes in other orientations can be found by using a Mohr diagram or Equation 1.7 to find the normal stress and Equation 7.1 to find from this the shearing resistance. The shear stress for various planes halfway through the loading is calculated directly from the shear stress equation 1.7. The two curves for this stage of the experiment are labeled II in Figure B.15. The curves of Figure 7.32 are labelled III. Notice that faulting finally occurs when the curves for shear stress and shearing resistance touch.

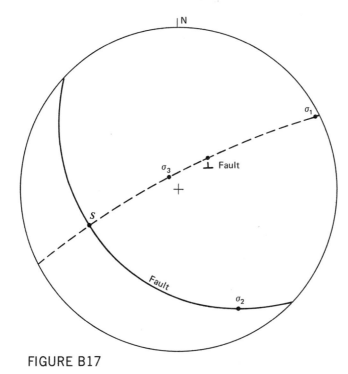

FIGURE B17

9. Faults 1 and 2 are plotted as great circles in stereographic projection (Fig. B.16). σ_2 is taken as parallel to the line of intersection of the faults. A great circle is then drawn with σ_2 as pole and this great circle contains the σ_1 and σ_3 directions. σ_1 is assumed to lie in the smaller angle (58°) between the fault planes. The trends and plunges are then read off the projection as follows: σ_1 trends to 191°, plunges 69°; σ_2 trends to 32°, plunges 20°; σ_3 to 300°, plunges 8°. Notice that the coefficient of internal friction must have been $\tan\left[90° - (58°/2)\right] = 1.8$.

10. The fault plane and striations (S) are plotted in stereographic projection (Fig. B.17). The σ_2 direction can be plotted immediately, at 90° from S in the fault plane. A great circle (dashed) is then drawn with σ_2 as pole, and this great circle will contain the σ_1 and σ_3 directions. σ_1 lies at 30° to S. There are two positions in the dashed great circle that satisfy this requirement, but only one of these is consistent with the observed reverse character of the fault. σ_1 is plotted in this position in Figure B.17. σ_3 is then plotted 90° from σ_1 in the dashed great circle. The trends and plunges are: σ_1 trends to 64°, plunges 2°; σ_2 trends to 153°, plunges 11°; σ_3 trends to 323°, plunges 79°.

REFERENCES

Agron, S. L., 1964. Tectonic boulders in Carboniferous conglomerate, Rhode Island. Geol. Soc. Am. Special Paper 76, p. 3.

Allen, C. R., 1962. Circum-Pacific faulting in the Philippines — Taiwan region. J. Geophysical Res., 67, pp. 4795-4812.

Allen, J. R. L., 1968. Current ripples — their relation to patterns of water and sediment motion. North Holland Publishing Co., Amsterdam, 433pp.

Allen, J. R. L., 1970. Physical processes of sedimentation. Earth Science Series No. 1, J. Sutton and J. V. Watson (eds.). George Allen & Unwin Ltd., London, 248pp.

Allen J. R. L., and N. R. Banks, 1972. An interpretation and analysis of recumbent-folded deformed cross-bedding. Sedimentology, 19, pp. 257-283.

Ampferer, O., and W. Hammer, 1911. Geologischer Querschnitt durch die Ostalpen von Allgäu zum Gardasee. Jahrb. Geol. Reichsandstalt, 61 (3-4), pp. 531-710.

Anderson, D. E., 1971. Kink bands and major folds, Broken Hill, Australia. Geol. Soc. Am. Bull., 82, pp. 1841-1862.

Anderson, E. M., 1948. On lineation and petrofabric structure and the shearing movement by which they have been produced. Geol. Soc. London, Quart. J., 104, pp. 99-132.

Anderson, E. M., 1951. The dynamics of faulting. Oliver and Boyd, Edinburgh, 206pp.

Andreatta, C., 1934. Analisi strutturali di rocce metamorfiche, 5, Olivinite. Period. Mineral., 5, pp. 237-253.

Ardell, A. J., J. M. Christie, and J. A. Tullis, 1974 (in preparation). Dislocation substructures in deformed quartz rocks.

Argand, E., 1911. Les Nappes de recouvrement des Alpes pennines et leurs prolongements structuraux. Matériaux p. la Carte Géol de la Suisse, n.s., livre, 31, pp. 1-26.

Argand, E., 1916. Sur l'arc des Alpes Occidentales. Eclogae Geol. Helv., 14, pp. 145-191.

Artemjev, M. E., and E. V. Artyushkov, 1971. Structure and isostasy of the Baikal rift and the mechanism of rifting. J. Geophysical Res., 76 (5), pp. 1197-1210.

Atwater, G. I., and M. J. Forman, 1959. Nature of growth of southern Louisiana salt domes and its effect on petroleum accumulation. Am. Assoc. Petrol. Geol. Bull., 43 (11), pp. 2592-2622.

Atwater, T., 1970. Implications of plate tectonics for the Cenozoic tectonic evolution of western North America. Geol. Soc. Am. Bull., 81, pp. 3513-3536.

Atwater, T., 1972. Test of new global tectonics: discussion. Am. Assoc. Petrol. Geol. Bull., 56 (2), pp. 385-392.

Atwater, T., and H. W. Menard, 1970. Magnetic lineations in the northeast Pacific. Earth Planetary Sci. Letters, 7, pp. 445-450.

Atwater, T. and J. D. Mudie, 1968. Block faulting on the Gorda Rise. Science, 159 (3816), pp. 729-731.

Aubouin, J., 1965. Geosynclines. Elsevier Publishing Co., New York, 335pp.

Ave'Lallemant, H. G. and N. L. Carter, 1971. Pressure dependence of quartz deformation lamellae orientations. Am. J. Sci., 270, pp. 218-235.

Badgley, P. C., 1965. Structural and tectonic principles. Harper & Row, New York, 521pp.

Badoux, H., 1963. Les bélemnites tronconnées de Leytron (Valais). Bull. Laboratoires de Géologie, Minéralogie, Géophysique et du Musée Géologique, Université de Lausanne Bull., 138, pp. 1-7. Also in Bull. de la Société vaudoise des Sciences Naturells, 68 (311), 1963.

Badoux, H., 1970. Les oolites déformées du Vélar (Massif de Morcles). Eclogae Geol. Helvet., 63, pp. 539-548.

Bailey, E., and W. J. McCallien, 1961. Structure of the Northern Apennines. Nature, 191, pp. 1136-1137.

Bailey, E. B., 1935. Tectonic essays, mainly Alpine. Oxford.

Bailey, E. H., W. P. Irwin, and D. L. Jones, 1964. Franciscan and related rocks, and their significance in the geology of Western California. Calif. Div. of Mines and Geol. Bull., 183, 177pp.

Baird, A. K., 1962. Superposed deformation in the Central Sierra Nevada foothills east of the Mother Lode. Univ. Calif. Publ., Geol. Sci., 42, pp. 1-70.

Baker, D. W., H.-R. Wenk, and J. M. Christie, 1969. X-ray analysis of preferred orientation in fine-grained quartz aggregates. J. Geol., 77, pp. 144-172.

Balk, R., 1936. Structural and petrologic studies in Dutchess County, New York. Geol. Soc. Am. Bull., 47, pp. 685-774.

Balk, R., 1937. Structural behaviour of igneous rocks. Geol. Soc. Am. Memoir 5, 177pp.

Balk, R., 1946. Gneiss dome at Shelburne Falls, Massachusetts. Geol. Soc. Am. Bull., 57, pp. 125-160.

Balk, R., 1952. Fabric of quartzites near thrust faults. J. Geol., 60, pp. 415-435.

Bankwitz, P. von, 1965. Uber Klufte, I. Beobachtungen in Thuringishen Schiefergebirge. Geologie. Zeitschrift fur des Gesamtegebiet der Geologie und der Mineralogie, 14 (3), pp. 241-253.

Bankwitz, P. von, 1966. Uber Klufte, II. Die Bildung der Kluftfläche und eine Systematik ihrer Strukturen. Geologie, 15, pp. 896-941.

Barazangi, M., and J. Dorman, 1969. World seismicity map compiled from ESSA Coast and Geodetic Survey epicenter data, 1961-1967. Bull. Seismol. Soc. Am., 59, pp. 359-380.

Barrett, C. S., and L. H. Leverson, 1940. The structure of aluminum after compression. Trans. Am. Inst. Metall. Engr., 137, pp. 112-126.

Barrett, C. S., and T. B. Massalski, 1966. Structure of metals, McGraw-Hill, New York, 654pp.

Bartram, J. G., 1929. Elk basin oil and gas field, Park County, Wyoming and Carbon County, Montana. Structure of typical American oil fields, 2, pp. 577-88. Am. Assoc. Petrol. Geol., 780pp.

Baumeister, T., and L. S. Marks, 1967. Standard handbook for mechanical engineers, 7th ed., McGraw-Hill, New York.

Bayly, M. B., 1964. A theory of similar folding in viscous materials. Am. J. Sci., 262, pp. 753-766.

Bayly, M. B., 1965. A correlation between cleavage fan angle and bed thickness. Geol. Mag., 102 (3), pp. 246-251.

Bayly, M. B., 1971. Similar folds, buckling, and great circle patterns. J. Geol., 79, pp. 110-118.

Beck, P. A., and H. Hu, 1966. The origin of recrystallization textures. In H. Margolin (coordinator), Recrystallization, grain growth and textures. Seminar Am. Soc. Metals, pp. 393-433.

Becke, F., 1913. Über Mineralbestand und Struktur der krystallinischen Schiefer: Akademie der Wissenschaften Wien, Denkschriften, Mathematisch-Naturwissenschaftliche Klasse, 75, pp. 1-53.

Becker, G. F., 1893. Finite homogeneous strain, flow and rupture of rocks. Geol. Soc. Am. Bull., 4, pp. 13-90.

Behre, C. H., 1933. Slate in Pennsylvania. Pennsylvania Geol. Survey Bull., 16, p. 400.

Bell, T. H., and M. A. Etheridge, 1973. Microstructure of mylonites and their descriptive terminology. Lithos, 6, pp. 337-348.

Bell, T. H., and M. A. Etheridge, 1974. The deformation and recrystallization of quartz in a mylonite zone, Central Australia. Tectonophysics, in press.

Beloussov, V. V., 1961. The origin of folding in the earth's crust. J. Geophysical Res., 66 (7), pp. 2241-54.

Beloussov, V. V., 1962. Basic problems in geotectonics. McGraw-Hill, New York, 809pp.

Beloussov, V. V., 1968. An open letter to J. Tuzo Wilson. Geotimes, 13 (10), pp. 17-19.

Beloussov, V. V., 1969. Interrelations between the earth's crust and upper mantle: pp. 698-712 in The Earth's Crust and Upper Mantle, P. Hart (ed.). Am. Geophysical Union, Geophysical Monograph 13, 735pp.

Beloussov, V. V., 1970. Against the hypothesis of ocean-floor spreading. Tectonophysics, 9, pp. 489-511.

Benioff, H., 1949. Seismic evidence for the fault origin of oceanic deeps. Geol. Soc. Am. Bull., 60, pp. 1837-56.

Benioff, H., 1954. Orogenesis and deep crustal structure: additional evidence from seismology. Geol. Soc. Am. Bull., 65, pp. 385-400.

Bertrand, M., 1884. Rapports de structure des Alpes de Glaris et du Bassin haviller du Nord. Soc. Geol. de France, Bull., 3eme Ser. 12.

Bienlawski, Z. T., 1967. Mechanism of brittle fracture of rock. Int. J. Rock Mech. Mineral Sci., 4, pp. 395-406.

Billings, M. P., 1972. Structural geology, 3rd ed., Prentice-Hall, Inc., Englewood Cliffs, N. J., 606pp.

Biot, M. A., 1957. Folding instability of a layered viscoelastic medium under compression. Roy. Soc. London Proc., Ser. A., 242, pp. 444-454.

Biot, M. A., 1959. The influence of gravity on the folding of a layered viscoelastic medium under compression. J. Franklin Inst., 267, pp. 211-228.

Biot, M. A., 1961. Theory of folding of stratified viscoelastic media and its implications in tectonics and orogenesis. Geol. Soc. Am. Bull., 72, pp. 1595-1620.

Biot, M. A. 1965. Mechanics of incremental deformations. Wiley, New York, 504pp.

Biot, M. A., H. Odé, and W. L. Roever, 1961. Experimental verification of the theory of folding of stratified viscoelastic media. Geol. Soc. Am. Bull., 72, pp. 1621-1630.

Birch, F., 1966. Compressibility, elastic constants: pp. 97-174 in Handbook of physical constants, S. P. Clark (ed.), Geol. Soc. Am. Memoir, 97, 587pp.

Bird, J. M., and J. F. Dewey, 1970. Lithosphere plate-continental margin tectonics and the evolution of the Appalachian orogen. Geol. Soc. Am. Bull., 81, pp. 1031-1060.

Bishop, D. G., 1972. Transposition structures associated with cleavage formation in the Otago schists. New Zealand J. Geol. and Geophys., 15 (3), pp. 360-371.

Bishop, J. F. W., 1953. A theoretical examination of the plastic deformation of crystals by glide. Phil. Mag., 44, pp. 51-64.

Bishop, J. F. W., 1954. A theory of the tensile and compressive textures of face-centered cubic metals. J. Mech. and Physics of Solids, 3, pp. 130-142.

Bishop, J. F. W., and R. Hill, 1951. A theory of plastic distortion of a polycrystalline aggregate under combined stresses. Phil Mag., 42, pp. 414-427.

Blacic, J. D., 1972. Effect of water on the experimental deformation of olivine. Geophysical Monograph 16, pp. 109-115.

Blake, M. C., and D. L. Jones,1974. Origin of Franciscan mélanges in Northern California: pp. 345-357 in Modern and ancient geosynclinal sedimentation, R. H. Dott and R. H. Shaver (eds.). Special Pub. 19, Soc. Econ. Paleontologists and Mineralogists.

Blake, M. C., W. P. Irwin, and R. G. Coleman, 1967. Upside-down metamorphic zonation, blueschist facies along a regional thrust in California and Oregon. U. S. Geol. Survey Prof. Paper 575-C, pp. Cl-19.

Blatt, H., G. Middleton, and R. Murray, 1972. Origin of sedimentary rocks. Prentice-Hall, Inc., Englewood Cliffs, N. J., 634pp.

Boland, J. N., A. C. McLaren, and B. E. Hobbs, 1971. Dislocations associated with optical features in naturally deformed olivine. Contr. Miner. Petrol., 30, pp. 53-63.

Bombolakis, E. G., 1968. Photoelastic study of initial stages of brittle fracture in compression. Tectonophysics, 6 (6), pp. 461-473.

Bombolakis, E. G., 1972. Study of crack behavior during compression. EOS Trans. Am. Geophysical Union, 53, p. 512.

Bonney, T. G., 1886. Anniversary address of the President. Geol. Soc. London Proc., 42, pp. 38-115.

Borg, I. Y., and J. Handin, 1966. Experimental deformation of crystalline rocks. Tectonophysics, 3, pp. 249-368.

Borg, I. Y., and J. Handin, 1967. Torsion of calcite single crystals. J. Geophysical Res., 72, pp. 641-670.

Borg, I. Y., and H. C. Heard, 1970. Experimental deformation of plagioclases. In Experimental and natural rock deformation, P. Paulitsch (ed.). Springer-Verlag, Berlin, pp. 375-403.

Bornhauser, M., 1958. Gulf coast tectonics. Am. Assoc. Petrol. Geol. Bull., 42, pp. 339-370.

Borradaile, G. J., 1973. Dalradian structure and stratigraphy of the Northern Lock Awe District, Argyllshire. Roy. Soc. Edinburgh Trans., 69, pp. 1-21.

Borradaile, G. J., and H. D. Johnson, 1973. Finite strain estimates form the Dalradian Dolomitic Formation, Islay, Argyll, Scotland. Tectonophysics, 18, pp. 249-259.

Boswell, P. G. H., 1961. Muddy sediments. W. Heffer & Sons, Cambridge, 149pp.

Bott, M. H. P., 1959. The mechanics of oblique slip faulting. Geol. Mag., 96, pp. 109-117.

Bott, M. H. P., 1971. The interior of the earth. E. Arnold, London, 316pp.

Bouma, A. H., 1959. Some data on turbidites from the Alpes Maritime France. Geologie en Mijnbouw, 21, pp. 223-227.

Brace, W. F., 1955. Quartzite pebble deformation in central Vermont. Am. J. Sci., 253, pp. 129-145.

Brace, W. F., 1961. Mohr construction in the analysis of large geologic strain. Geol. Soc. Am. Bull., 72, pp. 1059-1080.

Brace, W. F., 1964. Brittle fracture of rocks: pp. 111-180 in State of Stress in the Earth's Crust, W. R. Judd (ed.), American Elsevier Inc., New York, 732pp.

Brace, W. F., and E. G. Bombolakis, 1963. A note on brittle crack growth in compression. J. Geophysical Res., 68 (12), pp. 3709-3713.

Brace, W. F., and J. D. Byerlee, 1966. Stick slip as a mechanism for earthquakes. Science, 153 (3739), pp. 990-992.

Brace, W. F., and J. D. Byerlee, 1967. Recent experimental studies of brittle fracture in rocks: pp. 58-81 in Failure and Breakage of Rock, C. Fairhurst (ed.), Proc. 8th Symposium on Rock Mechanics, Univ. Minn., 1966.

Brace, W. F., and J. D. Byerlee, 1970. California earthquakes: why only shallow focus? Science, 168, pp. 1573-1575.

Brace, W. F., B. W. Paulding and C. Scholz, 1966. Dilatancy in the fracture of crystalline rocks. J. Geophysical Res., 71 (16), pp. 3939-3953.

Brace, W. F., E. Silver, K. Hadley, and C. Goetze, 1972. Cracks and pores: a closer look. Science,

178, pp. 162-164.

Braddock, W. A., 1970. The origin of slaty cleavage: evidence from Precambrian rocks in Colorado. Geol. Soc. Am. Bull., 81, pp. 589-600.

Breddin, H., 1957. Tectonische Fossil und Gesteins deformation im Gebeit von St. Goarhausen. Decheniana, 110, pp. 289-350.

Breddin, H., 1964. Die tectonische Deformation de Fossilien und Gesteinen der Molasse von St. Gallen (Schweiz). Geol. Mitt. Aachen, 4, pp. 1-68.

Bredehoeft, J. D., and B. B. Hanshaw, 1968. On the maintenance of anomalous fluid pressures: I. Thick sedimentary sequences. Geol. Soc. Am. Bull., 79, pp. 1097-1106.

Bridwell, R. J., 1974. Nonlinear in-plane bending of a plane stress-strain finite element model. J. Geophysical Res., 79 (11), pp. 1674-1678.

Brown, E. H., 1968. Metamorphic structures in part of the eastern Otago schists. New Zealand J. Geol. Geophysics, 11 (1), pp. 41-65.

Brown, J. McC., 1967. The Grenville front south of Coniston, Ontario. Guidebook. Geology of parts of Eastern Ontario and Western Quebec, S. E. Jenness (ed.). Geol. Assoc. Canada; Min. Assoc. Canada; Min. Assoc. Am., pp. 11-12.

Brown Monnet, V., 1948. Mississippian Marshal formation of Michigan. Am. Assoc. Petrol. Geol. Bull., 32, pp. 629-688.

Brune, J. N., 1968. Seismic moment, seismicity, and rate of slip along major fault zones. J. Geophysical Res., 73 (2), pp. 777-784.

Bryant, B., and J. C. Reed, Jr., 1969. Significance of lineations and minor folds near major thrust faults in the southern Appalachians and the British and Norwegian Caledonides. Geol. Mag., 106, pp. 412-429.

Bullard, E., J. E. Everett, and A. G. Smith, 1965. The fit of the continents around the Atlantic. Phil Trans. Roy. Soc. I, Series A. 258, pp. 41-51.

Burford, R. O., 1968. Evidence of recent strain in the San Andreas fault system from triangulation surveys: pp. 87-93 in Proceedings of a Conference on Geologic Problems of the San Andreas Fault System, W. R. Dickinson and A. Grantz (eds.). Stanford Univ. Publ. Geol. Sci., 11, 376pp.

Burke, K., and J. F. Dewey, 1973. Plume generated triple junctions: key indicators in applying plate tectonics to old rocks. J. Geol., 81, pp. 406-433.

Burke, K., W. S. F. Kidd, and J. T. Wilson, 1973. Relative and latitudinal motion of Atlantic hot spots. Nature, 245, pp. 133-137.

Busk, H. G., 1929. Earth flexures. Cambridge Univ. Press, London, 106pp.

Byerlee, J. D., 1967. Frictional characteristics of granite under high confining pressure. J. Geophysical Res., 72 (14), pp. 3639–3648.

Byerlee, J. D., 1970. The mechanics of stick-slip. Tectonophysics, 9, pp. 475-486.

Byerlee, J. D., and W. F. Brace, 1968. Stick slip, stable sliding, and earthquakes — effect of rock type, pressure, strain rate, and stiffness. J. Geophysical Res., 78 (18), pp. 6031-6037.

Byerlee, J. D., and W. F. Brace, 1972. Fault stability and pore pressure. Bull. Seismol. Soc. Am., 62 (2), pp. 657-660.

Byerly, P., 1926. The Montana earthquake of June 28, 1925, G. M. C. T. Bull. Seismol. Soc. Am., 16 (4), pp. 209-232.

Cady, W., 1972. Are the Ordovician northern Appalachians and the Mesozoic cordilleran system homologous? J. Geophysical Res., 77 (20), pp. 3806-3815.

Calnan, E. A., and C. J. B. Clews, 1950. Deformation textures in face-centered cubic metals. Phil. Mag., 41, pp. 1085-1100.

Calnan, E. A., and C. J. B. Clews, 1951. The development of deformation textures in metals — Part II. Body centred cubic metals. Phil. Mag., 42, pp. 616-635.

Carey, S. W., 1958. A tectonic approach to continental drift. pp. 177-355 in Continental drift: a symposium, S. W. Carey (ed.), Univ. of Tasmania, 375pp.

Carter, N. L., 1971. Static deformation of silica and silicates. J. Geophysical Res., 76 (23), pp.

5514-5540.

Carter, N. L., J. M. Cristie, and D. T. Griggs, 1964. Experimental deformation and recrystallization of quartz. J. Geol., 72, pp. 687-733.

Carter, N. L., and H. G. Ave' Lallemant, 1970. High temperature flow of dunite and peridotite. Geol. Soc. Am. Bull., 81, pp. 2181-2202.

Cebull, S. E., 1973. Concept of orogeny. Geology, 1 (3), pp. 101-102.

Chapman, C. A., 1939. Geology of the Mascoma Quadrangle New Hampshire. Geol. Soc. Am. Bull., 50, pp. 127-180.

Chapple, W. M., 1964. A mathematical study of finite amplitude rock folding (abs.). Trans. Am. Geophysical Union, 45, p. 104.

Chapple, W. M., 1968a. Buckling of an elastic, viscous or plastic layer. NSF Advanced Science Seminar in Rock Mechanics. Air Force Research labs, Bedford, Mass., 1, pp. 201-222.

Chapple, W. M., 1968b. The equations of elasticity and viscosity and their applications to faulting and folding. NSF Advanced Science Seminar in Rock Mechanics. Air Force Research Labs, Bedford, Mass., 1, pp. 225-284.

Chapple, W. M., 1968c. A mathematical theory of finite amplitude folding. Geol. Soc. Am. Bull., 79, pp. 47-68.

Chapple, W. M., 1969. Fold shape and rheology. The folding of an isolated viscous-plastic layer. Tectonophysics, 7, pp. 97-116.

Chin, G. Y., 1969. Deformation textures: pp. 51-80 in Textures in Research and Practice, J. Grewen and G. Wasserman (eds.). Springer-Verlag, Berlin.

Chin, G. Y., W. F. Hosford, and D. R. Mendorf, 1969. Accommodation of constrained deformation in f.c.c. metals by slip and twinning. Proc. Roy. Soc., A, 309, pp. 433–456.

Chin, G. Y., and W. L. Mammel, 1969. Generalizations and equivalence of minimum work (Taylor) and maximum work (Bishop-Hill) principles for crystal plasticity. Trans. Met. Soc., Amer. Inst. of Mech. Engr., 245, pp. 1211-1214.

Chin, G. Y., E. A. Nesbitt, and A. J. Williams, 1966. Anisotropy of strength in single crystals under plane strain compression. Acta Met., 14, pp. 467-476.

Chinnery, M. A., 1964. The strength of the earth's crust under horizontal shear stress. J. Geophysical Res., 69, pp. 2085-2089.

Chinnery, M. A., 1966. Secondary faulting. Canadian J. Earth Sci., 3, pp. 163-174.

Choukroune, P., 1971. Contribution à l'etude des mécanismes de la déformation avec schistosité grâce aux cristallisations syncimematiques dans les "zones abritées" ("pressure shadows"). Soc. Geol. de France, Bull., XIII (3-4), pp. 257-271.

Christensen, M. N., 1963. Structure of metamorphic rocks at Mineral King California. Univ. Calif. Publ., Geol. Sci., 42, pp. 159-198.

Christie, J. M. 1960. Mylonitic rocks of the Moine thrust zone in the Assynt region, northwest Scotland. Trans. Edinburgh Geol. Soc., 18, Pt. 1, pp. 79-93.

Christie, J. M., 1963. The Moine thrust zone in the Assynt region, northwest Scotland. Univ. Calif. Publ., Geol. Sci., 40, pp. 345-419.

Christie, J. M., and A. J. Ardell, 1974. The substructures of deformation lamellae in quartz. Geol., 2, pp. 405–408.

Christie, J. M., and H. W. Green, 1964. Several new slip mechanisms in quartz (abstract). EOS Trans., Am. Geophysical Union, 45, p. 103.

Christie, J. M., D. T. Griggs, and N. L. Carter, 1964. Experimental evidence for basal slip in quartz. J. Geol., 72, pp. 734-756.

Christie, J. M., D. T. Griggs, and N. L. Carter, 1966. Experimental deformation and recrystallization of quartz and Experimental evidence of basal slip in quartz: a reply. J. Geol., 74, pp. 368-371.

Clark, B. R., 1970. Origin of slaty cleavage in the Coeur d' Alene District, Idaho. Geol. Soc. Am. Bull., 81, pp. 3061-3072.

Clark, R. H., and D. B. McIntyre, 1951. The use of the terms pitch and plunge. Am. J. Sci., 249, pp. 591-599.

Cloos, E., 1932. "Feather joints" as indicators of the direction of movements on faults, thrusts, joints and magmatic contacts. Nat. Acad. Sci. Proc., 18, pp. 387-395.

Cloos, E., 1946. Lineation. Geol. Soc. Am. Memoir, 18, 122pp.

Cloos, E., 1947. Oolite deformation in the South Mountain Fold, Maryland. Geol. Soc. Am. Bull., 58, pp. 843-918.

Cloos, E., 1971. Microtectonics along the western edge of the Blue Ridge, Maryland and Virginia. Johns Hopkins Univ. Press, Baltimore, 234pp.

Cloos, E., and C. H. Broedel, 1940. Geologic map of Howard County. Scale 1:62,500. Maryland Geol. Survey.

Cloos, H., 1936. Einfuhrung in die Geologie, Berlin, 503pp.

Cloos, H., 1939. Heburn, Spaltung, Vulkanismus. Geol. Rundschau, 30, pp. 405-527.

Cobbold, P. R., J. W. Cosgrove, and J. M. Summers, 1971. Development of internal structures in deformed anisotropic rocks. Tectonophysics, 12, pp. 23-53.

Colette, B. J., 1958. On the origin of schistosity. Proc. Kon. Neder. Akad. Wetensch., 61, pp. 121-139.

Collet, L. W., 1936. The Structure of the Alps. London.

Coney, P. J., 1970. The geotectonic cycle and the new global tectonics. Geol. Soc. Am. Bull., 81, pp. 739-748.

Coney, P. J., 1973. Plate tectonics of marginal foreland thrust-fold belts. Geology, 1 (3), pp. 131-134.

Conybeare, C. E. B., and K. A. W. Crook, 1968. Manual of sedimentary structures. Commonwealth of Australia, Department of National Development, Bureau of Mineral Resources, Geol. and Geophys. Bull., 102, 327pp.

Corbett, K. D., 1973. Open-cast slump sheets and their relationship to sandstone beds in an Upper Cambrian flysch sequence, Tasmania, J. Sed. Petrol: 43, pp. 147-159.

Coulomb, C. A., 1773. Sur une application des règles de maximus et minimis a quelques problèmes de statique relatifs à l'architecture. Acad. Roy. des Sci., Memoires de Math. et de Physique par divers savans, 7, pp. 343-382.

Coulomb, J., 1972. Sea Floor Spreading and Continental Drift. D. Reidel, Dordrecht, Holland, 184pp.

Coward, M. P., 1973. Heterogeneous deformation in the development of the Laxfordian complex of South Uist, Outer Hebrides. Geol. Soc. London J., 129, pp. 139-160.

Cox, A., 1969. Geomagnetic reversals. Science, 163 (3864), pp. 237-245.

Cox, A. (ed.), 1973. Plate tectonics and geomagnetic reversals. W. H. Freeman and Co., San Francisco, 702pp.

Cox, A., R. R. Doell, and G. B. Dalrymple, 1963. Geomagnetic polarity epochs and Pleistocene geochronometry. Nature, 198, pp. 1049-1051.

Creer, K. M., 1964. A reconstruction of the continents for the upper Paleozoic from paleomagnetic data. Nature, 2031 (4950), pp. 1115-1120.

Crittenden, M. D., Jr., 1967. Viscosity and finite strength of the mantle as determined from water and ice loads. Geophys. J. Roy. Astron. Soc., 14, pp. 261-279.

Crowell, J. C., 1962. Displacement along the San Andreas Fault, California. Geol. Soc. Am. Spec. Paper 71, 61pp.

Dale, T. N. et al., 1914. Slate in the United States. U. S. Geol. Survey Bull., 586, 220pp.

Dallmeyer, R. D., 1972. Precambrian structural history of the Hudson Highlands near Bear Mountain, New York. Geol. Soc. Am. Bull., 83, pp. 895-904.

Daly, R. A., 1915. A geological reconnaissance between Golden and Kamloops, Brit. Columbia. Canada Geol. Survey Memoir 68, pp. 40-53.

Daly, R. A., 1947. The Vredefort ring-structure of South Africa. J. Geol., 55, pp. 125-145.

Dana, J., 1847. Geological results of the earth's contraction in consequence of cooling. Am. J. Sci., 3 (2), pp. 176-188.

Davies, A. F., and J. N. Brune, 1970. Regional and global fault slip rates from seismicity. Science, 229, pp. 101-107.

Davis, G. A., 1965. Role of fluid pressure in mechanics of overthrust faulting: discussion. Geol.

Soc. Am. Bull., 76, pp. 463-468.

Davis, G. A., and B. C. Burchfiel, 1973. Garlock fault: an inter-continental transform structure, Southern California. Geol. Soc. Am. Bull., 84, pp. 1407-1422.

DeBeaumont, E., 1852. Notice sur les systemes de montagnes. P. Bertrand, Paris, 3 vols.

De Caprariis, P., 1974. Stress-induced viscosity changes and the existence of dominant wavelengths in folds. Tectonophysics, 23, pp. 139-148.

Decker, R. W., and P. Einarsson, 1971. Rifting in Iceland. Trans. Am. Geophysical Union, 52 (4), p. 352.

DeJong, K. A., and R. Scholten (eds.), 1973. Gravity and tectonics. Wiley, New York, 502pp.

Den Hartog, J. P., 1948. Mechanics. McGraw-Hill, New York, 462pp. (Reprinted in 1961 by Dover Publications, New York.)

Dennis, J. G., 1967. International tectonic dictionary. Am. Assoc. Petrol. Geol.. Tulsa, 196pp.

Dennis, J. G., 1972. Structural geology. Ronald Press Co., New York, 532pp.

de Sitter, L. U., 1939. The principle of concentric folding and the dependence of tectonic structure on original sedimentary structure. Proc. Kon. Akad. Wetensch., Amsterdam, 42 (5), pp. 412-430.

de Sitter, L. U., 1956. Structural geology. McGraw-Hill, London, 552pp.

de Sitter, L. U., 1964. Structural geology, 2nd ed., McGraw-Hill, New York, 551pp.

Devore, G. W., 1969. Preferred mineral distributions of polymineralic rocks related to non-hydrostatic stresses as expressions of mechanical equilibria. J. Geol., 77, pp. 26-38.

Dewey, J. F., 1972. Plate tectonics. Sci. Am., 226 (5), pp. 56-72.

Dewey, J. F., and J. M. Bird, 1970a. Mountain belts and the new global tectonics. J. Geophysical Res., 75 (14), pp. 2625-2647.

Dewey, J. F., and J. M. Bird, 1970b. Plate tectonics and geosynclines. Tectonophysics, 10, pp. 625-638.

Dewey, J. F., and J. McManus, 1964. Superposed folding in the Silurian rocks of Co. Mayo, Eire, Geol. J. (Manch. Liver.), 4 (1), pp. 61-75.

Dewey, J. F., W. C. Pitman, III, W. B. F. Ryan, and J. Bonnin, 1973. Plate tectonics and the evolution of the Alpine system. Geol. Soc. Am. Bull., 84, pp. 3137-3180.

Dicke, R. H., 1962. The earth and cosmology. Science, 138, pp. 653-664.

Dickinson, W. R., 1966. Structural relationships of San Andreas Fault System, Cholame Valley and Castle Mountain Range, California. Geol. Soc. Am. Bull., 77, pp. 707-726.

Dickinson, W. R., 1970. Relations of andesites, granites, and derivative sandstones to arc-trench tectonics. Geophys. and Space Phys. Rev., 8 (4), pp. 813-860.

Dickinson, W. R., 1971. Plate tectonics in geologic history. Science, 174 (4005), pp. 107-113.

Dickinson, W. R., 1972. Evidence for plate-tectonic regimes in the rock record. Am. J. Sci., 272, pp. 551–576.

Dickinson, W. R., 1973. Width of modern arc-trench gaps proportional to past duration of igneous activity in associated magmatic arcs. J. Geophysical Res., 78 (17), pp. 3376-3389.

Dickinson, W. R., D. S. Cowen, and R. A. Schweickert, 1972. Test of new global tectonics: discussion. Am. Assoc. Petrol. Geol. Bull., 56 (2), pp. 375-384.

Dickinson, W. R., and A. Grantz (eds.), 1968. Proceedings of a conference on geologic problems of the San Andreas fault system. Stanford Univ. Publ., Geol. Sci., 11, 376pp.

Dieterich, J. H., 1969. Origin of cleavage in folded rocks. Am. J. Sci., 267, pp. 155-165.

Dieterich, J. H., 1970. Computer experiments on mechanics of finite amplitude folds. Canadian J. Earth Sci., 7, pp. 467-476.

Dieterich, J. H., and N. L. Carter, 1969. Stress-history of folding. Am. J. Sci., 267, pp. 129-154.

Dietz, R. S., 1961a. Continent and ocean basin evolution by spreading of the sea floor. Nature, 190 (4779), pp. 854-857.

Dietz, R. S., 1961b. Vredefort ring-structure: Meteorite impact scar. J. Geol., 69, pp. 499-516.

Dietz. R. S., 1963. Collapsing continental rises: an actualistic concept of geosynclines and mountain building. J. Geol., 71 (3), pp. 314-333.

Dietz, R. S. 1972. Geosynclines, mountains and continent-building, Scientif. Am, 226, pp. 30-38.

Dietz, R. S., and J. C. Holden, 1966. Miogeoclines in space and time. J. Geol. 74, pp. 566-583.

Dietz, R. S., and J. C. Holden, 1970. Reconstruction of Pangaea: breakup and dispersion of continents, Permian to present. J. Geophysical Res., 75, pp. 4939-4955.

Dietz, R. S., and J. C. Holden 1974. Collapsing continental rises: actualistic concept of geosynclines—a review pp. 14-25. In R. H. Dott and R. H. Shaver (eds.), Modern and Ancient Geosynclincal Sedimentation: Soc. Econ. Paleontologists and Mineralogists, Spec. Pub. 19.

Dillamore I. L., E. Butler and D. Green, 1968. Crystal rotations under conditions of imposed strain and the influence of twinning and cross-slip. Metal Sci. J., 2, pp. 161-167.

Doll, C. G., W. M. Cady, J. B. Thompson, and M. P. Billings, 1961. Centennial geologic map of Vermont. State of Vermont, United States of America.

Donath, F. A., 1961. Experimental study of shear failure in anistropic rocks. Geol. Soc. Am. Bull., 72, pp. 985-990.

Donath, F. A., and R. B. Parker, 1964. Folds and folding. Geol. Soc. Am. Bull., 75, pp. 45-62.

Dott, R. H., 1974. The geosynclinal concept: pp. 1-13 in R. H. Dott and R. H. Shaver (eds.), Modern and ancient geosynclinal sedimentation. Soc. Econ. Paleontologists and Mineralogists, Spec. Pub. 19.

Drake, C. L., M. Ewing, and J. Sutton, 1959. Continental margins and geosynclines; the east coast of North America, north of Cape Hatteras: pp. 110-198 in Physics and Chemistry of the Earth, 5, L. H. Ahrens, K. Rankama, and S. K. Runcorn (eds.). Pergamon Press, London.

Dunn, D. E., L. J. LaFountain and R. E. Jackson, 1973. Porosity dependence and mechanism of brittle fracture in sandstones. J. Geophysical Res., 78 (14), pp. 2403-2417.

Dunnet, D. (1969). A technique of finite-strain analysis using elliptical particles. Tectonophysics, 7, pp. 117-136.

Durney, D. W., 1972. Deformation history of the Western Helvetic Nappes, Valais, Switzerland. Ph. D. thesis, Univ. of London, London, 251pp.

Durney, D. W., and J. G. Ramsay, 1973. Incremental strains measured by syntectonic crystal growths: pp. 67-96 in Gravity and Tectonics, K. A. DeJong and R. Scholten (eds.). Wiley, New York, 502pp.

DŽUŁYŃSKI, S., and E. K. Walton, 1965. Sedimentary features of flysch and greywackes. Developments in sedimentology, 7, Elsevier Pub. Co., Amsterdam, 274pp.

Edmond, J. M., and M. S. Paterson, 1972. Volume changes during the deformation of rocks at high pressures. Int. J. Rock Mech. Min. Sci., 9, pp. 161-182.

Egyed, L., 1957. A new dynamic conception of the internal constitution of the earth. Geol. Rundschau, 46, pp. 101-121.

Eisbacher, G. H., 1970. Deformation mechanics of mylonitic rocks and fractured granites in Cobequid Mountains, Nova Scotia, Canada. Geol. Soc. Am. Bull., 81, pp. 2009-2020.

Elliott, D., 1970. Determination of finite strain and initial shape from deformed elliptical objects. Geol. Soc. Am. Bull., 81, pp. 2221-2236.

Elliott, D., 1972. Deformation paths in structural geology. Geol. Soc. Am. Bull., 83, pp. 2621-2638.

Elliott, D., 1973. Diffusion flow laws in metamorphic rocks. Geol. Soc. Am. Bull., 84, pp. 2645-2664.

Elter, P., and L. Trevisan, 1973. Olistostromes in the tectonic evolution of the northern Apennines: pp. 175-188 in Gravity and Tectonics, K. A. DeJong and R. Scholten (eds.). Wiley, New York, 502pp.

Engel, A. E. J., C. G. Engel, and R. G. Havens, 1965. Chemical characteristics of oceanic basalts and upper mantle. Geol. Soc. Am. Bull., 76, pp. 719-734.

Engelder, J. T., 1973. The influence of quartz fault-gouge on sliding mode and stick-slip stress drops. Trans. Am. Geophysical Union, 54 (4), p. 465.

Ernst, W. G., 1970. Tectonic contact between the Franciscan mélange and the Great Valley sequence—crustal expression of a late Mesozoic Benioff zone. J. Geophysical Res., 75 (5), pp. 886-901.

Ernst, W. G., Y. Seki, H. Onuki, and M. C. Gilbert, 1970. Comparative study of low-grade metamorphism in the California Coast Ranges and the outer Metamorphic Belt of Japan, Geol. Soc. Am. Memoir 124, 276pp.

Eskola, P. E., 1949. The problem of mantled gneiss domes. Geol. Soc. London, Quart. J., 104, pp. 461-476.

Etheridge, M. A., 1971. Experimental investigations of the mechanisms of mica preferred orientation in foliated rocks. Unpublished Ph. D. thesis, Australian Nat. Univ., Canberra.

Etheridge, M. A., 1973. Variation of the length/thickness ratio of biotite crystals of metamorphic rocks: a discussion. J. Geol., 81 (2), pp. 234-236.

Etheridge, M. A., and B. E. Hobbs, 1974. Chemical and deformational controls on recrystallization of mica. Contr. Miner. Petrol., 43, pp. 111-124.

Etheridge, M. A., B. E. Hobbs, and M. S. Paterson, 1973. Experimental deformation of single crystals of biotite. Contr. Miner. Petrol., 38, pp. 21-36.

Etheridge, M. A., M. S. Paterson, and B. E. Hobbs, 1974. Experimentally produced preferred orientation in synthetic mica aggregates. Contr. Miner. Petrol., 44, pp. 275-294.

Evans, D. M., 1966. The Denver area earthquakes and the Rocky Mountain Arsenal disposal well. The Mountain Geologist, 3 (1), pp. 23-36.

Ewing, M., and J. Ewing, 1970. Seismic reflection: pp. 1-52 in The Sea, A. E. Maxwell (ed.), 4, pt. I. Wiley-Interscience, New York.

Fairbairn, H. W., 1937. Structural petrology. Queens Univ., Kingston.

Fairbairn, H. W., 1939. Hypotheses of quartz orientation in tectonites. Geol. Soc. Am. Bull., 50, pp. 1475-1492.

Fairbairn, H. W., 1950. Pressure shadows and relative movements in a shear zone. Trans. Am. Geophysical Union, 31 (6), pp. 914-916.

Felkel, E., 1929. Gefugestudien in Kalktektonite. Geol. Jahrb. Bundesanst. Wien., 79, pp. 33-86.

Ferreira, M. P., and F. J. Turner, 1964. Microscopic structure and fabric of Yule Marble experimentally deformed at different strain rates. J. Geol., 72, pp. 861-875.

Fischer, A. G., B. C. Heezen, R. E. Boyce, D. Bukry, R. G. Douglas, R. E. Garrison, S. A. Kling, V. Krasheninnikov, A. P. Lisitzin, and A. C. Pimm, 1970. Geological history of the western North Pacific. Science, 168, pp. 1210-1214.

Fitch, T. J., 1972. Plate convergence, transcurrent faults, and internal deformation adjacent to southeast Asia and the western Pacific. J. Geophysical Res., 77 (23), pp. 4432-4460.

Fletcher, R. C., 1972. Application of a mathematical model to the emplacement of mantled gneiss domes. Am. J. Sci., 272, pp. 197-216.

Fleuty, M. J., 1964. The description of folds. Geol. Assoc. Proc., 75, Pt. 4, pp. 461-489.

Flinn, D., 1956. On the deformation of the Funzie Conglomerate, Fetlar, Shetland. J. Geol., 64, pp. 480-505.

Flinn, D., 1962. On folding during three-dimensional progressive deformation. Geol. Soc. London, Quart. J., 118, pp. 385-433.

Flinn, D., 1965. Deformation in metamorphism: pp. 46-72 in Controls of Metamorphism, W. S. Pitcher and G. W. Flinn (eds.). Oliver & Boyd, Edinburgh, 368pp.

Forman, D. J., 1971. The Arltunga Nappe Complex MacDonnell Ranges, Northern Territory, Australia. Geol. Soc. Australia J., 18, pp. 173-182.

Forristall, G. Z., 1972. Stress distributions and overthrust faulting. Geol. Soc. Am. Bull., 83, pp. 3073-3082.

Francis, P. W., 1972. The pseudotachylyte problem. Comments on Earth Science Geophys., 3, pp. 35-53.

Freedman, J., D. U. Wise, and R. D. Bentley, 1964. Pattern of folded folds in the Appalachian Piedmont along Susquehanna River. Geol. Soc. Am. Bull., 75, pp. 621-638.

Friedman, M., 1972. Residual elastic strain in rocks. Tectonophysics, 15, pp. 297-330.

Friedman, M., and J. M. Logan, 1970. Microscopic feather fractures. Geol. Soc. Am. Bull., 81, pp. 3417-3420.

Friedman, M., R. D. Perkins, and S. J. Green, 1970. Observation of brittle-deformation features at the maximum stress of Westerly granite and Solenhofen limestone. Int. J. Rock Mech. Min. Sci., 7, pp. 297-306.

Fyson, W. K., 1964. Repeated trends of folds and crossfolds in Paleozoic rocks, Parrsboro, Nova Scotia. Canadian J. Earth Sci., 1, pp. 167-183.

Gansser, A., 1964. Geology of the Himalayas. Wiley-Interscience, New York, 289pp.

Garg, S. K., and A. Nur, 1973. Effective stress laws for fluid-saturated porous rocks. J. Geophysical Res., 78 (26), pp. 5911-5921.

Gay, N. C., 1968a. The motion of rigid particles embedded in a viscous fluid during pure shear deformation of the fluid. Tectonophysics, 5 (2), pp. 81-88.

Gay, N. C., 1968b. Pure shear and simple shear deformation of inhomogeneous viscous fluids. 1. Theory. Tectonophysics, 5 (3), pp. 211-234.

Gay, N. C., 1968c. Pure shear and simple shear deformation of inhomogeneous viscous fluids. 2. The determination of the total finite strain in a rock from objects such as deformed pebbles. Tectonophysics, 5 (4), pp. 295-302.

Gay, N. C., 1969. The analysis of strain in the Barberton Mountain Land, Eastern Transvaal, using deformed pebbles. J. Geol., 77, pp. 377-396.

Gay, N. C, 1970. The formation of step structures on slickensided shear surfaces. J. Geol., 78, pp. 523-532.

George, T. N., 1965. The geological growth of Scotland. In The geology of Scotland, G. Y. Craig (ed.). Oliver and Boyd, Edinburgh, 556pp.

Ghosh, S. K., 1966. Experimental tests of buckling folds in relation to strain ellipsoid in simple shear deformations. Tectonophysics 3, pp. 169-185.

Ghosh, S. K., 1968. Experiments of buckling of multilayers which permit interlayer gliding. Tectonophysics, 6, pp. 207-250.

Gill, J. E., 1971. Continued confusion in the classification of faults. Geol. Soc. Am. Bull., 82, pp. 1389-1392.

Gillespie, P., A. C. McLaren, and J. N. Boland, 1971. Operating characteristics of an ion-bombardment apparatus for thinning non-metals for transmission electron microscopy. J. Materials Sci., 6, pp. 87-89.

Gilluly, J., 1963. The tectonic evolution of the western United States. Geol. Soc. London, Quart. J., 119, pp. 133-174.

Gilluly, J., 1972. Tectonics involved in the evolution of mountain ranges: pp. 406-435 in The nature of the solid earth, E. C. Robertson (ed.). McGraw-Hill New York, 677pp.

Gilluly, J., 1973. Steady plate motion and episodic orogeny and magmatism. Geol. Soc. Am. Bull., 84, pp. 499-514.

Goldsmith, H. L., and S. G. Mason, 1967. The micro-rheology of dispersions In Rheology Theory and Applications, F. R. Eirich (ed.), 4. Academic Press, New York and London.

Gordon, R. B., and C. W. Nelson, 1966. Anelastic properties of the earth. Geophysical Rev., 4, pp. 457-474.

Gough, D. I., 1969. Incremental stress under a two-dimensional artificial lake. Canadian J. Earth Sci., 6, pp. 1067-1075.

Gramberg, J., 1965. Axial Clevage Fracturing, a significant process in mining and geology. Engineering Geol., I pp. 31-72.

Green, H. W., 1970. Diffusional flow in polycrystalline materials. J. Appl. Phys., 41, pp. 3899-3902.

Green, H. W., 1968. Syntectonic and annealing recrystallization of fine-grained quartz aggregates. Unpublished Ph. D. Thesis, Univ. of Calif., Los Angeles, 203pp.

Green, H. W., 1967. Extreme preferred orientation produced in quartz by annealing: an analog of "cube texture." Science, 157 (3795), pp. 1444-1447.

Green, H. W., D. T. Griggs, and J. M. Christie, 1970. Syntectonic and annealing recrystallization of fine-grained quartz aggregates: pp.272-335 in Experimental and Natural Rock Deformation, P. Paulitsch (ed.). Springer-Verlag, Berlin-Heidelberg.

Greenly, E., 1919. The geology of Anglesey. Geol. Survey England and Wales Mem., 1, 388pp.

Greenly, E., 1930. Foliation and its relations to folding in the Mono Complex at Rhoscolyn (Anglesey). Geol. Soc. London, Quart. J., 86, pp. 169-190.

Gregory, J. W., 1921. The rift valleys and geology of East Africa, Seeley Service and Co., London, 479pp.

Griffith, A. A., 1924. Theory of rupture. Proc. 1st Int. Cong. Applied Mech., Delft, pp. 55-63.

Griggs, D. T., 1939. A theory of mountain building. Am. J. Sci., 237 (9), pp. 611-650.

Griggs, D. T., 1940. Experimental flow of rocks under conditions favoring recrystallization. Geol. Soc. Am. Bull., 51, pp. 1001-1022.

Griggs, D. T., 1967. Hydrolytic weakening of quartz and other silicates. Geophys. J. Roy. Astron. Soc., 14, pp. 19-31.

Griggs, D. T., 1972. The sinking lithosphere and the focal mechanism of deep earthquakes: pp. 361-384 in The Nature of the Solid Earth, E. C. Robertson (ed.). McGraw-Hill, New York 677pp.

Griggs, D. T., and J. D. Blacic, 1965. Quartz: anomalous weakness of synthetic crystals. Science, 147, pp. 292-295.

Griggs, D. T. and J. Handin, 1960. Observations on fracture and a hypothesis of earthquakes: pp. 347-373 in D. T. Griggs and J. Handin (eds.), Rock Deformation., Geol. Soc. Am., Memoir, 79, 382pp.

Griggs, D. T., M. S. Paterson, H. C. Heard, and F. J. Turner, 1960. Annealing recrystallization in calcite crystals and aggregates: pp. 21-37 in Rock Deformation, D. T. Griggs and J. Handin (eds.) Geol. Soc. Am. Memoir, 79, 382pp.

Griggs, D. T., F. J. Turner, and H. C. Heard, 1960. Deformation of rocks at 500° to 800°C: pp. 39-104 in Rock Deformation, Griggs and Handin (eds.). Geol. Soc. Am. Memoir, 79, 382pp.

Grim, R. E., 1962. Applied clay mineralogy. McGraw-Hill, New York, 422pp.

Grindley, G. W., 1963. Structure of the Alpine schists of South Westland, Southern Westland, Southern Alps. New Zealand. New Zealand J. Geol. and Geophys., 6 (5), pp. 872-930.

Groves G. W., and A. Kelly, 1969. Change of shape due to dislocation climb. Phil. Mag., 19, pp. 977–986.

Gutenberg, B., and C. F. Richter, 1951. Evidence from deep-focus earthquakes: pp. 305-313 in Internal Constitution of the Earth, 2nd ed., G. Gutenberg (ed.). Dover Publications, 439pp.

Gwinn, V. E., 1964. Thin-skinned tectonics in the plateau and northwestern Valley and Ridge provinces of the Central Appalachians. Geol. Soc. Am. Bull., 75, pp. 863-900.

Haarman, H., 1930. Die Oszillationstheorie Eine Erhlarung der Krustenbewegungen von Erde und Mond. Ferdinand Enke Verlag, Stuttgart, 260pp.

Hall, A. L., and G. A. F. Molengraaf, 1925. The Vredefort Mountain Land in the southern Transvaal and the northern Orange Free State. Verh. K. on Akad. Wetensch, Amsterdam, 24 (3), 183pp.

Hall, W. D. M., 1972. The structural history and metamorphic history of the Lower Pennine Nappes, Valle di Bosco, Ticino, Switzerland. Unpublished Ph. D. Thesis, Imperial College, London, 220pp.

Haller, J., 1956. Der zentrale metamorphe Komplex von NE Grønland Teil I, Mededel. om Grønland, 73.

Haller, J., 1962. Structural control of regional metamorphism in the east Greenland Caledonides. Geol Soc. London Proc. No 1594, pp. 21-25.

Haller, J., 1971. Geology of the east Greenland Caledonides. Wiley- Interscience, New York, 413pp.

Hamblin, W. K., 1965. Origin of "reverse drag" on the downthrown side of normal faults. Geol. Soc. Am. Bull., 76, pp. 1145-1164.

Hamilton, W., 1969. Mesozoic California and the underflow of Pacific mantle. Geol. Soc. Am. Bull., 80, pp. 2409-2429.

Handin, J., 1966. Strength and ductility: pp.223-289 in Handbook of Physical Constants, S. P. Clark (ed.). Geol. Soc. Am. Memoir, 97, 223pp.

Handin, J., 1968. Experimental evidence for the effects of pore water pressure on the strength and ductility of rocks: pp. 285-316 in NSF advanced science seminar in rock mechanics for college teachers of structural geology, R. E. Riecker (ed.). Terrestrial Sciences Lab., Air Force Cambridge Research Laboratories, Bedford, Mass., 594pp.

Handin, J., 1969. On the Coulomb-Mohr failure criterion. J. Geophysical Res., 74 (22) pp. 5343-5348.

Handin, J., R. V. Hager, Jr., M. Friedman, and J. N. Feather, 1963. Experimental deformation of sedimentary rocks under confining pressure: pore pressure tests. Am. Assoc. Petrol. Geol. Bull., 47 (5), pp. 718-755.

Handin, J., M. Friedman, J. M. Logan, L. J. Pattison, and H. S. Swolfs, 1972. Experimental folding of rocks under confining pressure: buckling of single-layer rock beams: pp. 1-28 in flow and Fracture of rocks, H. C. Heard, I. Y. Borg, N. L. Carter, and C. B. Raleigh (eds.). Am. Geophysical Union, Geophys. Monograph 16, 352pp.

Hansen, E., 1971. Strain facies. Springer-Verlag, New York, 207pp.

Hanshaw, B. B., and J. D., Bredehoeft 1968. On the maintenance of anomalous fluid pressures: II. Source layer at depth. Geol. Soc. Am. Bull., 79, pp. 1107-1122.

Hara, I., 1971. An ultimate steady-state pattern of c-axis fabric of quartz in metamorphic tectonites. Geol. Rundschau, 60 (3), pp. 1142-1173.

Hara, I., K. Takeda, T. Kimura, 1973. Preferred lattice orientation of quartz in shear deformation. Hiroshima Univ., J. Sci., Ser. C, 7 (1), pp. 1-10.

Harker, A., 1932. Metamorphism. Methuen & Co., London, 362pp.

Harper, C. T., 1967. The geological interpretation of potassium argon ages of metamorphic rocks from the Scottish Caledonides. Scot. J. Geol., 3 (1), pp. 46-66.

Hartman, P., and E. den Tex, 1964. Piezocrystalline fabrics of olivine in theory and nature. pp. 54-114 in Rock Deformation and Tectonics. Int. Geol. Cong., 22nd, India, 1964. Proc. Sec. 4.

Hayes, D. E., and M. Ewing, 1968. Pacific boundary structure: pp. 29-72 in The Sea, 4, pt. 2, A. E. Maxwell (ed.). Wiley-Interscience, New York, 664pp.

Healy, J. H., W. W. Rubey, D. T. Griggs, and C. B. Raleigh, 1968. The Denver Earthquakes. Science, 161, pp. 1301-1310.

Heard, H. C., 1960. Transition from brittle fracture to ductile flow in Solenhofen limestome as a function of temperature confining pressure, and interstitial fluid pressure: pp. 193–226 in D. T. Griggs and J. Handin (eds.). Rock Deformation Geol. Soc. Am. Memoir, 79, 382pp.

Heard, H. C., 1963. Effect of large changes in strain rate in the experimental deformation of Yule marble. J. Geol., 71, pp. 162-195.

Heard, H. C., 1968. Experimental deformation of rocks and the problem of extrapolation to nature; pp. 439-508 in NSF advanced science seminar in rock mechanics for college teachers of structural geology, 1967, R. E. Riecker (ed.) Terrestrial Science Lab, Air Force Cambridge Research Laboratories, Bedford, Mass., 594pp.

Heard, H. C., 1972. Steady-state flow in polycrystalline halite at pressures of 2 kilobars: pp. 191-210 in flow and Fracture of rocks. H. C. Heard, I. Y. Borg, N. L. Carter, and C. B. Raleigh (eds.). Am. Geophysical Union, Geophys. Monograph 16, 352pp.

Heard, H. C., and N. L. Carter, 1968. Experimentally induced "natural" intergranular flow in quartz and quartzite. Am. J. Sci., 266, pp. 1-42.

Heard, H. C., and C. B. Raleigh, 1972. Steady-state flow in marble at 500° to 800°C. Geol. Soc. Am. Bull., 83, pp. 935-956.

Heard, H. C., and W. W. Rubey, 1966. Tectonic implications of gypsum dehydration. Geol. Soc. Am. Bull., 77, pp. 741-760.

Heezen, B. C., 1960. The rift in the ocean floor. Sci. Am., 203, pp. 98-110.

Heezen, B. C., 1969. The world rift system: an introduction to the symposium. Tectonophysics, 8, pp. 268-279.

Heezen, B. C., and M. Ewing, 1963. Mid-oceanic ridge: pp. 388-410 in The Sea, III, M. N. Hill (ed.),. Wiley, New York, 963pp.

Heier, K. S., and J. M. Rhodes, 1966. Thorium, uranium and potassium concentrations in granites and gneisses of the Rum Jungle complex, Northern Territory, Australia. Econ. Geology, 61, pp. 563-571.

Heim, A., 1878. Untersuchungen über den Mechanismus der Gebirgsbildung, B. Schwabe, Basel, 246pp.

Heim, A., 1900. Gneissfaltenlung in alpinem Central Massive: Vierteljahrscrift der Naturf. Gesellsch. in Zurich. Jahrg., 45.

Heim, A., 1919. Geologie der Schweiz. C. H. Tauchnitz, Leipsig.

Heim, A., 1932. Bergsturz und Menschenleben. Fretz und Wasmuth Verlag, 218pp.

Heirtzler, J. R., 1968. Sea-floor spreading. Scientif. Am., 219 (6), pp. 60-70.

Heirtzler, J. R., G. O. Dickson, E. M. Herron, W. C. Pitman, and X. Le Pichon, 1968. Marine magnetic anomalies, geomagnetic field reversals, and motions of the ocean floor and continents. J. Geophysical Res., 73 (6), pp. 2119-2136.

Helm, D. G., and A. W. B. Siddans, 1971. Deformation of a slaty, lapillar tuff in the English Lake District: discussion. Geol. Soc. Am. Bull., 82 (2), pp. 523-531.

Heanley, K. J., 1970. The structural and metamorphic history of the Sulitjelma region, Norway, with special reference to the nappe hypothesis. Norsk Geol. Tidsk., 50 (2).

Heritsch, F., 1929. The Nappe Theory in the Alps. Methuen and Co., London, 228pp.

Hess, H. H., 1962. History of ocean basins: pp. 599-620 in Petrologic studies: a volume in honor of A. F. Buddington, A. E. J. Engel, H. L. James, and B. F. Leonard (eds.). Geol. Soc. Am. Bull.

Heuer, A. H., R. F. Firestone, J. D. Snow, H. W. Green, R. G. Howe, and J. M., Christie 1971. An improved ion-thinning device. Sci. Instrum. Rev., 42, pp. 1177-1184.

Hietanen, A., 1938. On the petrology of the Finnish quartzites. Comm. Geol. Finlande Bull., 122.

Higgins, M. W., G. W. Fisher, I. Zietz, 1973. Aeromagnetic discovery of a Baltimore Gneiss Dome in the Piedmont of northwestern Delaware and southeastern Pennsylvania. Geology, 1, pp. 41-43.

Hill, M. L., and T. W. Dibblee, Jr. 1953. San Andreas, Garlock and Big Pine faults, California. Geol. Soc. Am. Bull., 64, pp. 443-458.

Hills, E. S., 1953. Outlines of Structural Geology. Methuen and Co., London, 182pp.

Hills, E. S., 1963. Elements of Structural Geology, 1st ed. Methuen and Co., London, 483pp.

Hills, E. S., 1972. Elements of Structural Geology, 2nd ed., Wiley, New York, 502pp.

Hirth, J. P., and J. Lothe, 1968. Theory of dislocations. McGraw-Hill, New York, 780pp.

Hobbs, B. E., 1965. Structural analysis of the rocks between the Wyangala Batholith and the Copperhannia thrust, New South Wales, Geol. Soc. Australia J., 12, pp. 1-24.

Hobbs, B. E., 1966a. Microfabric of tectonites from the Wyangala Dam area, New South Wales, Australia. Geol. Soc. Am. Bull., 77, pp. 685-706.

Hobbs, B. E., 1966b. The structural environment of the northern part of the Broken Hill orebody. Geol. Soc. Australia J., 13, pp. 315-338.

Hobbs, B. E., 1968. Recrystallization of single crystals of quartz. Tectonophysics, 6, pp. 353-401.

Hobbs, B. E., 1971. The analysis of strain in folded layers. Tectonophysics, 11, pp. 329-375.

Hobbs, B. E., and T. P. Hopwood, 1969. Structural features of the Central Tablelands: pp. 200-216 in The geology of New South Wales, G. H. Packham (ed.). Geol. Soc. Australia J., 16, 654pp.

Hobbs, B. E., A. C. McLaren, and M. S. Paterson, 1972. Plasticity of single crystals of synthetic quartz pp. 29-53 in Flow and Fracture of Rocks, H. C. Heard, I. Y. Borg, N. I., Carter, and C. B. Raleigh (eds.). Am. Geophysical Union, Geophys. Monograph Ser., 16,.

Hobbs, B. E., and J. L. Talbot 1966. The analysis of strain in deformed rocks. J. Geol. 74, pp. 500-513.

Hobbs, B. E., and J. L. Talbot, 1968. The relationship of metamorphic differentiation to other structural features at three localities. J. Geol., 76, pp. 581-587.

Hodgson, R. A., 1961. Classification of structures on joint surfaces. Am. J. Sci., 259, pp. 493-502.

Hoeppener, R., 1956. Zum problem der Bruchbildung, Schieferung und Faltung. Geol. Rundschau. 45, pp. 247-283.

Hoek, E., 1965. Rock fracture under static stress conditions. Nat. Mech. Eng. Res. Inst., Council for Science and Industrial Research, Pretoria, C.S.I.R. Rept. MEG, 383, pp. 70-75.

Hoffman, P., J. F. Dewey and K. Burke 1974. Aulocogens and their genetic relation to geosynclines, with a Proterozoic example from Great Slave Lake, Canada: pp. 38-55 in Modern and Ancient geosynclinal Sedimentation, R. H. Dott and R. H. Shaver (eds.). Soc. Econ. Paleontologists and Mineralogists, Spec. Pub. 19.

Holcombe, R. J., 1973. Mesoscopic and microscopic analysis of deformation and metamorphism near Ducktown, Tennessee. Ph. D. thesis, Stanford Univ., California, 225pp.

Holmes, A., 1928. The Nomenclature of Petrology, 2nd ed., Thos. Murby and Co., London, 284pp.

Holmes, A., 1931. Radioactivity and earth movements. Trans. Geol. Soc. Glasgow, 18, pp. 559-606.

Holmes, A., 1944. Principles of physical geology. Thomas Nelson and Sons, London, 532pp.

Holmes, A., and D. L Reynolds, 1954. The superpositon of Caledonian folds on older fold systems in the Dalradians of Malin Head, Co. Donegal. Geol. Mag., 91, pp. 417-444.

Honda, H., 1962. Earthquake mechanisms and seismic waves. Tokyo Univ., Geophys. Notes Supp., 15, pp. 1-97.

Hooper, P. R., 1968. The "a" lineation and the trend of the Caledonides of northern Norway Norsk Geologisk Tidsskrift, 48, pp. 261-268.

Hopwood, T. P., 1968. Derivation of a coefficient of degree of preferred orientation from contoured fabric diagrams. Geol. Soc. Am. Bull., 79, pp. 1651-1654.

Hosford, W. F., 1965. Axially symmetric flow of single crystals. Trans. Met. Soc., Am. Inst. Mech. Eng., 233. pp. 329-333.

Hosford, W. F., 1966. Plane-strain compression of aluminum crystals. Acta Met., 14, pp. 1085-1094.

Hossack, J. R., 1968. Pebble deformation and thrusting in the Bygdin area (Southern Norway). Tectonophysics, 5, pp. 315-339.

Hsu, K. J., 1955. Granulites and mylonites of the region about Cucamonga and San Antonio Canyons, San Gabriel Mountains, California. Univ. of Calif. Publs. Geol. Sci., 30, pp. 223-352.

Hsu, K. J., 1969. Role of cohesive strength in the mechanics of overthrust faulting and of landsliding. Geol. Soc. Am. Bull., 80, pp. 927-952.

Hsu, K. J., 1971. Franciscan melanges as a model for eugeosynclinal sedimentation and underthrusting tectonics. Jour. Geophys. Res., 76 (5), pp. 1162-1170.

Hsu, K. J., 1972. The concept of the geosyncline, yesterday and today. Leicester Lit. and Philos. Soc. Trans., 66, pp. 26-48.

Hsu, K. J., 1974. Mélanges and their distinction from olistostromes: pp. 321-333 in Modern and Ancient geosynclinal sedimentation, R. H. Dott and R. H. Shaver (eds.), Soc. Econ. Paleontologists and Mineralogists, Spec. Pub. 19.

Hsu, K. J., and R. Ohrbom, 1969. Mélanges of San Francisco Penninsula: Geologic reinterpretation of type Franciscan. Am. Assoc. Petrol. Geol. Bull., 53, pp. 1348-1367.

Hsu, T. C., 1966. The characteristics of coaxial and non-coaxial strain paths. J. Strain Anal; 1, pp. 216-222.

Hu, H., 1963. Annealing of silicon-iron single crystals. In L. Himmel (ed.), Recovery and recrystallization of metals. Wiley-Interscience, New York, pp. 311-362.

Hubbert, M. K. and W. W. Rubey, 1959. Role of fluid pressure in mechanics of overthrust faulting. Geol. Soc. Amer. Bull., 70, pp. 115-206.

Hudleston, P. J., 1973a. An analysis of "single-layer" folds developed experimentally in viscous media. Tectonophysics, 16, pp. 189-214.

Hudleston, P. J., 1973b. The analysis and interpretation of minor folds developed in the Moine Rocks of Monar, Scotland. Tectonophysics, 17, pp. 89-132.

Hudleston, P. J. and O. Stephansson, 1973. Layer shortening and fold-shape development in the buckling of single layers. Tectonophysics, 17, pp. 299-321.

Hurley, P. M. and J. R. Rand, 1969. Pre-drift continental nuclei. Science, 164 (3885), pp. 1229–1242.

Ilhan, E., 1971. Earthquakes in Turkey: pp.431-442 in Geology and History of Turkey, A. S. Campbell (ed.). Petrol. Exploration Soc., Libya, 511 pp.

Illies, J. H., 1970. Graben tectonics as related to crust-mantle interaction: pp. 4-27 in Graben Problems, J. H, Illies and St. Mueller (eds.). Intl. Upper Mantle Proj., Sci. Report 27, E. Schweitzerbart'sche Verlagsbuchandlung, Stuttgart, 316pp.

Ingerson, E., 1938. Summary of an article by Bruno Sander: "Über Zusammenhange zwischen Teilbewegung und Gefuge in Gesteinen." Rep. Comm. Structural Petrology, Bull. Nat. Res. Coun., pp. 23-31.

Irvine, T. N., 1965. Sedimentary structures in igneous intrusions with particular reference to the Duke Island Ultramafic Complex: pp. 220-232 in Primary sedimentary structures and their hydrodynamic interpretation, G. V. Moddleton (ed.). Soc. Econ. Palae. and Mineralogists, Spec. Pub. 12, 265pp.

Isacks, B. and P. Molnar, 1969. Mantle earthquake mechanisms and the sinking of the lithosphere. Nature, 223, pp. 1121-1124.

Isacks, B., and P. Molnar, 1971. Distribution of stresses in the descending lithosphere from a global survey of focal-mechanism solutions of mantle earthquakes. Geophys. and Space Phys. Rev.,9 (1), pp. 103-174.

Isacks, B., J. Oliver and L. R. Sykes, 1968. Seismology and the new global tectonics. J. Geophysical Res., 78 (18), pp. 5855-5899.

Isacks, B., L. R. Sykes, and J. Oliver, 1969. Focal mechanisms of deep and shallow earthquakes in the Tonga-Kermadec region and the tectonics of island arcs. Geol. Soc. Am. Bull., 80, pp. 1443-1470.

Jackson, E. D., E. A. Silver and G. B. Dalrymple, 1972. Hawaiian-Emperor chain and its relation to Cenozoic circumpacific tectonics. Geol. Soc. Am. Bull., 83, pp. 601-618.

Jaeger, J. C., 1969. Elasticity, fracture and flow with engineering and geological applications. Methuen & Co., London, 268pp.

Jaeger, J. C., 1970. Shear fracture of anisotropic rocks. Geol. Mag., 97 (1), pp. 65-72.

Jaeger, J. C., 1971. Friction of rocks and stability of rock slopes. Geotechnique, 21 (2), pp. 97-134.

Jaeger, J. C., and N. G. W. Cook, 1969. Fundamentals of Rock Mechanics. Methuen & Co., London, 513pp.

Jahns, R. H., 1943. Sheet structure in granites: its origin and use as a measure of glacial erosion in New England. J. Geol., 51 (2), pp. 71-98.

James, A. V. G., 1920. Factors producing columnar structure in lavas and its occurrence near Melbourne, Australia. J. Geol., 28, pp. 458-471.

Jeffrey, G. B., 1923. The motion of ellipsoidal particles immersed in a viscous fluid. Roy. Soc. London Proc., Ser. A, 102, pp. 161-177.

Johnson, A. M. 1970. Physical Processes in Geology. W. H. Freeman & Co., San Francisco, 577pp.

Johnson, M. R. W., 1957. The structural geology of the Moine thrust zone in the Coulin Forest, wester Ross. Geol. Soc. London, Quart, J., 113, pp. 241-270.

Johnson, M. R. W., 1960. The structural history of the Moine thrust zone at Lochcarron, wester Ross. Roy. Soc. Edinburgh Trans., 64, pp. 139-168.

Johnson, M. R. W., 1963. Some time relations of movement and metamorphism in the Scottish Highlands. Geologie en Mijnbouw, 42, pp. 121-142.

Johnson, M. R. W., 1967. Mylonite zones and mylonite banding. Nature, 213 (5073), pp. 246-247.

Johnson, M. R. W., 1970a. Torridonian and Moinian. In The Geology of Scotland, G. Y., Craig (ed.). Oliver and Boyd, Edinburgh, 556pp.

Johnson, M. R. W., 1970b. Dalradian. In The Geology of Scotland, G. Y. Craig (ed.). Oliver and Boyd, Edinburgh, 556pp.

Kallend, J. S., and G. J. Davies, 1972. A simulation of texture development in metals. Phil. Mag., 25, pp. 471-490.

Kamb, W. B., 1959a. Ice petrofabric observations from Blue Glacier, Washington, in relation to theory and experiment. J. Geophysical Res., 64, pp. 1891-1909.

Kamb, W. B., 1959b. Theory of preferred orientation developed by crystallization under stress. J. Geol., 67, pp. 153-170.

Kamb, W. B., 1961. The thermodynamic theory of nonhydrostatically stressed solids. J. Geophysical Res., 66, pp. 259-271.

Kanamori, H., and F., Press, 1970. How thick is the lithosphere? Nature, 226, pp. 330-331.

Karig, D. E., 1971. Origin and development of marginal basins in the western Pacific. J. Geophysical Res., 76 (11), pp. 2542-2561.

Karig, D. E., and Mammerickx, J., 1972. Tectonic framework of the New Hebrides island arc. Marine Geology, 12, pp. 187-205.

Katsumata, M., and L. R. Sykes, 1969. Seismicity and tectonics of the western Pacific: Izu-Mariana-Caroline and Ryukyu—Taiwan Regions. J. Geophysical Res., 74 (25), pp. 5923-5948.

Kawachi, Y., 1968. Large-scale overturned structure in the Sanbagawa metamorphic zone in central Shikoku, Japan. J. Geol. Soc. Japan, 74 (12), pp. 607-616.

Kay, G. M., 1951. North American Geosynclines. Geol. Soc. Am. Memoir, 48, 143pp.

Kelley, V. C., and N. J. Clinton, 1960. Fracture systems and tectonic elements of the Colorado Plateau. New Mexico Univ. Publ., Geology (6), 104pp.

Kelling, G., and E. K. Walton, 1957. Load-cast structures: the relationship to upper-surface structures and their mode of formation. Geol. Mag., 94 (6), pp. 481-490.

Kelly, A., and G. W. Groves, 1970. Crystallography and crystal defects. Longmans, London, 428pp.

Kelvin, W., and P. G. Tait, 1883. A Treatise on Natural Philosophy, 2nd ed., Cambridge. Republished, 1962, by Dover, New York, in two parts, 1035pp.

Kelvin, W., 1911. On homogeneous division of space: pp. 333-349 in Mathematical and Physical Papers of Sir William Thomson, J. Larmor (ed.). Cambridge Univ. Press, 5.

Kennedy, W. A., 1946. The Great Glen Fault. Geol. Soc. London, Quart. J., 102, pp. 41-76.

Kennedy, W. Q., 1955. The tectonics of the Morar anticline and the problem of the north-west Caledonian front. Geol. Soc. London, Quart, J., 110, pp. 357-382.

Khattri, K., 1973. Earthquake focal mechanism studies—a review. Earth Sci. Rev., 9, pp. 19-63.

Kidd, W. S. F., K. Burke, and T. J. Wilson, 1973. The present plume population. EOS, Trans. Am. Geophysical Union, 54 (4), p 238.

King, B. C., and N. Rast 1955. Tectonic styles in the Dalradians and Moines of parts of the Central Highlands of Scotland. Geol. Assoc. London Proc., 66, pp. 243-269.

King, P. B., 1950 Tectonic Framework of Southeastern United States. Bull. Am. Assoc. Petrol. Geol., 34, pp. 635–671.

Knill, J. L., 1959. The tectonic pattern in the Dalradian of the Craignish-Kilmelfort District, Argyllshire. Geol. Soc. London, Quart. J., 115, pp. 339-364.

Knill, J. L., 1960. A classification of cleavages, with special references to the Craignish district of Scottish highlands. 21st Int. Geol. Cong. Report, pt. 18, pp. 317-325.

Knopf. E. B., 1931. Retrogressive metamorphism and phyllonitization. Am. J. Sci., 221 (121), pp. 1-27.

Knopf, E. B., and E. Ingerson, 1938. Structural Petrology. Geol. Soc. Am. Memoir, 6, 270pp.

Kobe, H. W., 1966. Struktur des Gebietes zwischen Gresso und Passo della Garina, Tessin, (Mit Deutungsversuch der tectonischstructurellen Verhaltnisse des Gebietes vom Valle di Vergeletto bis zum untersten Val Verzasca.) Eclogae Geol. Helv., 59 (2), pp. 789-802.

Kocks, U. F., 1970. The relation between polycrystal deformation and single crystal deformation. Metall. Trans., 1, pp. 1121-1143.

Krumbein, W. C., 1942. Criteria for subsurface recognition of unconformities. Am. Assoc. Petrol. Geol. Bull., 26, pp. 36-62.

Krumbein, W. C., and L. L. Sloss, 1963. Stratigraphy and sedimentation, 2nd ed. W. H., Freeman and Co., San Francisco, 660pp.

Kuenen, Ph. H., 1953. Significant features of graded bedding. Am. Assoc. Petrol. Geol. Bull., 37, pp. 1044-1066.

Kühn-Velten, H., 1955. Subagiratische Rutschungen in hoheren Oberderon des Sauerlandes. Geol. Rundschau, 44, pp. 3-25.

Kumazawa, M., 1963. A fundamental thermodynamic theory on non-hydrostatic field and on the stability of mineral orientation and phase equilibrium. Nagoya Univ., J. Earth Sci., 11, pp. 145-217.

Kuno, H., 1959. Lateral variation of basalt magma types across continental margins and island arcs. Bull. Vulcan., 29, pp. 195-222.

Kupfer, D. H., 1968. Relationship of internal to external structure of salt domes: pp. 79-89 in Diapirs and diapirism, J. Braunstein and G. D. O'Brien (eds.). Am. Assoc. Petrol. Geol. Memoir, 8, Tulsa, Oklahoma, 444pp.

Kvale, A., 1948. Petrologic and structural studies in the Bergsdalen Quadrangle, Western Norway, Part II: structural geology. Bergens Mus. Arb., 1946-7. Naturv. rekke 1, pp. 1-255.

Kvåle, A., 1953. Linear structures and their relation to movement in the Caledonides of Scandinavia and Scotland. Geol. Soc. London, Quart. J., 109, pp. 51-73.

Lachenbruch, A. H., and G. A. Thompson, 1972. Oceanic ridges and transform faults: their intersection angles and resistance to plate motion. Earth Planetary Sci. Letters 15, pp. 116-122.

Landis, C. A., and D. G. Bishop 1972. Plate tectonics and regional stratigraphic-metamorphic relations in the southern part of the New Zealand Geosyncline. Geol. Soc. Am. Bull., 83 (8), pp. 2267-2284.

Landis, C. A., and D. S. Coombs, 1967. Metamorphic belts and orogenesis in southern New Zealand. Tectonophysics, 4 (4-6), pp. 501-518.

Langseth, M. G., and R. P. Von Herzen, 1970. Heat flow through the floor of the world oceans: pp. 299-352 in The Sea 4, pt. 1, A. E. Maxwell (ed.). Wiley-Interscience, New York.

Lajtai, E. Z., 1969a. Shear strength of weakness planes in rock. Int. J. Rock Mech. Min. Sci., 6, pp. 499-515.

Lajtai, E. Z., 1969b. Mechanics of second-order faults and tension gashes. Geol. Soc. Am. Bull., 80, pp. 2253-2272.

Lajtai, E. Z., 1971. A theoretical and experimental evaluation of the Griffith theory of brittle fracture. Tectonophysics, 11, pp. 129-156.

Lapré, J. F., 1965. Minor structures in the Upper Vicdessos valley (Aston Massif, France). Leidse Geol. Mededel. 33, pp. 255-274.

Lapworth, C., 1885. The highland controversy in British geology; its causes, course and consequences. Nature, 32, pp. 558–559.

Laubscher, H. P., 1972. Some overall aspects of Jura dynamics. Am. J. Sci., 272, pp. 293-304.

Leeman, E. R., 1964. The measurement of stress in rock: I. The principles of rock stress measurement; II. Borehole rock stress measuring instruments; III. The results of some rock stress investigations. J. S. Africa, Inst. Min. Metall., 65, pp. 45-114; 65, pp. 254-284.

Lees, G. M., 1952. Foreland folding. Geol. Soc. London, Quart. J., 108, pp. 1-34.

Leith, C. K., 1905. Rock cleavage. U. S. Geol. Survey Bull., 239, pp. 1-216.

Lensen, G. J., 1961. Principal horizontal stress directions as an aid to the study of crustal deformation: pp. 389-397 in A symposium on Earthquake Mechanism. Publs. of the Dominion Observatory, 14 (10).

Le Pichon, X., 1968. Sea-floor spreading and continental drift. J. Geophysical Res., 73 (12), pp. 3661-3697.

Lewis, B. R., P. S. Forward, and J. B. Roberts, 1965. Geology of the Broken Hill Lode, Reinterpreted: pp. 319-332 in Geology of Australian Ore Deposits, J. McAndrew (ed.). 8th Commonwealth Min. and Met. Cong., 1.

Lindström, M., 1958a. Tectonic transports in three small areas in the Caledonides of Swedish Lapland. Lunds Universitets Arsskrift. N. F. Ard 2, 54, Nr. 3, pp. 5-85.

Lindström, M., 1958b. Tectonic transports in the Caledonides of Northern Scandinavia East and South of the Rombak-Sjangeli Window. Publs. Inst. Min. Palaeont. Quart. Geol., 43, Lund.

Link, T. A., 1949. Interpretations of foothills structures, Alberta. Am. Assoc. Petrol. Geol. Bull., 33, pp. 1175-1501.

Lister, G. S., 1974. The theory of deformation fabrics. Unpublished Ph.D thesis — Australian National University 2 Vols. 463pp.

Lister, G. S., and Hobbs, B. E., 1974. The influence of deformation path on fabric development in quartzites (in preparation.)

Lister, G. S., and Paterson, M. S., 1974. Fabric transitions in quartzites (in preparation).

Lister, G. S., and Hobbs, B. E., 1974. Computer simulation of preferred orientation development by slip.

Lombard, A. E., 1948. Appalachian and Alpine structures—a comparative study. Am. Assoc. Petrol. Geol. Bull., 32, pp. 709-744.

Lomer, W. M., and J. F. Nye, 1952. A dynamical model of a crystal structure. IV. Grain boundaries. Roy. Soc. London, Proc., A, 212, pp. 576-584.

Lugeon, M., 1901. Les grandes nappes de recouvrement des Alpes du Chablais et de Suisse. Soc. Geol. de France Bull., 4eme. Ser. 1, pp. 723-825.

Lugeon, M., 1902. Sur la coupe geologique du massif du Simplon. C. R. Acad. Sci., Paris, 134, pp. 726-727.

Magnée, I. D., 1935. Observations sur l'origine des gisements de pyrite du sud de l'Espagne et du Portugal. Int. Geol. Cong. VII, Report, Paris, 95.

Malone, E. J., 1962. Darwin, N. T., 1:250000 Geological Series. Bur. Min. Resources, Australia, expln. notes D/52-4.

March, A, (1932). Mathematische Theorie der Regelung nach der Korngestalt bei Affiner Deformation. Z. Krist., 81, pp. 285-297.

Marjoribanks, R. W., 1974. The structural and metamorphic geology of the Ormiston Region, Central Australia. Unpubl. Ph. D. thesis, Australian Nat. Univ., Canberra, 311pp.

Martin, R. J. III, 1972. Time-dependent crack growth in quartz and its application to the creep of rocks. J. Geophysical Res., 77 (8), pp. 1406-1419.

Mayer, G., and W. A. Backofen, 1968. Constrained deformation of single crystals. Trans. metal. Soc., Am. Inst. Mech. Engr., 242, pp. 1587-1594.

Maxwell, A. E., R. P. Von Herzen, K. J. Hsü, J. E. Andrews, T. Saito, S. F. Percival, E. D. Milow, and R. E. Bovce, 1970. Deep sea drilling in the south Atlantic. Science, 168, pp. 1047-1059.

Maxwell, D. T., and J. Hower, 1967. High-grade diagenesis and low grade metamorphism of illite in the preCambrian Belt Series. Am. Min., 52, pp. 843-857.

Maxwell, J. C., 1962. Origin of slaty and fracture cleavage in the Delaware Water Gap Area, New Jersey and Pennsylvania: pp. 281-311 in Petrologic Studies. Geol. Soc. Am. (Buddington Volume).

Maxwell, J. C., 1968. Continental drift and a dynamic earth. Am. Scientist, 56, pp. 35-51.

McClintock, F. A., and A. S. Argon, 1966. Mechanical behavior of materials. Addison-Wesley, Reading, Mass., 770pp.

McClintock, F. A., and L. B. Walsh, 1962. Friction on Griffith cracks under pressure. Fourth U.S. Nat. Congress of Appl. Mech., Proc., pp. 1015-1021.

McConnell, R. B., 1972. Geological development of the rift system of East Africa. Geol. Soc. Am. Bull., 83, pp. 2549-2572.

McDougall, I., and D. H., Tarling, 1963. Dating of polarity zones in the Hawaiian Islands. Nature, 200, pp. 54-56.

McIntyre, D. B., 1951. The tectonics of the area between Grantown and Tomintoul (mid Strathsprey). Geol. Soc. London, Quart J., 107, pp. 1-22.

McIntyre, D. B., 1954. The Moine thrust: its discovery, age and tectonic significance. Geologists Assoc. Proc., 65, pp. 203-219.

McIntyre, D. B. and L. E. Weiss, 1956. Construction of block diagrams to scale in orthographic projection. Geologists Assoc., Proc., 67, pts. 1 and 2, pp. 142-155.

McKee, E. D., M. A. Reynolds, and C. H. Baker, Jr., 1962a. Laboratory studies on deformation in unconsolidated sediment. United States Geol. Surv. Professional Papers 450D, pp. 151-155.

McKee, E. D., M. A. Reynolds, and C. H. Baker, Jr., 1962b. Experiments on intraformational recumbent folds in cross-bedded sand. United States Geol. Surv. Professional Papers. 450D, 155-160.

McKenzie, D. P., 1969a. The relation between fault plane solutions for earthquakes and the directions of the principal stresses. Bull. Seis. Soc. Amer., 59 (2), pp. 591-601.

McKenzie, D. P., 1969b. Speculations on the consequences and causes of plate motions. Geophys. J. Roy. Astron. Soc., 18, pp. 1-32.

McKenzie, D. P., 1970. Plate tectonics of the Mediterranean region. Nature, 226, pp. 239-243.

McKenzie, D. P., 1972. Plate tectonics: pp. 323-360 in The Nature of the Solid Earth, E. C., Robertson (ed.) McGraw-Hill, New York, 677pp.

McKenzie, D. P., and J. N. Brune, 1972. Melting on fault planes during large earthquakes. Geophys. J. Roy. Astron. Soc., 29, pp. 65-78.

McKenzie, D. P., and W. J. Morgan, 1969. Evolution of triple junctions. Nature, 224, pp. 125-133.

McKenzie, D. P., and R. L. Parker, 1967. The north Pacific: an example of tectonics on a sphere. Nature, 216 (5122), pp. 1276-1280.

McKenzie, D. P., and J. G. Sclater, 1969. Heat flow in the eastern Pacific and sea floor spreading. Bull. Volcanol., 33, pp. 101-118.

McKinstry, H. C., 1953. Shears of second order. Am. J. Sci., 251, pp. 401-414.

McLaren, A. C., and P. P. Phakey, 1965. Dislocations in quartz observed by transmission electron microscopy. *J. Applied Phys.,* 36 pp. 3244-3246.

McLaren, A. C., J. A. Retchford, D. T. Griggs, and J. M. Christie, 1967. Transmission electron microscope study of Brazil twins and dislocations experimentally produced in natural quartz. Phys. Stat. Sol., 19, pp. 631-644.

McLaren, A. C., C. F. Osborne, and L. A. Saunders, 1971. X-ray topographic study of dislocations in synthetic quartz. Phys. Stat. Sol., (a) 4, pp. 235-247.

McLaren, A. C., R. G. Turner, J. N. Boland, and B. E. Hobbs, 1970. Dislocation structure of the deformation lamellae in synthetic quartz; a study by electron and optical microscopy. Contr. Min. Petrol., 29, pp. 104-115.

McLaren, A. C., and B. E., Hobbs, 1972. Transmission electron microscope investigation of some naturally deformed quartzites. Am. Geophysical Union, Geophys. Monograph, Ser. 16, pp. 55-66.

McLean, D., 1957. Grain Boundaries in Metals. Oxford Univ. Press, 346pp.

McLean, D., 1973. Montreal summer school on grain boundaries. Met. Sci. J., 7, pp. 211-212.

McLeish, A. J., 1971. Strain analysis of deformed Pipe Rock in the Moine Thrust zone, Northwest Scotland. Tectonophysics, 12, pp. 469-504.

Means, W. D., 1963. Mesoscopic structures and multiple deformation in the Otago schist. New Zealand J. Geol. and Geophys., 6 (5), pp. 801-816.

Means, W. D., 1966. A macroscopic recumbent fold in schist near Alexandra, Central Otago. New Zealand J. Geol. and Geophys., 9, pp. 173-194.

Means, W. D., and M. S. Paterson, 1966. Experiments on preferred orientation of platy minerals. Contr. Miner. and Petrol., 13, pp. 108-133.

Means, W. D., and P. F. Williams, 1972. Crenulation cleavage and faulting in an artificial salt-mica schist. J. Geol., 80, pp. 569-591.

Menard, H. W., 1964. Marine geology of the Pacific. McGraw-Hill, New York, 271pp.

Menard, H. W., 1967. Extension of northeastern Pacific fracture zones. Science, 155, pp. 72-74.

Menard, H. W., 1969. The deep-ocean floor. Scientif. Am., Sept. 1969, pp. 127-142.

Menard, H. W., 1973. Epeirogeny and plate tectonics. Trans. Am. Geophysical Union, 54 (12), pp. 1244-1255.

Menard, H. W., and T. M. Atwater, 1968. Changes in direction of sea floor spreading. Nature, 219, pp. 463-467.

Meservey, R., 1969. Topological inconsistency of continental drift on the present-sized earth. Science, 166, pp. 609-611.

Meyerhoff, A. A., 1970. Continental drift: implications of paleomagnetic studies, meteorology, physical oceanography, and climatology. J. Geol., 78 (1), pp. 1-50.

Meyerhoff, A. A., and H. A. Meyerhoff, 1972. The new global tectonics — major inconsistencies. Am. Assoc. Petrol. Geol. Bull., 56 (2), pp. 269-336.

Milnes, A. G., 1965. Structure and history of the Antigorio Nappe (Simplon Group, North Italy). Schweiz. Mineral. Petrog. Mitt., 45 (1), pp. 167-177.

Milnes, A. G., 1971. A model for analyzing the strain history of folded competent layers in deeper parts of orogenic belts. Eclogae Geol. Helv., 64, pp. 335-342.

Misch, P., 1970. Paracrystalline microboudinage in a metamorphic reaction sequence. Geol. Soc. Am. Bull., 81, pp. 2483-2486.

Mitchell, A. H., and H. G. Reading, 1969. Continental margins, geosynclines, and ocean floor spreading. J. Geol., 77, pp. 629-646.

Mitchell, A. H., and H. G. Reading, 1971. Evolution of island arcs. J. Geol., 79 (3), pp. 253-284.

Mitronovas, W., and B. L. Isacks, 1971. Seismic velocity anomalies in the upper mantle beneath the Tonga-Kermadec island arc. J. Geophysical Res., 76 (29), pp. 7154-7180.

Miyashiro, A., 1973. Metamorphism and Metamorphic Belts. William Clowes and Sons, Ltd., London, 492pp.

Miyashiro. A., 1974. Volcanic rock series in island arcs and active continental margins. Am. J. Sci., 274, pp. 321-355.

Moench, R. H., 1966. Relation of S_2 schistosity to metamorphosed clastic dikes, Rangeley-Phillips area, Maine. Geol. Soc. Am. Bull., 77 (12), pp. 1449-1462.

Moffatt, W. G., G. W. Pearsall, and J. Wulff, 1964. The Structure and Properties of Materials. Structure. Wiley, New York, 1, 236pp.

Mogi, K., 1972. Fracture and flow of rocks. Tectonophysics, 13, pp. 541-568.

Molnar, P., and T. Atwater, 1973. Magnetic anomalies and evolution of the south Pacific: implications for fixed hot spots. Trans. Am. Geophysical Union, 54, p. 240.

Monroe, J. N., 1969. Slumping structures caused by organically derived gases in sediments. Science, 164 (3886), pp. 1394-1395.

Moody, J. D., and M. J. Hill, 1956. Wrench-fault tectonics. Geol. Soc. Am. Bull., 67, pp. 1207-1246.

Moon, C. F., 1972. The microstructure of clay sediments. Earth Sci. Rev., 8, pp. 303-321.

Moore, J. C., and J. E. Geigle, 1972. Incipient axial plane cleavage: deep sea occurrence. Geol. Soc. Am. (Abstracts) '72. Annual Mtg., p. 600.

Moore, J. G., 1960. Curvature of normal faults in the Basin and Range province of the western United States: pp. B409-B411. in Short Papers in the Geological Sciences, U. S. Geol. Survey Prof. Paper 400B.

Morgan, W. J., 1968. Rises, trenches, great faults, and crustal blocks. J. Geophysical Res., 73, pp. 1959-1982.

Morgan, W. J., 1972a. Deep mantle convection plumes and plate motions. Am. Assoc. Petrol. Geol. Bull., 56 (2), pp. 203-213.

Morgan, W. J., 1972b. Plate motions and deep mantle convection: pp. 7-22. in Studies in Earth and Space Sciences, R. Shagam et al. (eds.), Geol. Soc. Am. Memoir, 132, 683pp.

Morgenstern, N. R., and J. S. Tchalenko, 1967. Microscopic structures in kaolin subjected to direct shear. Geotechnique, 17, pp. 309-328.

Morley, L. W., 1963. Plate tectonics and geomagnetic reversals. Unpublished manuscript. See excerpt reproduced on p. 224 in Plate Tectonics and Geomagnetic Reversals, A. Cox (ed.). W. H. Freeman and Co. San Francisco, 702pp.

Morris, T. O., and W. G. Fearnsides, 1926. The stratigraphy and structure of the Cambrian slate belt of Nantlle (Caernavonshire). Geol. Soc., Quart. J., 82, pp. 250-303.

Muehlberger, W. R., 1961. Conjugate joint sets of small dihedral angle. J. Geol., 69, pp. 211-219.

Muehlberger, W. R., and P. S. Clabaugh, 1968. Internal structure and petrofabrics of gulf coast salt domes: pp. 90-98 in Diapirs and Diapirism, J. Braunstein and G. D. O'Brien (eds.). Am. Assoc. Petrol. Geol., Memoir 8, Tulsa, Oklahoma., 444pp.

Mukhopadhyay, D., 1972. Deformation of a slaty, lapillar tuff in the Lake District, England: discussion. Geol. Soc. Am. Bull., 83, pp. 547-548.

Naha, K., and Chaudhuri, A. K., 1968. Large scale fold interference in a metamorphic-migmatitic complex. Tectonophysics, 6, pp. 127-142.

Naha, K., and S. K. Ray, 1972. Structural evolution of the Simla Klippe in the Lower Himalayas. Geol. Rundschau, 61 (3), pp. 1050-1086.

Naha, K., and R. V. Halyburton, 1974. Late stress systems deduced from conjugate folds and kink bands in the "Main Raialo Syncline," Udaipur District, Rajasthan, India. Geol. Soc. Am. Bull., 85 (2), pp. 251-256.

Naylor, R. S., 1960. Origin and regional relationships of the core-rocks of the Oliverian Domes: pp. 231-240 in Studies of Appalachian Geology: Northern and Maritime, E-an Zen, W. S. White, J. B. Hadley, and J. B. Thompson (eds.). Wiley-Interscience, New York, 475pp.

Nesbitt, R. W., A. D. T. Goode, A. C. Moore, and T. P. Hopwood, 1970. The Giles Complex, Central Australia: a stratified sequence of mafic and ultramafic intrusions. Special Publ., Symposium on the Bushveld Igneous Complex and other layered intrusions. Geol. Soc, S. Africa (1), pp. 547-564.

Nettleton, L. L., 1934. Fluid mechanics of salt domes. Am. Assoc. Petrol. Geol. Bull., 18 (9), pp. 1175-1204.

Neumann, E. R., 1969. Experimental recrystallization of dolomite and comparison of preferred orientations of calcite and dolomite in deformed rocks. J. Geol., 77, pp. 426-438.

Nevin, S. M., 1949. Principles of structural geology (3rd ed.). Wiley, New York, 320pp.

Nickelsen, R. P., 1966. Fossil distortion and penetrative rock deformation in the Appalachian Plateau, Pennsylvania. Jour. Geol., 74, pp. 924-931.

Nickelsen, R. P., and V. N. D. Hough, 1967. Jointing in the Appalachian Plateau of Pennsylvania. Geol. Soc. Amer. Bull., 78, pp. 609-630.

Norris, D. K., 1967. Structural analysis of the Queensway folds, Ottawa, Canada. Can. Jour. Earth Sci., 4, pp. 299-321.

Norris, D. K., and K. Barron, 1969. Structural analysis of features on natural and artificial faults. Geol. Survey Canada Paper 68-52, pp. 136-157.

Nur, A., and J. D. Byerlee, 1971. An exact effective stress law for elastic deformation of rock with fluids. Jour. Geophys. Res., 76(26), pp. 6414-6419.

Nye, J. F., 1964. Physical properties of crystals. Clarendon, Oxford, 322pp.

O'Brien, G. D., 1968. Survey of diapirs and diapirism: pp. 1-9 in Diapirs and Diapirism, J. Braunstein and G. D. O'Brien (eds.). Am. Assoc. Petrol. Geol. Memoir 8, Tulsa, Oklahoma, 444pp.

O'Brien, N. R., 1970. The fabric of shale—an electron microscope study. Sedimentology, 15, pp. 229-246.

Ocamb, R. D., 1961. Growth faults of South Louisiana. Gulf Coast Assoc. Geol. Societies, Trans., 11, pp. 139-176.

O'Driscoll, E. S., 1962. Experimental patterns in superposed similar folding. J. Soc. Petrol. Geol., Alberta, 10, pp. 145-167.

O'Driscoll, E. S., 1964. Interference patterns from inclined shear fold systems. Canadian Petrol. Geol. Bull., 12, pp. 279-310.

Oertel, G., 1970. Deformation of a slaty, lapillar tuff in the Lake District, England. Geol. Soc. Am. Bull., 81, pp. 1173-1188.

Oertel, G., 1971. Deformation of a slaty, lapillar tuff in the English Lake District: reply. Geol. Soc. Am. Bull., 82 (2), pp. 533-536.

Olesen, N. Ø., E. S. Hansen, L. H. Kristensen, and T. Thyrsted, 1973. A preliminary account on the geology of the Selbu-Tydal area, the Trondheim Region, Central Norwegian Caledonides. Leidse Geol. Mededel., 49(2), pp. 259-277.

Orowan, E., 1954. Dislocations in metals. Am. Inst. Metall. Eng., New York, 183pp.

Orowan, E., 1967. Seismic damping and creep in the mantle. Roy. Astron. Soc. Geophys. J., 14, pp. 191-218.

Owens, W. H., 1973. Strain modification of angular density distributions. Tectonophysics, 16, pp. 249-261.

Pakiser, L. C., J. P. Eaton, J. H. Healy and C. B. Raleigh, 1969. Earthquake prediction and control. Science, 166, pp. 1467-1474.

Park, C. F., and R. A. MacDiarmid, 1964. Ore Deposits. W. H. Freeman and Co., San Francisco, 522pp.

Park, R. G., 1969. Structural correlation in metamorphic belts. Tectonophysics, 7 (4), pp. 323-338.

Parker, J. M., 1942. Regional systematic jointing in slightly deformed sedimentary rocks. Geol. Soc. Am. Bull., 53, pp. 381-408.

Parrish, D. K., 1973. A nonlinear finite element fold model. Am. J. Sci., 273, pp. 318-334.

Paterson, M. S., 1958. Experimental deformation and faulting in Wombeyan marble. Geol. Soc. Am. Bull., 69, pp. 465-476.

Paterson, M. S., 1969. The ductility of rocks: pp. 377-392 in Physics of Strength and Plasticity, The Orowan 65th Anniversary Volume, A. S. Argon (ed.). The M. I. T. Press.

Paterson, M. S., 1970a. A high-pressure, high-temperature apparatus for rock deformation. Int. J. Rock Mech. Min Sci., 7, pp. 517-526.

Paterson, M. S., 1970b. Experimental deformation of minerals and rocks under pressure: pp. 197-235 in Mechanical Behaviour of Materials under Pressure, H. Ll. D. Pugh (ed.). Elsevier Pub. Co., New York 785pp.

Paterson, M. S., 1973. Nonhydrostatic thermodynamics and its geologic applications. Geophys. and Space Phys., Rev., 11, pp. 355-389.

Paterson, M. S., and F. J. Turner, 1970. Experimental deformation of constrained crystals of calcite in extension: pp. !09-141 in Experimental and Natural Rock Deformation, P. Paulitsch (ed.). Springer-Verlag, Berlin.

Paterson, M. S., and L. E. Weiss 1961. Symmetry concepts in the structural analysis of deformed rocks. Geol. Soc. Am. Bull., 72, pp. 841-882.

Paterson, M. S., and L. E. Weiss, 1966. Experimental deformation and folding in phyllite. Geol. Soc, Am. Bull., 77, pp. 343-374.

Paterson, M. S., and L. E. Weiss, 1968. Folding and boudinage of quartz-rich layers in experimentally deformed phyllite. Geol. Soc. Am. Bull., 79, pp. 795-812.

Pettijohn, F. J., and P. J. Potter, 1964. Atlas and glossary of primary sedimentary structures. Springer-Verlag, Berlin, 370pp.

Pettijohn, F. J., P. E. Potter, and R. Siever, 1972. Sand and Sandstone. Springer-Verlag, Berlin, 618pp.

Phillips, F. C., 1937. A fabric study of some Moine Schists and associated rocks. Geol. Soc., London, Quart. J., 93, pp. 581-620.

Phillips, F. C., 1945. The microfabric of Moine schists. Geol. Mag., 82, pp. 205-220.

Phillips, F. C., 1971. The Use of Stereographic Projection in Structural Geology. Edward Arnold, London, 90pp.

Phillips, F. C., and G. Windsor, 1970. The study of fabrics of geological bodies by Professor Bruno Sander. Pergamon Press, New York, 641pp.

Pitcher, W. S., 1969. Northeast-trending faults of Scotland and Ireland and chronology of displacements: pp. 724-733 in North Atlantic Geology and Continental Drift, M. Kay (ed.) Am. Assoc. Petrol. Geol., Tulsa, Oklahoma, 1082pp.

Pitman, W. C. III, and J. R. Heirtzler 1966. Magnetic anomalies over the Pacific-Antarctic ridge. Science, 154, pp. 1164-1171.

Pitman, W. C. III, E. M. Herron, and J. R. Heirtzler, 1968. Magnetic anomalies in the South Pacific Ocean and sea floor spreading. J. Geophysical Res., 73 pp. 2069-2085.

Pitman, W. C. III, and M. Talwani, 1972. Sea-floor spreading in the North Atlantic. Geol. Soc. Am. Bull., 83, pp. 619-646.

Plafker, G., 1972. Alaskan earthquake of 1964 and Chilean earthquake of 1960: implications for arc tectonics. J. Geophysical Res., 77 (5), pp. 901-925.

Platt, L. B., 1962. Fluid pressure in thrust faulting, a corollary. Am. J. Sci., 260, pp. 107-114.

Plessman, W. von, 1964. Gesteinslösung, ein Hauptfaktor beim Scheiferungsprozess. Geol. Mitt., 4 (1), pp. 69-82.

Potter, P. E., and F. J. Pettijohnn 1963. Pcurrents and Basin Analysis. Springer-Verlag, Berlin, 296pp.

Press, F., 1973. The gravitational instability of the lithosphere: pp. 7-16 in Gravity and Tectonics, K. A. DeJong and R. Scholten (eds.). Wiley, New York, 502pp.

Price, N. J., 1966. Fault and Joint Development in Brittle and Semi-brittle Rock. Pergamon Press, New York, 176pp.

Price, R. A., 1973. Large-scale gravitational flow of supracrustal rocks, Southern Canadian Rockies: pp. 491-502 in Gravity and Tectonics, K. A. DeJong and R. Scholten (eds.). Wiley, New York, 502pp.

Price, R. A., and E. W. Mountjoy, 1970. Geologic structure of the Canadian Rocky Mountains between Bow and Athabasca Rivers—a progress report: pp. 7-26, in Geol. Assoc. Canada, Spec. Paper 6, J. O. Wheeler (ed.) 166pp.

Purdue, A. H., 1909. The slates of Arkansas with a bibliography of the geology of Arkansas. Geological Survey of Arkansas.

Raleigh, C. B., 1965. Glide mechanisms in experimentally deformed minerals. Science, 150, pp. 739-741.

Raleigh, C. B., 1968. Mechanisms of plastic deformation of olivine. J. Geophysical Res., 73, pp. 5391-5406.

Raleigh, C. B., and D. T. Griggs, 1963. Effect of the toe in the mechanics of overthrust faulting. Geol. Soc. Am. Bull., 74, pp. 819-830.

Raleigh, C. B., and S. H. Kirby, 1970. Creep in the upper mantle. Min. Soc. Am., Spec. Paper 3, pp. 113-121.

Raleigh, C. B., and M. S. Paterson, 1965. Experimental deformation of serpentinite and its tectonic implications. J. Geophysical Res., 70, pp. 3965-3985.

Raleigh, C. B., J. H. Healy and J. D. Bredehoeff, 1972. Faulting and crustal stress at Rangely, Colorado: pp. 275-284 in Flow and Fracture of Rocks, H. C. Heard, I. Y. Borg, N. L. Carter, and C. B. Raleigh (eds.). Am. Geophysical Union, Geophys. Monograph 16, 352pp.

Ramberg, H., 1955. Natural and experimental boudinage and pinch-and-swell structures. J. Geol., 63, pp. 512-526.

Ramberg, H., 1959. Evolution of ptygmatic folding. Norsk. Geol. Tiddskr., 39, pp. 99-151.

Ramberg, H., 1961. Contact strain and folding instability of a multilayered body under compression. Geol. Rundschau, 51, pp. 405-439.

Ramberg, H., 1963a. Evolution of dragfolds. Geol. Mag., 100 (2), pp. 97-106.

Ramberg, H., 1963b. Fluid dynamics of viscous buckling applicable to folding of layered rocks. Am. Assoc. Petrol. Geol. Bull., 47, pp. 484-505.

Ramberg, H., 1964. Selective buckling of composite layers with contrasted rheological properties. A theory for simultaneous formation of several orders of folds. Tectonophysics, 1, pp. 307-341.

Ramberg, H., 1967. Gravity, Deformation and the Earth's Crust as studied by Centrifuged Models. Academic Press, New York, 241pp.

Ramberg, H., 1968. Instability of layered systems in the field of gravity. Phys. Earth Planet. Inter. 1, pp. 427-447.

Ramberg, H., 1972. Theoretical model of density stratification and diapirism in the earth. J. Geophysical Res., 77 (5), pp. 877-889.

Ramsay, D. M., and B. A. Sturt, 1970. Polyphase deformation of a polymict Silurian conglomerate from Mageröy Norway. J. Geol., 78, pp. 264-280.

Ramsay, D. M., and B. A. Sturt, 1973a. An analysis of noncylindrical and incongruous fold pattern from the Eo-Cambrian rocks of Sorøy, Northern Norway. I. Noncylindrical, incongruous and aberrant folding. Tectonophysics, 18, pp. 81-107.

Ramsay, D. M., and B. A. Sturt, 1973b. An analysis of noncylindrical and incongruous fold pattern from the Eo-Cambrian rocks of Sorøy, Northern Norway. II. The significance of synfold stretching lineation in the evolution of noncylindrical folds. Tectonophysics, 18, pp. 109-121.

Ramsay, J. G., 1958a. Superimposed folding at Loch Monar, Inverness-shire and Ross-shire. Geol. Soc. London, Quart. J., 113, pp. 271-308.

Ramsay, J. G., 1958b. Moine-Lewisian relations at Glenelg, Inverness-shire. Geol. Soc. London, Quart. J., 113, pp. 487-523.

Ramsay, J. G., 1960. The deformation of early linear structures in areas of repeated folding. J. Geol., 68, pp. 75-93.

Ramsay, J. G., 1962a. The geometry and mechanics of formation of "similar" type folds. J. Geol., 70, pp. 309-327.

Ramsay, J. G., 1962b. Interference patterns produced by the superposition of folds of "similar" type. J. Geol., 60, pp. 466-481.

Ramsay, J. G., 1963. Stratigraphy, structure and metamorphism in the Western Alps. Geol. Assoc. Proc., 74, pp. 357-392.

Ramsay, J. G., 1964. The uses and limitations of beta-diagrams and pi-diagrams in the geometrical analysis of folds. Geol. Soc. London Quart. J., 120, pp. 435-454.

Ramsay, J. G., 1965. Structural investigations in the Barberton Mountain Land, Eastern Transvaal. Geol. Soc. S. Africa Trans., 66, pp. 353-401.

Ramsay, J. G., 1967. Folding and Fracturing of Rocks. McGraw-Hill, New York, 568pp.

Ramsay, J. G., and R. H. Graham, 1970. Strain variation in shear belts. Canadian J. of Earth Sci., 7, pp. 786-813.

Ramsay, J. G., and D. S. Wood, 1973. The geometric effects of volume change during deformation processes. Tectonophysics, 16, pp. 263-277.

Ransom, D. M., 1968. The relationship of lode shape to wall-rock structure in the southern half of the Broken Hill orebody. Geol. Soc. Australia, J., 15, (1), pp. 57-64.

Ransom, D. M., 1971. Host control of recrystallized quartz grains. Mineralog. Mag., 38, pp. 83-88.

Rast, N., 1963. Structure and metamorphism of the Dalradian rocks of Scotland: pp. 123-175 in The British Caledonides, M. R. W. Johnson and F. H. Stewart (eds.). Oliver and Boyd, London, 280pp.

Ratcliffe, N. M., 1971. The Ramapo fault system in New York and adjacent northern New Jersey: a case of tectonic heredity. Geol. Soc. Am. Bull., 82, pp. 125-142.

Ray, S. K., 1974. Inversion of fold-hinge in superposed folding: an example from the Precambrian of Central Rajasthan, India. Precambrian Research 1, pp. 157-164.

Rettger, R. E., 1935. Experiments on soft rock deformation. Am. Assoc. Petrol. Geol. Bull., 19, pp. 271-292.

Reynolds, D. L., and A. Holmes, 1954. The superposition of Caledonoid folds on an older fold-system in the Dalradians of Malin Head, Co. Donegal. Geol. Mag., 91 pp. 417-444.

Richards, J. R., H. Berry, and J. M. Rhodes, 1966. Isotopic and lead-alpha ages of some Australian zircons. Geol. Soc. Australia, J., 13, pp. 69-96.

Richter, F. M., 1973. Dynamical models for sea floor spreading. Geophys. and Space Phys. Rev., 11 (2), pp. 223-287.

Rickard, M. J., 1961. A note on cleavages in crenulated rocks. Geol. Mag., 98 (4), pp. 324-332.

Rickard, M. J., 1972. Fault classification: discussion. Geol. Soc. Am. Bull., 83, pp. 2545-2546.

Riecker, R. E., 1972. Fracture mechanics and earthquake source mechanisms. Geotimes, 17 (4), pp. 15-18.

Roberts, D., 1972. Tectonic deformation in the Barents Sea region of Varanger Peninsula, Finmark. N. G. U., Bull., 10 (282), pp. 1-39.

Roberts, D., and K. E. Strömgård, 1972. A comparison of natural and experimental strain patterns around fold hinge zones. Tectonophysics, 14, pp. 105-120.

Roberts, J. C., 1961. Feather-fracture and the mechanics of rock-jointing. Am J. Sci., 259, pp. 481-492.

Roberts, J. L., 1966. The formation of similar folds by inhomogeneous plastic strain, with reference to the fourth phase of deformation affecting the Dalradian rocks in the southwest highlands of Scotland. J. Geol., 74 (6), pp. 831-855.

Roberts, J. L., 1972. The mechanics of overthrust faulting: a critical review. 24th Int. Geol. Cong., Sec. 3, pp. 593-598.

Robinson, P., 1958. The structural and metamorphic geology of the Brighton-Taieri mouth area, East Otago, New Zealand. Unpubl. M.Sc. thesis, Univ. Otago, New Zealand.

Rodgers, J., 1950. Mechanics of Appalachian folding as illustrated by Sesquatchie anticline, Tennessee and Alabama. Am. Assoc. Petrol. Geol. Bull., 34, pp. 672-681.

Rodgers, J., 1953. The folds and faults of the Appalachian Valley and Ridge province. Kentucky Geol. Survey, Ser. 9, Spec. Publ. 1, pp. 150-166.

Rodgers, J., 1964. Basement and no-basement hypotheses in the Jura and the Appalachian valley and ridge: pp. 71-80 in Tectonics of the Southern Appalachians, Virginia Polytech. Inst., Dept. Geol. Sci. Memoir, 1.

Rodgers, J., 1970. The Tectonics of the Appalachians. Wiley-Interscience, New York, 271pp.

Ross, J. V., 1973. Mylonitic rocks and flattened garnets in the Southern Octanogan of British Columbia. Canadian J. of Earth Sci., 10, pp. 1-17.

Sahama, T. G., 1936. Die Regelung von Quarz und Glimmer in den Gesteinen der Finnishch-Lapplandichen Granulitformation. Comm. Géol. de Finlande, Bull. (113), pp. 1-110.

Sander, B., 1911. Über Zusammenhange Zwischen Teilbewegung und Gefüge in Gesteinen. Tschermaks Mineral. Petrogr. Mitt., 30, pp. 381–384.

Sander, B., 1912. Über tektonishche Gesteinsfazies. Verhandlugen der Geologischen Reichsanstalt in Wien, Nr. 10.

Sander, B., 1930. Gefügekunde der Gesteine. Springer, Vienna, 352pp.

Sander, B., 1932. Zur Kinematik passiver Gefügeregelungen. Zeits. Krist., 81, pp. 298-308.

Sander, B., 1948; 1950. Einführung in die Gefügekunde der Geologischen Körper. Springer, Vienna, I, 1948, 215pp; II, 1950, 409pp.

Sander, B., 1970. An Introduction to the Study of Fabrics of Geological Bodies. Translated by F. C., Phillips and G. Windsor. Pergamon Press, New York, 641pp.

Sanders, J. E., 1960. Origin of convoluted laminae. Geol. Mag., 97, pp. 409-421.

Savage, J. C., and R. O. Burford, 1973. Geodetic determination of relative plate motion in central California. J. Geophysical Res., 78 (5), pp. 832-845.

Savage, J. F., 1965. Terrestrial Photogrammetry for Geological purposes. ITC Publ., Delft Spec. Publ., 2, pp. 41-53.

Savage, J. F., 1967. Tectonic analysis of Lechada and Curavacas synclines, Yuso Basin, Léon, Northwest Spain. Leisde Geol. Mededel., 39, pp. 193-247.

Schardt, H., 1904. Note sur le profil géologique et la tectonique de massif du Simplon. Eclogae Geol. Helv., 8, pp. 173-200.

Schmid, E., and W. Boas, 1935. Kristallplastizitat. Springer-Verlag, Berlin.

Schmidt, W., 1925. Gefugestatistik. Min. Petrol. Mitt., 38, p. 392.

Schmidt, W., 1932. Tectonik und Verformungslehre. Borntraeger, Berlin, 208pp.

Schneegans, D., 1938. La géologie des nappes de l'Ubaye-Embrunaise entre la Durance et l'Ubaye. Mémoires pour servir à l'explication de la Carte Géologique Détaillée de la France, 339pp.

Scholz, C. H., 1968. Microfracturing and the inelastic deformation of rock in compression. J. Geophysical Res., 73 (4), pp. 1417-1432.

Scholz, C. H., 1972. Static fatigue of quartz. J. Geophysical Res., 77 (11), pp. 2104-2114.

Scholz, C. H., and T. J. Fitch, 1969. Strain accumulation along the San Andreas fault. J. Geophysical Res., 74 (27), pp. 6649-6666.

Scholz, C. H., L. R. Sykes, and Y. P. Aggarwal, 1973. Earthquake prediction: a physical basis. Science, 181 (4102), pp. 803-810.

Scholz, C. H., M. Wyss and S. W. Smith, 1969. Seismic and aseismic slip on the San Andreas fault. J. Geophysical Res., 74 (8), pp. 2049-2069.

Schwerdtner, W. M., 1970. Hornblende lineations in Trout Lake area, Lac la Ronge map sheet, Saskatchewan. Canadian J. Earth. Sci., 7, pp. 884-899.

Sclater, J. G., 1972. Heat flow and elevation of the marginal basins of the western Pacific. J. Geophysical Res., 72 (29), pp. 5705-5719.

Sclater, J. G., R. N. Anderson, and M. L. Bell, 1971. The elevation of ridges and the evolution of the central eastern Pacific. J. Geophysical Res., 76, pp. 7888-7915.

Sclater, J. G., and C. E. Corry, 1967. Heat flow, Hawaiian area. J. Geophysical Res., 72, pp. 3711-3715.

Sclater, J. G., R. N. Anderson and M. L. Bell, 1971. The elevation of ridges and the evolution of the central eastern Pacific. J. Geophysical Res., 76, pp. 7888-7915.

Scott, J. S., and H. I. Drever, 1953. Frictional fusion along a Himalayan thrust. Roy. Soc. Edinburgh, Proc., Sec. b, 2, pp. 121-140.

Scott, W. H., E. Hansen, and R. J. Twiss, 1965. Stress analysis of quartz deformation lamellae in a minor fold. Am. J. Sci., 263, pp. 729-746.

Secor, D. T., 1965. Role of fluid pressure in jointing. Am J. Sci., 263, pp. 633-646.

Secor, D. T., 1969. Mechanics of natural extension fracturing at depth in the earth's crust. Canada Geol. Survey Paper 68-52, pp. 3-48.

Sedgwick, A., 1835. Remarks on the structure of large mineral masses, and especially on the chemical changes produced in the aggregation of stratified rocks during different periods after their deposition. Geol. Soc. London Trans., Ser. 2, 3, pp. 461-486.

Seifert, K. W., 1965. Deformation bands in albite. Am. Min., 50, pp. 1469-1472.

Selley, R. C., 1970. Ancient Sedimentary Environments. Chapman and Hall, London, 237pp.

Shackleton, R. M., 1958. Downward facing structures of the Highland Border, Geol. Soc. London, Quart. J., 113, pp. 361-392.

Shand, J., 1916. The pseudotachylyte of Parijs (Orange Free State). Geol. Soc. London, Quart. J., 72, pp. 198-221.

Shelley, D., 1971. The origin of cross-girdle fabrics of quartz. Tectonophysics, 11, pp. 61-68.

Sherwin, J. A., and W. M. Chapple, 1968. Wavelengths of single layer folds: a comparison between theory and observation. Am. J. Sci., 266, pp. 167-179.

Shonle, J. I., 1965. Resource letter CM-1 on the teaching of angular momentum and rigid body motion. Am J., Phys., 33 (11), pp. 879-887.

Shonle, J. I., 1966. Correction to resource letter CM-1. Am. J. Phys., 34 (3), p. 273.

Shrock, R. R., 1948. Sequence in Layered Rocks. McGraw-Hill Book Co., New York, 507pp.

Siddans, A. W. B., 1972. Slaty cleavage—a review of research since 1815. Earth Sci. Rev., 8, pp. 205-232.

Singh, R. P., 1968. Experimental tests of buckling folds in relation to strain ellipsoid in simple shear deformations: a discussion. Tectonophysics 5, pp. 341-343.

Smith, C. S., 1948. Grains, phases and interfaces: an interpretation of microstructure. Trans. Am. Inst. Mech. Engr., 175, pp. 15-51.

Smith, C. S., 1964. Some elementary principles of polycrystalline microstructure. Met. Rev., 9, pp. 1-48.

Smith, B., and T. N., George, 1948. British Regional Geology, North Wales, 2nd ed., revised by T. N., George. Dept. Sci. and Industrial Research. His Majesty's Stationary Office, London, p. 89.

Smith, G., and A. Hallam, 1970. The fit of the southern continents. Nature, 225, pp. 139-144.

Smith, J. G., 1965. Orogenesis in western Papua and New Guinea. Tectonophysics, 2 (1), pp. 1-27.

Smith, W. K., 1951. Quitman field: pp. 315-319 in Occurrence of oil and gas in northeast Texas. Univ. Texas Publ., 5116.

Sorby, H. C., 1853. On the origin of slaty cleavage. Edinburgh New Phil. J., 55, pp. 137-148.

Sorby, H. C., 1856a. On Slaty Cleavage as exhibited in the Devonian Limestones of Devonshire. Phil. Mag., J. Sci., Ser. 4, 11.

Sorby, H. C., 1856b. On the theory of the origin of slaty cleavage. Phil. Mag., Ser. 4, 12, pp. 127-129.

Spiegel, M. R., 1959. Vector analysis and an introduction to tensor analysis. McGraw-Hill, New York, 225pp.

Spry, A., 1962. The origin of columnar jointing, particularly in basalt flows. Geol. Soc. Australia, J., 8, pt. 2, pp. 191-216.

Spry, A., 1969. Metamorphic Textures. Pergamon Press, Oxford, 350pp.

Stanley, D. J., 1963. Variability in Annot sandstone turbidites. J. Sed. Petrol., 33, pp. 783-788.

Stanton, R. L., 1972. Ore Petrology. McGraw-Hill, New York, 713pp.

Stanton, R. L., and H. G., Willey, 1970. Experimental modification of naturally deformed galena crystals and their grain boundaries. Min. Mag., 37 (291), pp. 852-857.

Stauder, W., 1962. The focal mechanism of earthquakes. In Advances in Geophysics, H. E., Landsberg and J. Van Mieghem (eds.) Academic Press, London, 374pp.

Stauder, W., 1968. Tensional character of earthquake foci beneath the Aleutian trench with relation to sea-floor spreading. J. Geophysical Res., 73 (24), pp. 7693-7701.

Stauder, W., 1972. Fault motion and spatially bounded character of earthquakes in Amchitka and the Delarof Islands. J. Geophysical Res., 77 (11), pp. 2072-2080.

Stauffer, M. R., 1964. The geometry of conical folds. New Zealand J. Geol. and Geophys., 7 (2), pp. 340-347.

Stauffer, M. R., 1967. Tectonic strain in some volcanic, sedimentary and intrusive rocks near Canberra, Australia: a comparative study in deformation fabrics. New Zealand J. Geol. and Geophys., 10, pp. 1079-1108.

Stearns, D. W., 1968. Certain aspects of fractures in naturally deformed rocks: pp. 97-118 in NSF advanced science seminar in rock mechanics for college teachers of structural geology, R. E. Riecker (ed.). Terrestrial Sciences Laboratory, Air Force Cambridge Research Laboratories, Bedford, Mass., 594pp.

Stephansson, O., and H. Berner, 1971. Finite element method in tectonic processes. Phys. Earth Planetary Inter. 4, pp. 301-321.

Stirewalt, G. L., and D. E. Dunn, 1973. Mesoscopic fabric and structural history of Brevard zone and adjacent rocks, North Carolina. Geol. Soc. Amer. Bull., 84, pp. 1629-1650.

Strand, T., 1961. The Scandinavian Caledonides: a review. Am. J. Sci., 259, pp. 161-172.

Strand, T., and O. Kulling, 1972. Scandinavian Caledonides. Wiley-Interscience, New York, London.

Sturt, B. A., 1961. Preferred orientation of nepheline in deformed nepheline syenite gneisses, Sorøy, Northern Norway. Geol. Mag., 98, pp. 464-466.

Sturt, B. A., and J. Taylor, 1971. The timing and environment of emplacement of the Storelv Gabbro. Sorøy. Norges Geologiske Undersokelse, 272, pp. 1-34.

Suppe, J., 1970. Offset of late Mesozoic basement terrains by the San Andreas fault system. Geol. Soc. Am. Bull., 81, pp. 3253-3258.

Sugden, W., 1950. The influence of water films adsorbed by mineral grains upon the compaction of natural sediments and notes on allied phenomena. Geol. Mag., 87, pp. 26-40.

Sutton, J., 1969. Rates of change within orogenic belts: pp. 239-250 in Time and Place in Orogeny, P. E. Kent, G. E., Satterthwaite and A. M. Spencer (eds.). Geol. Soc. London, Spec. Publ. 3, 311pp.

Sutton, J., and J. Watson, 1959. Structures in the Caledonides between Loch Duich and Glenelg, Northwest Highlands. Geol. Soc. London, Quart. J., 114, pp. 231-258.

Sutton, J., and J. Watson, 1962a. Further observations on the margin of the Laxfordian complex of the Lewisian near Loch Laxford, Sutherland. Roy. Soc. Edinburgh Trans., 65, pp. 89-106.

Sutton, J., and J. Watson, 1962b. An interpretation of Moine-Lewisian relations in Central Ross-shire. Geol. Mag., 99, pp. 527-541.

Sullivan, C. J., and R. S. Matheson, 1952. Uranium-copper deposits, Rum Jungle, Australia. Econ. Geology, 47, pp. 751-758.

Sykes, L. R., 1966. The seismicity and deep structure of island arcs. J. Geophysical Res., 71 (12), pp. 2981-3006.

Sykes, L. R., 1968. Seismological evidence for transform faults, sea floor spreading and continental drift: pp. 120-150 in The History of the Earth's Crust, R. A., Phinney (ed.). Princeton Univ. Press, New Jersey, 244pp.

Sykes, L. R., 1971. Aftershock zones of great earthquakes, seismicity gaps, and earthquake prediction for Alaska and the Aleutians. J. Geophysical Res., 75, pp. 8021-8041.

Sykes, L. R., and M. Sbar, 1973. Intraplate earthquakes, lithospheric stresses and the driving mechanism of plate tectonics. Nature, 245, pp. 298-302.

Sylvester, A. G., and J. M. Christie, 1968. The origin of crossed-girdle orientations of optic axes in deformed quartzites. J. Geol., 76, pp. 571-580.

Syme Gash, P. J., 1971. A study of surface features relating to brittle and semibrittle fracture. Tectonophysics, 12, pp. 349-391.

Synge, J. L., 1960. Classical dynamics: pp. 1-225 in Encyclopedia of Physics, III (1), S. Flugge (ed.). Springer-Verlag, Berlin, 902pp.

Takeuchi, H., S. Uyeda, and H. Kanamori, 1970. Debate About the Earth. Freeman, Cooper and Co., San Francisco, 281pp.

Talbot, J. L., 1965. Crenulation cleavage in the Hunsrückschiefer of the Middle Moselle region. Geol. Rundschau, 54, pp. 1026-1043.

Talwani, M., C. C. Windisch, and M. G. Langseth, 1971. Reykjanes ridge crest: a detailed geophysical study. J. Geophysical Res., 76 (2), pp. 473-517.

Taylor, B. J., I. C. Burgess, D. H. Land, D. A. C. Mills, D. B. Smith, and P. T. Warren, 1971. British Regional Geology, Northern England, 4th ed. Natural Environment Research Council Institute of Geological Sciences, Her Majesty's Stationary Office, London, 121pp.

Taylor, G. I., 1923. The motion of ellipsoidal particles in a viscous fluid. Proc. Roy. Soc. Lond., Series A, Vol. 103, pp. 58-61.

Taylor, G. I., 1938. Plastic strain in metals. Jour. Inst. Metal. 62, pp. 307-324.

Taylor, G. I., 1955. Strains in crystalline aggregates. Proc. Colloquium on Deformation and Flow of Solids (Madrid, 1955). Springer, Berlin, pp. 3-12.

Tchalenko, J. S., 1970. Similarities between shear zones of different magnitudes. Geol. Soc. Amer. Bull., 81, pp. 1625-1640.

Tchalenko, J. S., and N. N. Ambraseys, 1970. Structural analysis of the Dasht-e Bayaz (Iran) earthquake fractures. Geol. Soc. Amer. Bull., 81, pp. 41-60.

Ten Haaf, E., 1959. Graded bedding of the Northern Appenines. Doctoral thesis, Univ. of Groningen, 102pp.

Terzaghi, K., and R. B. Peck, 1948. Soil Mechanics in Engineering Practice. Wiley, New York, 566pp.

Theodore, T. G., and J. M. Christie, 1969. Mylonites and mylonitic gneisses: correct use of the terms. Geol. Soc. Am. (Abstracts 1969), Cordilleran section, pt. 3, pp. 69-70.

The Symbols Committee of the Royal Society, 1971. Quantities, Units, and Symbols. Roy. Soc. Great Britain, 48pp.

Thompson, G. A., 1966. The rift system of the western United States: pp. 280-290 in The World Rift System, T. N. Irvine (ed.). Canada Geol. Survey Paper 66-14. Canada Dept. Mines and Tech. Surveys, Ottawa, 471pp.

Thompson, J. B., 1950. A gneiss dome in southeastern Vermont. Unpubl. Ph. D. thesis, Massachusetts Institute of Technology, 149pp.

Thompson, J. B., P. Robinson, T. N. Clifford, and N. J. Trask, 1968. Nappes and Gneiss Domes in west central New England: pp. 203-218 in Studies of Appalachian Geology: Northern and Maritime, E-An Zen, W. S. White, J. B. Hadley, and J. B. Thompson (eds.). Wiley-Interscience, New York, 475pp.

Thomson, W., and P. G. Tait, 1912. Principles of mechanics and dynamics, Part I. Constable and Co., 508pp. Reprinted in 1962 by Dover Publications.

Timur, A., W. B., Hempkins, and R. M. Weinbrandt, 1971. Scanning electron microscope study of pore systems in rocks. J. Geophysical Res., 76 (20), pp. 4932-4948.

Tjia, H. D., 1964. Slickensides and fault movements. Geol. Soc. Am. Bull., 75, pp. 683-686.

Tobisch, O. T., 1965. Observations on primary deformed sedimentary structures in some metamorphic rocks from Scotland. J. Sed. Petrol., 35, pp. 415-419.

Tobisch, O. T., 1966. Large scale basin and dome pattern resulting from the interference of major folds. Geol. Soc. Am. Bull., 77 (4), pp. 393-408.

Tocher, D., 1960. Creep on the San Andreas fault: creep rate and related measurements at Vineyard, California. Bull. Seis. Soc. Am., 50, pp. 396-415.

Torrance, K. E., and D. L. Turcotte, 1971. Structure of convection cells in the mantle. J. Geophysical Res., 76, pp. 1113-1138.

Treagus, S. H., 1973. Buckling stability of a viscous single-layer system, oblique to the principal compression. Tectonophysics, 19, pp. 271-289.

Trener, G. B., 1906. Geologische Aufnahme in nordlichen Abhange der Presanella-gruppe. Geol. Reichsanstalt Jahrb., 56, pp. 453-470.

Trouw, R. A. J., 1973. Structural geology of the Marsfjällen area, Caledonides of Västerbotten, Sweden. Sveriges Geol. Undersökning, Ser. C. (689), 115pp.

Trumpy, R., 1960. Paleotectonic Evolution of the Central and Western Alps. Geol. Soc. Am. Bull., 71, pp. 843-908.

Trumpy, R., 1973. The timing of orogenic events in the Central Alps: pp. 229-251 in Gravity and Tectonics, K. A. DeJong and R. Scholten (eds.). Wiley, New York, 502pp.

Trüstedt, O.; 1907. Die Erzlagerstätten von Pitkäranta am Ladoga-See. Comm. Geol. de Finlande Bull., 19, p. 333.

Tullis, J., J. M. Christie, and D. T. Griggs, 1973. Microstructures and preferred orientations of experimentally deformed quartzites. Geol. Soc. Am. Bull., 84, pp. 297-314.

Tullis, T. E., 1971. Experimental development of preferred orientation of mica during recrystallization, Ph. D. thesis, Univ. of Calif. at Los Angeles, 262pp.

Turcotte, D. L., and E. R. Oxburgh, 1973. Mid-plate tectonics. Nature 244, pp. 337-339.

Turner, F. J., 1941. The development of pseudo stratification by metamorphic differentiation in the schists of Otago, New Zealand. Am. J. Sci., 239, pp. 1-16.

Turner, F. J., 1957. Lineation, symmetry and internal movement in monoclinic tectonite fabrics. Geol. Soc. Am. Bull., 68, pp. 1-17.

Turner, F. J., D. T. Griggs, and H. C. Heard, 1954. Experimental deformation of calcite crystals. Geol. Soc. Am. Bull., 65, pp. 883-934.

Turner, F. J., and L. E. Weiss, 1963. Structural analysis of metamorphic tectonites. McGraw-Hill, New York, 545pp.

Tuttle, O. T., 1949. Structural petrology of planes of liquid inclusions. J. Geol., 57, pp. 331-355.

Twenhofel, W. H., 1936. Marine unconformities, marine conglomerates and thicknesses of strata. Am. Assoc. Petrol. Geol. Bull., 20, pp. 677-703.

Tyler, J. H., 1972. Pigeon Point Formation: an upper Cretaceous shoreline succession, central California coast. J. Sed. Petrol., 42, pp. 537-557.

Vacquier, V., 1972. Geomagnetism in marine geology. Elsevier Pub. Co., New York, 185pp.

van Bemmelan, R. W., 1960. New views on East-Alpine orogenesis. Rept. 21st Int. Geol. Cong., Copenhagen, 18, pp. 99-116.

van Veen, J., 1965. The tectonic and stratigraphic history of the Cardano Area, Cantabrian Mountains, Northwest Spain, Leidse Geol. Mededel., 35, pp. 45-104.

Vernon, R. H., 1968. Microstructures of high-grade metamorphic rocks at Broken Hill, Australia. J. Petrol., 9, pp. 1-22.

Vernon, R. H., 1969. The Willyama Complex, Broken Hill area: pp. 20-55 in The Geology of New South Wales, G. H. Packham (ed.). Geol. Soc. Australia, J., 16, 654pp.

Vernon, R. H., 1970. Comparative grain-boundary studies of some basic and ultrabasic granulites, nodules and cumulates. Scottish Geol. J., 6, pp. 337-351.

Vernon, R. H., and D. M. Ransom, 1971. Retrograde schists of the amphibolite facies at Broken Hill, New South Wales. Geol. Soc. Australia, J., 18, pp. 267-278.

Vistelius, A. B., 1966. Structural Diagrams, Pergamon Press, New York, 178pp.

Vine, F. J., and D. H. Matthews, 1963. Magnetic anomalies over oceanic ridges. Nature, 199, pp. 947-949.

Voll, G., 1960. New work on petrofabrics. Geol. J., (Liverpool and Manchester), 2 (3), pp. 503-597.

Walcott, R. I., 1970. Flexural rigidity, thickness and viscosity of the lithosphere. J. Geophysical Res., 75, pp. 3941-3954.

Waters, A. C., and C. O. Campbell, 193t. Mylonites from the San Andreas fault zone. Am. J. Sci., 5th Ser., 29 pp. 473-503.

Weertman, J., 1968. Dislocation climb theory of steady-state creep. Am. Soc. Metals. Trans., 61, pp. 681-694.

Weertman, J., 1970. The creep strength of the earth's mantle. Geophys. and Space Plhys. Rev., 8, pp. 145–168.

Wegmann, C. E., 1929. Beispiele tektonischer Analysen des Grundgebirges in Finnland. Comm. Géol. de Finlande Bull., 87, pp. 98-127.

Wegmann, C. E., 1932. Note sur le Boudinage. Soc. Geol. de France Bull., pp. 471-491.

Weiss, L. E., 1954. A study of tectonic style—structural investigation of a marble-quartzite complex in southern California. Univ. Calif. Publ., Geol. Sci., 30 (1), pp. 1-102.

Weiss, L. E., 1959a. Geometry of superposed folding. Geol. Soc. Am. Bull., 70, pp. 91-106.

Weiss, L. E., 1959b. Structural analysis of the Basement System at Turoka, Kenya. Overseas Geol. and Min. Resources, 7 (1-2), pp 3-35, 123-153, respectively.

Weiss, L. E., 1969. Flexural-slip folding of foliated model materials: pp. 294-357 in Proc. Conference on Research in Tectonics, A. J. Baer and D. K. Morris (eds.). Canada Geol. Survey, Ottawa, Paper 68-52.

Weiss, L. E., and D. B. McIntyre, 1957. Structural geometry of Dalradian rocks at Loch Leven, Scottish Highlands. J. Geol., 65, pp. 575-602.

Weissel, J. K., and D. E. Hayes, 1971. Asymmetric seafloor spreading south of Australia. Nature, 231, pp. 518-521.

Wellman, H. W., 1955. New Zealand quarternary tectonics. Geol. Rundschau, 43, pp. 248-257.

Wenk, H.-R., 1973. The structure of the Bergell Alps. Eclogae Geol. Helv., 66 (2), pp. 255-291.

Wesson, P. S., 1970. The position against continental drift. Roy. Astron. Soc., Quart. J., 2, pp. 312-340.

Wesson, P. S., 1972. Objections to continental drift and plate tectonics. J. Geol., 80, pp. 185-197.

Wettstein, A., 1886. Über die Fischfauna des Tertiären Glarner Schiefers, Schweiz. Paläont. Ges. Abh., 13, pp. 1-101.

Weymouth, J. H., and W. O. Williamson, 1953. The effects of extrusion and some other working processes on the micro structures of clay. Am. J. Sci., 251, pp. 89-108.

Wheeler, J. O., J. D. Aitken, M. J. Berry, H. Gabrielse, W. W. Hutchinson, W. R. Jacoby, J. W. H. Monger, E. R. Niblett, D. K. Norris, R. A. Price, and R. A. Stacey, 1972. The Cordilleran structural province: pp. 1-82 in Variations in Tectonic Styles in Canada, R. A. Price and R. J. W. Douglas (eds.). Geol. Assoc. Canada, Spec. Paper 11.

Whitcomb, J. H., J. D. Garmany, and D. L. Anderson, 1973. Earthquake prediction: variation of seismic velocities before the San Francisco earthquake. Science, 180, pp. 632-635.

White, D. A., D. H. Roeder, T. H. Nelson, and J. C. Crowell, 1970. Subduction. Geol. Soc. Am. Bull., 81, pp. 3431-3432.

White, W. S., 1949. Cleavage in East Central Vermont. Trans. Am. Geophysical Union, 30 (4), pp. 587-594.

Whitten, C. A., 1956. Crustal movement in California and Nevada. Trans. Am. Geophysical Union, 37, pp. 393-398.

Wickens, A. J., 1971. Variations in lithospheric thickness in Canada. Canadian J. of Earth Sci., 8, pp. 281-284.

Wickham, J. S., 1973. An estimate of strain increments in a naturally deformed carbonate rock. Am. J. Sci., 273, pp. 23-47.

Williams, E., 1960. Intra-stratal flow and convolute folding. Geol. Mag., 97, pp. 208-214.

Williams, P. F., 1967. Structural analysis of the Little Broken Hill area of New South Wales. Geol. Soc. Australia, J., 14, pp. 317-332.

Williams, P. F., 1970. A criticism of the use of style in the study of deformed rocks. Geol. Soc. Am. Bull., 81, pp. 3283-3296.

Williams, P. F., 1971. Structural analysis of Bermagui Area, New South Wales. Geol. Soc. Australia, J., 18, pp. 215-228.

Williams, P. F., 1972. Development of metamorphic layering and cleavage in low-grade metamorphic rocks at Bermagui, Australia. Am. J. Sci., 272, pp. 1-47.

Williams, P. F., A. R. Collins, and R. G. Wiltshire, 1969. Cleavage and penecontemporaneous deformation structures in sedimentary rocks. J. Geol., 77, pp. 415-425.

Williams, P. F., B. E. Hobbs, R. H. Vernon, and D. E. Anderson, 1971. The structural and metamorphic geology of basement rocks in the McMurdo sound area, Antarctica. Geol. Soc. Australia, J., 18, pp. 127-142.

Wilson, C. J. L., 1973. The prograde microfabric in a deformed quartzite sequence, Mount Isa, Australia. Tectonophysics, 19, pp. 39-81.

Wilson, G., 1953. Mullion and rodding structures in the Moine series of Scotland Geol. Assoc. Proc., 64, pp. 118-151.

Wilson, G., 1961. The tectonic significance of small-scale structures and their importance to the geologist in the field. Ann. Soc. Géol. de Belgique 84 pp. 424-548.

Wilson, J. T., 1963a. Evidence from islands on the spreading of the ocean floor. Nature, 197, pp. 536-538.

Wilson, J. T., 1963b. Continental drift. Scientif. Am., 208, pp. 86-100.

Wilson, J. T., 1963c. Hypothesis of earth's behavior. Nature, 198, (4884), pp. 925-929.

Wilson, J. T., 1965. A new class of faults and their bearing on continental drift. Nature, 207, (4995) pp. 343-347.

Wilson, J. T., 1966. Did the Atlantic close and then re-open? Nature, 211 (5050), pp. 676-681.

Wilson, J. T., 1968a. A reply to V. V. Beloussov. Geotimes, 13 (10), pp. 20-22.

Wilson, J. T., 1968b. Static or mobile earth: the current scientific revolution. Am. Phil. Soc. Proc., 112 (5), pp. 309-320.

Wise, D. U., 1964. Microjointing in basement, middle Rocky Mountains of Montana and Wyoming. Geol. Soc. Am. Bull., 75, pp. 287-306.

Wonsiewicz, B. C., and G. Y. Chin, 1970. Inhomogeneity of plastic flow in constrained deformation. Metall. Trans., 1, pp. 57-61.

Wood, B. L., 1963. Structure of the Otago Schists. New Zealand J. Geol. and Geophys., 6 (5), pp. 641-680.

Wood, D. S., 1973. Patterns and magnitudes of natural strain in rocks. Roy. Soc. London, Philos. Trans., A274, pp. 373-382.

Wood, D. S., 1974. Current views of the development of slaty cleavage. Ann. Rev.: Earth Sci., 2, pp. 1-35.

Woodring, W. P., and R. Stewart, 1934. Geologic map of Kettleman Hills, California. U. S. Geol. Survey.

Woodring, W. P. et al., 1941. Geology of the Kettleman Hills Oil Field, California. U. S. Geol. Survey Prof. Paper 195.

Wright, J. B., 1968. South Atlantic continental drift and the Benue trough. Tectonophysics, 6, (4), pp. 301-310.

Wyllie, P. J., 1971. The Dynamic Earth: Textbook in Geosciences. Wiley, New York, 416pp.

Yagishita, K., 1971. On microfabrics of slump fold of the Saikawa Anticline in northern Fossa Magna, Central Japan. Geol. Soc. Japan, J., 77 (12), pp. 779-790.

Yar Khan, M., 1972. The structure and microfabric of a part of the Arltunga Nappe Complex, Central Australia. Unpubl. Ph. D. thesis, Australia Nat. Univ., Canberra, 113pp.

Zingg, T., 1935. Beitrag zur Schotteranalyse. Schweiz, Min. Petrog, Mitt. 15, pp. 39-140.

Zwart, H. J., 1958. La Faille de Mérens dans les Pyrénées Ariégeoises. Soc. Geol. de France Bull., 8, pp. 794-796.

Zwart, H. J., and J. A. Oele, 1966. Rotated magnetite crystals from the Rocroi-Massif (Ardennes). Geologie en Mijnbouw, 45 (3), pp. 70-74.

INDEX